Electroanalysis of Biochemistry and Material Chemistry—2nd Edition

Electroanalysis of Biochemistry and Material Chemistry—2nd Edition

Guest Editor

Guangjin Wang

MDPI

Basel • Beijing • Wuhan • Barcelona • Belgrade • Novi Sad • Cluj • Manchester

Guest Editor
Guangjin Wang
School of Materials and
Energy
Foshan University
Foshan
China

Editorial Office
MDPI AG
Grosspeteranlage 5
4052 Basel, Switzerland

This is a reprint of the Special Issue, published open access by the journal *Molecules* (ISSN 1420-3049), freely accessible at: https://www.mdpi.com/journal/molecules/special_issues/9SXCIK4O9X.

For citation purposes, cite each article independently as indicated on the article page online and as indicated below:

Lastname, A.A.; Lastname, B.B. Article Title. *Journal Name* **Year**, *Volume Number*, Page Range.

ISBN 978-3-7258-7857-4 (Hbk)
ISBN 978-3-7258-7858-1 (PDF)
https://doi.org/10.3390/books978-3-7258-7858-1

Contents

About the Editor

Guangjin Wang

Guangjin Wang has been a Full Professor at the School of Materials and Energy, Foshan University, China, since 2019. He received his B.Eng. degree in Inorganic Nonmetallic Materials and Engineering from the School of Materials Science and Engineering, Jingdezhen Ceramics Institute, in 2009, and his Ph.D. in New Energy Materials from the State Key Laboratory of Advanced Technology for Materials Synthesis and Processing, Wuhan University of Technology, in 2014. From 2012 to 2014, he was a visiting student at the Ningbo Institute of Materials Technology & Engineering, Chinese Academy of Sciences. From 2014 to 2019, he joined the School of Chemistry and Materials Science, Hubei Engineering University, where he served as a Lecturer and then Associate Professor. In 2018, he went to the Center for Functional Nanomaterials, Brookhaven National Laboratory, as a visiting scholar. His research focuses on the design and fabrication of electrocatalysts for a range of electrochemical reactions, including oxygen reduction/evolution, hydrogen oxidation/evolution, carbon dioxide reduction, and biomass conversion reactions. His work also involves exploring the reaction mechanisms of these electrochemical reactions via first-principles calculation methods. Recently, his research interest has extended to the detection of contaminants in seawater (uranyl ions) and food (heavy metal ions and pesticide residues) matrices by means of electrochemical sensing technology.

molecules

Editorial

Electroanalysis of Biochemistry and Material Chemistry—2nd Edition

Guangjin Wang

School of Materials and Energy, Foshan University, Foshan 528051, China; wgj501@163.com

1. Introduction

Building on the success of the first edition of the Special Issue, entitled "Electroanalysis of Biochemistry and Material Chemistry", and with the aim of collecting more recent research advances in biochemistry and materials chemistry, the second edition of the Special Issue has been launched in *Molecules*. Through the joint efforts of reviewers and editors, 11 papers—including eight research articles and three reviews—have been published in this Special Issue. Herein, a brief review of the latest findings of the abovementioned published papers has been summarized in the following section.

2. An Overview of the Published Articles

Owing to their outstanding physicochemical properties, including high hardness, high melting point and excellent chemical and electrochemical corrosion resistance, transition metal carbides have been employed as reinforcing phases in coating materials to enhance the mechanical properties of metallic substrates [1]. In Contribution 1, Jiang et al. systematically reviewed the preparation methods, structural characteristics and application performance of five typical transition metal carbide-reinforced coatings, including those strengthened with WC, TiC, NbC, $Ti_{n+1}AlC_n$ (MAX phases) and Cr_3C_2-TaC. Subsequently, Jiang et al. (Contribution 2) investigated the impact of Y_2O_3 usage on the hardness, wear resistance and electrochemical corrosion resistance of Fe60/WC/Y_2O_3 coating prepared by laser cladding on 42CrMo steel substrates. It was found that the introduction of Y_2O_3 accelerated the decomposition of metal carbides to form gaseous products, thereby promoting the formation of a porous structure within the coating matrix. When the Y_2O_3 content reached 2.5 wt.%, the coating achieved a hardness of 861.97 HV, a friction coefficient of 0.675 and a wear volume of 1.8×10^{-3} mm^3. This sample also exhibited excellent electrochemical corrosion resistance, with a corrosion potential of -0.704 V and a corrosion current density of 0.013 mA cm^{-2}. The usage of additives, as well as their distribution in matrices, impacts material physicochemical properties [2]. Wang et al. (Contribution 3) reported the relationship between the distribution of Tb and Ga in sintered Nd-Fe-B magnet matrices and their magnetic properties. It was found that the introduction of Ga promoted the diffusion of Tb within the magnet matrices, thereby enhancing their coercivity, and the simultaneous introduction of Tb and Ga into the magnet matrices induced the formation of a $(Nd,Tb)_2Fe_{14}B$ shell–core structure, improving the matrices' coercivity by 53.15%.

Post-processing time is another critical factor affecting the physicochemical properties of materials [3]. Yan et al. (Contribution 4) discussed the effect of chromizing time on the composition and microstructure of chromized layers coated on GCr15 bearing steel by powder pack chromizing. It was found that extending the chromizing treatment time hardly altered the distribution profiles of Cr, Fe, and C, or the main crystalline phases (i.e., $(Cr,Fe)_{23}C_6$ and $(Cr,Fe)_7C_3$) of the coatings, but significantly increased their thickness. The

https://doi.org/10.3390/molecules31091474

hardness and brittleness of the coatings increased in tandem with the chromizing time, whereas the adhesion strength between the chromized coating and the GCr15 bearing steel substrate deteriorated. As compared to the bare GCr15 bearing steel substrate, the wear resistance of the chromized sample was improved by approximately five times under the optimal chromizing time due to the high hardness of the coatings and the robust metallurgical bonding between the chromized coating and the GCr15 substrate. In addition, Niu et al. (Contribution 5) applied Nb as a strengthening additive to regulate the microstructure of NiTi alloy by the formation of dual-phase networks, thereby enhancing the electrochemical performance of NiTi alloy bipolar plates for proton exchange membrane fuel cells. Electrochemical analysis revealed that the corrosion resistance of the Nb-doped NiTi alloy was significantly improved. Under typical simulated operating conditions for proton exchange membrane fuel cells, its corrosion current density and interfacial contact resistance accounted for only 74% and 71% of those of the NiTi alloy, respectively.

On the other hand, doping with heterogeneous elements is also one of the most important methods to improve the application performance of materials [4]. Wu et al. (Contribution 6) reported the physicochemical and electrochemical properties of Bi^{3+}-doped Ti/Sb-SnO_2/PbO_2 electrode prepared by electrodeposition. It was found that Bi^{3+} doping endowed the electrode with excellent hydrophilicity (a contact angle of 21.6°), a long service life of approximately 2000 h, a high electrochemical active area corresponding to a total voltammetric charge of 21.20 C cm^{-2}, and outstanding electrocatalytic activity for the oxygen evolution with an onset potential as low as 1.80 V. Its superior electrochemical performance in zinc electrowinning was attributed to its high electron transfer capability, reflected by the lowest charge transfer resistance of 7.30 Ω. In addition, first-principles calculations were employed to reveal the underlying mechanism responsible for the enhanced electrochemical performance of Bi^{3+}-doped electrodes. The calculated results demonstrated Bi^{3+} doping reduced the band gap of the as-prepared electrode, thereby improving its electronical conductivity. Moreover, the formation of OOH determined the reaction rate of the oxygen evolution reaction on the surface of the electrode, with a free energy barrier of 2.23 eV. In addition, hydrophilicity is employed to evaluate the performance of electrode materials in a redox flow battery [5]. For example, Mei et al. (Contribution 7) prepared highly hydrophilic nanorod-like NiMoS-modified carbon felt composites using one-step hydrothermal method, with a contact angle of 140°. Electrochemical measurements revealed that the charge transfer resistance of the as-prepared composites was 524 mΩ. When employed as the anode in a polysulfide-ferricyanide redox flow battery, the composites delivered an energy efficiency of 70% and a voltage efficiency of 87% in the first cycle, at a current density of 40 mA cm^{-2}. Under the same current density, the assembled battery maintained an energy efficiency of 40% after 2500 charge–discharge cycles, with a retained average coulombic efficiency as high as 99.9%.

Due to their large specific surface area, high porosity, well-defined crystal structure and tunable functionalized surfaces and good biological compatibility, metal–organic frames (MOFs) and their derivatives have attracted considerable attention in energy storage and electrochemical sensing applications [6]. Therefore, Suganthi et al. (Contribution 8) systematically reviewed the synthetic strategies of MOFs and MOF/MXene hybrids, as well as their supercapacitive performance and electrochemical detection performance for various hazardous substances, including heavy metal ions and picric acid. Meanwhile, based on the self-dissociation feature of metallic Ni in terephthalic acid solution, Li et al. (Contribution 9) prepared Ni-MOF microbelt arrays and applied them as the electrode materials for supercapacitors. It was found that the unique array structure endowed the as-prepared microbelt arrays with excellent electrochemical properties, achieving a

https://doi.org/10.3390/molecules31091474

capacitance of 1124 F g^{-1} at 1 A g^{-1}, while still retaining 590 F g^{-1} even at a high current density up to 10 A g^{-1}.

Apart from the above-mentioned contributions, Li et al. (Contribution 10) investigated the effect of the corona discharge treatment on the microstructure and chemical properties of polyethylene using various spectroscopic characterization techniques. Prolonging the increased the thickness of the corona degradation layer was found, which led to a decrease in the S parameter and an increase in the W parameter. This phenomenon was attributed to the reduction in free volume elements and the formation of oxygen-containing groups, including hydroxyl, carbonyl and ester groups. Furthermore, Ma et al. (Contribution 11) systematically reviewed the research progress of enzymatic electrochemical sensors for the detection of organophosphorus pesticides, with a focus on carbon-based, polymer-based, metal-based, metal compound-based, MOF-based and covalent organic framework-based electrochemical active materials, as well as their key performance indicators, including sensitivity, selectivity, storage stability, repeatability, reproducibility, anti-interference ability and recovery in real samples.

3. Conclusions

This Special Issue has collected important research findings across the following areas: (1) the influence of additive usage and distribution, postprocessing duration and second-phase formation on the physicochemical properties of alloy materials; (2) the effect of heterogeneous element doping and microstructure modulation on the electrochemical corrosion resistance of polycompounds; (3) the synthetic strategies and the supercapacitive performance of MOF-based materials; (4) the impact of corona discharge treatment on the physicochemical properties of polyethylene; and (5) the recent advances in the active materials of enzymatic electrochemical sensors for the detection of organophosphorus pesticides. The findings provide valuable references for researchers working in electrochemistry and materials chemistry. Currently, the third edition of the Special Issue entitled "Electroanalysis of Biochemistry and Material Chemistry" has been launched in Molecules. Researchers are welcome to submit their latest original work to this new Special Issue.

Acknowledgments: We express our sincere appreciation to the authors who contributed to this Special Issue and to the reviewers who made the effort to provide valuable feedback on the articles.

Conflicts of Interest: The author declares no conflicts of interest.

List of Contributions:

1. Jiang, D.; Wang, G.; Dong, W.; Hong, X.; Guo, C. Recent Advance on Metal Carbides Reinforced Laser Cladding Coatings. *Molecules* **2025**, *30*, 1820.
2. Jiang, H.; Jiang, D.; Guo, C.; Hong, X. The Influence of Y_2O_3 Dosage on the Performance of Fe60/WC Laser Cladding Coating. *Molecules* **2025**, *30*, 4598.
3. Wang, L.; Li, W.; Wang, X.; Deng, Z.; Gao, S. Enhanced Coercivity and Tb Distribution Optimization of Sintered Nd-Fe-B Magnets by TbF_3 Grain Boundary Diffusion Facilitated by Ga. *Molecules* **2025**, *30*, 594.
4. Yan, D.; Wei, C.; Tang, P.; Huang, S.; Lin, S.; Shi, Q.; Hong, X. Effect of Pack Chromizing on Microstructure and Tribological Properties of GCr15 Bearing Steel. *Molecules* **2025**, *30*, 3690.
5. Niu, Z.; Li, Y.; Li, Y.; Wang, X.; Pan, Y.; He, Z.; Zhang, G.; Wang, Z.; Zhou, Q. Improving Anti-Corrosion and Conductivity of NiTi Alloy Bipolar Plate Used for PEMFCs via Nb Alloying. *Molecules* **2025**, *30*, 3658.
6. Wu, J.; Kang, X.; Xu, S.; Wei, Z.; Xu, S.; Liu, K.; Feng, Q.; Jia, B.; Wang, Y. Influence of Bi^{3+} Doping on Electrochemical Properties of Ti/Sb-SnO_2/PbO_2 Electrode for Zinc Electrowinning. *Molecules* **2024**, *29*, 4062.

7. Mei, D.; Liu, B.; Ma, H.; Zhang, Z.; Wu, F.; Chen, Y.; Ali, J.; Xing, F.; Xiong, L. NiMoS-Modified Carbon Felt Electrode for Improved Efficiency and Stability in a Neutral S/Fe Redox Flow Battery. *Molecules* **2025**, *30*, 1219.
8. Suganthi, S.; Ahmad, K.; Oh, T.H. Progress in MOFs and MOFs-Integrated MXenes as Electrode Modifiers for Energy Storage and Electrochemical Sensing Applications. *Molecules* **2024**, *29*, 5373.
9. Li, H.; Li, Y.; Song, S.; Tian, Y.; Feng, B.; Li, B.; Liu, Z.; Zhang, X. Facile Growing of Ni-MOFs on Ni Foam by Self-Dissociation Strategy for Electrochemical Energy Storage. *Molecules* **2025**, *30*, 513.
10. Li, J.; Shen, Z.; Tie, L.; Long, T.; Zhong, Q.; Chen, X.; Yin, C.; Liufu, L.; Huang, X.; Xiong, B.; et al. Surface Microstructure Study on Corona Discharge-Treated Polyethylene Using Positron Annihilation Spectroscopy. *Molecules* **2024**, *29*, 4147.
11. Ma, S.; Chen, Z.; Yang, F.; Yao, T.; Wang, S.; Yu, Y.; Xiong, L.; Hong, X.; Wang, G. Advances in Active Materials of Enzymatic Electrochemical Sensors for Detecting Organophosphorus Pesticides. *Molecules* **2026**, *31*, 717.

References

1. Zhou, X.; Li, Y.; Sun, C.; Chen, L.; Fang, F.; Jiang, J. Optimization on mechanical properties of transition metal carbides: A combined experimental and theoretical study. *Mater. Chem. Phys.* **2022**, *282*, 125955. [CrossRef]
2. Moghimian, P.; Poirié, T.; Habibnejad-Korayem, M.; Zavala, J.A.; Kroeger, J.; Marion, F.; Larouche, F. Metal powders in additive manufacturing: A review on reusability and recyclability of common titanium, nickel and aluminum alloys. *Addit. Manuf.* **2021**, *43*, 102017. [CrossRef]
3. Malakizadi, A.; Mallipeddi, D.; Dadbakhsh, S.; M'Saoubi, R.; Krajnik, P. Post-processing of additively manufactured metallic alloys—A review. *Int. J. Mach. Tools Manuf.* **2022**, *179*, 103908. [CrossRef]
4. Xu, X.; Cui, T.; McConohy, G.; Jagad, H.D.; Lee, S.S.; Wang, S.; Melamed, C.; Yang, Y.; Barks, E.; Kaeli, E.; et al. Heterogeneous doping via nanoscale coating impacts the mechanics of Li intrusion in brittle solid electrolytes. *Nat. Mater.* **2026**, *25*, 627–634. [CrossRef] [PubMed]
5. Zhang, L.; Feng, R.; Wang, W.; Yu, G. Emerging chemistries and molecular designs for flow batteries. *Nat. Rev. Chem.* **2022**, *6*, 524–543. [CrossRef] [PubMed]
6. Abánades Lázaro, I.; Chen, X.; Ding, M.; Eskandari, A.; Fairen-Jimenez, D.; Giménez-Marqués, M.; Gref, R.; Lin, W.; Luo, T.; Forgan, R.S. Metal–organic frameworks for biological applications. *Nat. Rev. Methods Primers* **2024**, *4*, 42. [CrossRef]

https://doi.org/10.3390/molecules31091474

molecules

MDPI

Review

Recent Advance on Metal Carbides Reinforced Laser Cladding Coatings

Dazhi Jiang [1], Guangjin Wang [2], Wei Dong [3], Xiaodong Hong [2,*] and Chenguang Guo [1,*]

[1] School of Mechanical Engineering, Liaoning Technical University, Fuxin 123000, China
[2] School of Materials and Energy, Foshan University, Foshan 528000, China
[3] School of Materials Science and Engineering, Liaoning Technical University, Fuxin 123000, China
* Correspondence: hongxiaodong@lntu.edu.cn (X.H.); gchg_neu@163.com (C.G.)

Abstract: The laser cladding technique can be adapted to fabricate composite coatings on the surface of the metal substrate, which not only effectively improves the surface properties of materials, but also greatly expands their application range. Metal carbides exhibit extremely high hardness, melting point, and outstanding chemical stability. The hardness of carbides is much higher than that of general metal materials. Therefore, various metal carbides serve as reinforcing agents for enhancing the overall performance of metal-based coatings. To date, there is no special review about metal carbide-reinforced laser cladding coatings. In view of the outstanding performance and wide application of metal carbides in laser cladding coatings, herein, recent advances in various metal carbide-reinforced metal coatings are highlighted. According to the type of metal carbides, the whole review is classified into five sections: WC-reinforced coatings, TiC-reinforced coatings, NbC-reinforced coatings, $Ti_{n+1}AlC_n$ (MAX) reinforced coatings, and Cr_3C_2, TaC-reinforced coatings. The preparation method, microstructure feature, and application performance of various carbide-reinforced composite coatings are summarized. At last, some prospects are put forward on the current issues and future development directions, aiming to provide comprehensive and in-depth references for the research and application in the field of composite coatings.

Keywords: laser cladding coating; metal carbides; reinforcing agent; composite coating

1. Introduction

With rapid technological development, the requirements for material performance in various industrial fields are becoming increasingly stringent. Traditional materials often struggle to meet the high-performance requirements of wear resistance, corrosion resistance, and high-temperature resistance under complex working conditions. Therefore, material surface modification technology has emerged as a key strategy to improve the comprehensive performance of materials. Laser cladding, as an advanced surface modification technology, has received much attention in recent years. It can accurately prepare coatings with specific properties on low-cost substrate materials, which not only effectively improves the surface properties of materials, but also greatly expands their application range [1]. When a high-energy laser beam is irradiated onto the surface of the substrate material, a huge amount of energy is instantly released, causing a very small area of the substrate surface to rapidly heat up to a molten state. At the same time, the alloy powder is transported to the laser action area through a powder feeding device, and the alloy powder quickly melts at high temperatures and is fully mixed with the melted matrix material. With the movement of the laser beam, the molten pool rapidly cools and solidifies, forming a layer of fusion coating on the surface of the substrate that is firmly metallurgically bonded to the substrate. This metallurgical bonding method ensures good bonding strength between the cladding layer and the substrate, effectively transmitting stress and jointly bearing external loads [2]. With the continuous advancement of laser technology, from the initial simple preparation of single component cladding layers to the development of complex multi-element alloy systems and gradient functional coatings today, laser cladding technology has become increasingly important in the field of material surface modification, continuously driving the technological upgrading and development of related industries.

The performance of laser cladding coatings is closely related to their composition, which is complex and diverse. Different components work together to determine the various characteristics of the coating [3]. Laser cladding coatings are mainly composed of substrate, cladding material, and additives. The substrate material is the fundamental support for laser cladding coatings, which not only provides physical support for the coating, but also has a significant impact on the bonding strength between the coating and the substrate. Usually, the substrate material is the workpiece itself that needs surface strengthening or modification, commonly including various metal materials such as steel, aluminum alloy, titanium alloy, etc. Cladding materials are the core part of laser cladding coatings, and they come in various types. Different types of cladding materials can be selected according to different usage requirements, such as metal powder, ceramic powder, and composite material powder. During a laser cladding process, special additives, such as deoxidizer, alloying element, and activator, are often added to improve the quality and performance of the coating [4].

Ceramic materials have the advantages of high hardness, high melting point, high-temperature resistance, wear resistance, corrosion resistance, etc. Therefore, ceramic powders have also been widely served as reinforcing agents for preparing high-performance composite coatings. Commonly used ceramic powders include aluminum oxide (Al_2O_3), zirconium oxide (ZrO_2), tungsten carbide (WC), etc. Among the existing ceramic powders, metal carbides exhibit extremely high hardness, high melting point, and good chemical stability. The hardness of some carbides such as WC and TiC is much higher than that of general metal materials. When these carbides are added to the laser cladding coating, the overall hardness of the coating is significantly improved [5]. In addition, in the actual process of friction and wear, hard carbide particles act as a hard shield to resist the cutting and plowing effects of abrasive particles. Even under harsh wear conditions, carbide particles still maintain relative stability, effectively protecting the coating substrate and greatly reducing the wear rate of the coating. This makes metal carbide-enhanced laser cladding coatings have significant advantages in applications that require high wear resistance, such as mining machinery, mold manufacturing, etc. In fact, metal carbides in laser cladding coatings play three functions: diffusion strengthening effect, refining grain effect, and improving interface integration. Thus, metal carbides are widely used to enhance the performance of laser cladding coatings, and metal carbide-enhanced laser cladding coatings have become one of the research hotspots in the field of materials science.

To date, there is no special review about metal carbide-reinforced metal composite coatings. In view of the outstanding performance and wide application of metal carbides in laser cladding coatings, herein, the research progress of various metal carbide-reinforced composite coatings is highlighted. According to the type of commonly used metal carbides, five sections are divided, including WC-reinforced coatings, TiC-reinforced coatings, NbC-reinforced coatings, $Ti_{n+1}AlC_n$ (MAX)-reinforced coatings, and Cr_3C_2- and TaC-reinforced coatings, as shown in Figure 1. The preparation method, microstructure feature, and application performance of various carbide-reinforced metal composite coatings are summarized in detail.

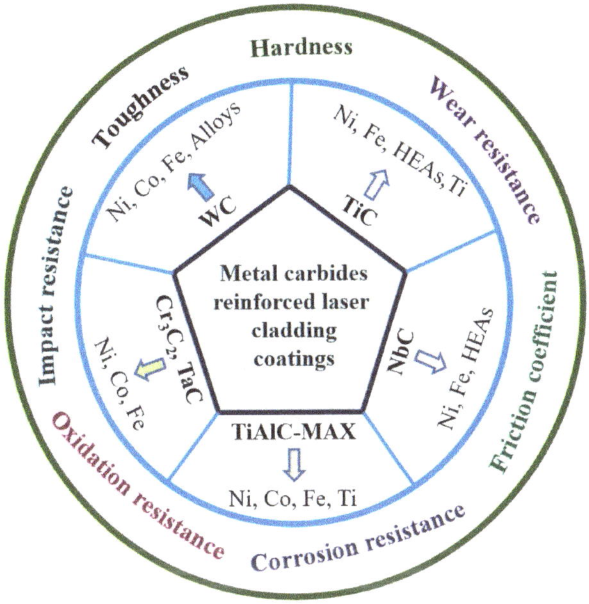

Figure 1. The scheme of various metal carbide-reinforced laser cladding coatings.

2. WC-Reinforced Composite Coatings

The Mohs hardness of WC is between 8.5 and 9.5, second only to diamond, which makes it perform well in various wear environments. Its high hardness enables WC-containing coatings to effectively resist abrasive wear, adhesive wear, and fatigue wear. Furthermore, the coatings containing WC exhibit superior corrosion resistance, high-temperature resistance, good bonding strength, and surface smoothness [6]. Thus, WC-reinforced composite coatings have been widely used in fields such as mechanical components, cutting tools, and molds.

2.1. WC/Ni Composite Coatings

As an important means of material surface protection and strengthening, Ni-based coatings have numerous significant advantages. Excellent corrosion resistance is the most prominent characteristic of Ni-based coatings. Ni itself has good chemical stability. In different corrosive environments, Ni-based coatings can form dense and stable passivation films, effectively preventing external corrosive media from contacting the substrate material and greatly extending the service life of the substrate material. Ni-based coatings also have excellent high-temperature resistance [7]. Under high-temperature conditions, Ni-based coatings can maintain good mechanical properties and chemical stability. In high-temperature working environments, Ni-based coatings not only protect the substrate material from oxidation but also maintain the structural integrity and mechanical properties of components, ensuring the normal operation of equipment. Good wear resistance is also a major advantage of Ni-based coatings. By designing the coating composition and microstructure reasonably, Ni-based coatings can have high hardness and toughness. During the process of friction and wear, Ni-based coatings can effectively resist wear, reduce the wear rate of material surfaces, and improve the wear resistance life of components. In addition, Ni-based coatings usually have good adhesion with the substrate material, which can firmly adhere to the surface of the substrate and is not easy to peel off. This ensures the reliability and stability of the coating during long-term service. Ni-based coatings play an indispensable role in modern industry due to their many advantages [8]. All kinds of Ni/WC composite coatings have been reported to boost the mechanical performance of metal materials. To develop high-performance Ni/WC composite coatings, some basic issues have been explored, including the influence of WC content and particle size, the optimization of process parameters, the addition of WC and other ceramic phases, and so on.

To reveal the influence of WC content, Wang et al. [9] prepared Ni60A/WC coatings on the substrate of Inconel 718 alloy and revealed the influence of WC content on the WC/matrix micro-interface evolution, microhardness, and corrosion resistance. The microstructures of different samples are provided in Figure 2a,b. The result indicated that the coating with 35% WC presented the highest microhardness of 375 $HV_{0.5}$. While the coating of Ni60A + 15% WC exhibited the best corrosion resistance. Hu et al. [10] fabricate Ni/WC composite coatings on 304 steel through a high-speed laser cladding method, as described in Figure 2c. During the thermal processing, WC particles decomposed and generated the WC, W_2C, and $Cr_{23}C_6$ strengthening phases. The increasing WC fraction would refine the coating structure. When the WC content was less than 30%, the wear resistance and hardness of the coating were gradually improved with an increasing WC content. Liu et al. [11] fabricated Ni50-WC coatings and investigated the influence of WC content on morphology, corrosion, and mechanical performance. Through comparison, the coating of Ni50-20WC showed a higher microhardness (817.9 HV) and a lower friction coefficient than that of Ni50-10WC coating, due to the generation of fine and compact structure. Furthermore, the Ni50-20WC coating presented an outstanding corrosion resistance. Li et al. [12] prepared Ni45/WC coatings on the substrate of 42CrMoA to enhance the impact resistance and wear resistance. When the WC content was 12 wt%, the coating exhibited a lower wear rate (91.59%) than that of the Ni45 coating. However, the coating containing 12 wt% WC presented a poor impact resistance. Regarding the performance of the coating with a high content of WC, Shen et al. [13] fabricated three NiCrSiBC/WC composite coatings on the carbon steel with a WC ratio of 40%, 50%, and 60%. When the content of WC was enhanced to 50% from 40%, the tensile stress of coating increased and the fracture toughness significantly reduced, causing an increase in crack susceptibility. When it reached 50%, coating hardness would not increase with an increase in WC fraction, while the cracking susceptibility increased, which was attributed to more serious cracks generated.

The influence of WC particle size on the performance of Ni/WC coating on 304 steel was disclosed by Tan [14]. Three kinds of particle sizes were selected as fillers, including 100–145 μm, 65–100 μm, and 30–65 μm. The content of WC was fixed as 40% and the matrix was Ni1725 powder. Through comparison, the Ni/WC coating made by the size of 30–65 μm exhibited the highest microhardness, the lowest friction coefficient, and wear volume, among the three coatings fabricated by different WC particles.

Except for the study on the WC content and particle size, the structure design of WC-reinforced coating is considered a crucial factor for determining the performance. In this respect, Wang et al. [15] fabricated Ni-WC gradient composite coating on the substrate of Q345R steel. The first layer was C276/10WC, and the subsequent layers were Ni60/10WC, Ni60/30WC, and Ni60/50WC. Gradient composition coating effectively reduced the cracking tendency. The microhardness of the resulting gradient coating reached 1053.5 $HV_{0.2}$, about four times that of the Q345R. Furthermore, the wear loss was 82.7% of the Q345R.

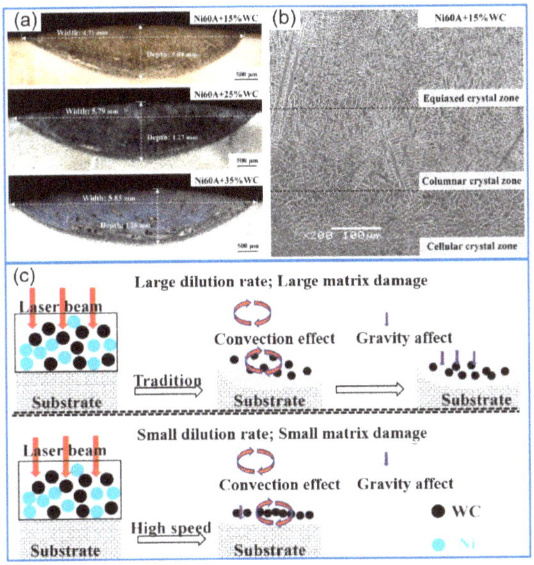

Figure 2. Effect of WC content on the composite coatings (**a**) and microstructure distribution (**b**) [9]; Copyright (2024) Elsevier. Convection and gravity model of Ni/WC composite coating during solidification, traditional laser cladding and HSLC (**c**) [10]; Copyright (2022) Elsevier.

In addition to the influence of coating composition, the effect of process parameters on the performance was also investigated. For example, Bian et al. [16] fabricated Ni60-WC composite coatings on H13 steel and discussed the influence of powder feeding rate on the microhardness of coating. The result demonstrated that the feeding rate affected the generation of new compounds in composite coating. When operated at 25 g/min, more compounds were produced, and the resulting coating presented the maximum microhardness of 936HV, which was 2.4 times that of the substrate H13.

To overcome the shortage of single WC, other ceramic phases or rare-earth activators have been introduced to further enhance the mechanical performance of laser cladding coatings. Li et al. [17] designed Ni62/WC+ TiC composite coatings by cladding on 65Mn steel. The ratio of WC/TiC was fixed as 3:1, and the mass fraction of the reinforcing phase was 10%, 20%, and 30%. The coating containing 20 wt% reinforcing phases had the highest microhardness of 1000 HV1, three times higher than that of 65Mn steel. Moreover, this coating also presented the lowest wear rate. In addition, Cr_3C_2 and WC acted as reinforcing phases for fabricating Ni/WC + Cr_3C_2 coating on the surface of Q235 steel [18]. When 5% Cr_3C_2 was introduced into Ni/WC, the plastic deformation of the Ni-3 coating was prohibited. Furthermore, this coating exhibited the lowest weight loss and the highest microhardness of 831$HV_{0.1}$. In Chen's work [19], in situ-generated NbC enhanced Ni-WC composite coating was prepared on AISI 1045 steel, under the presence of Nb, C, WC, and Ni50A powder. Through adjusting the ratio of Nb and C, different amounts of NbC were in situ-synthesized in composite coating. The result confirmed that the optimum Nb/C ratio was 1: 2, and the resulting coating possessed the best wear resistance and hardness. They also fabricated (Nb, M)C (M = Ti, V, and Zr) reinforced Ni-WC coatings by introducing other metal carbides [20]. A core-shell structure was formed in $(Nb_{0.5}Ti_{0.5})C$ coating with W_pC-shell and $(Nb_{0.5}Ti_{0.5})C$-core, and this coating possessed the best wear resistance and hardness. In addition to the reinforcing agents, rare-earth La_2O_3 can be used to improve the corrosion resistance and wear of the WC-containing coatings. In this respect, Zhao et al. [21] prepared La_2O_3-doped Ni60/WC coatings on the substrate of AISI 304 steel.

Compared with the coating of Ni60/WC+ nano-La_2O_3, the micron-La_2O coating had a higher microhardness, better corrosion resistance, and wear resistance, due to the fine grain strengthening effect.

2.2. WC/Co, WC/Fe Composite Coatings

In addition to the WC/Ni-based composite coatings, WC-reinforced Co-based and Fe-based coatings have been widely reported in recent years. For example, Jiang et al. [22] prepared the WC-Co composite coatings on Ti-6Al-4 V alloy substrate by laser cladding. In their work, WC particles were uniformly wrapped by the Co phase and denoted as WC-17Co, with a diameter of 80 μm. Thanks to the synergistic role of fine microstructure and metallurgical bonding, the microhardness of coating material reached 1536 $HV_{0.5}$, and the wear rate was as low as 1.5 g/h. Lv et al. [23] fabricated WC/Co-based coatings on 42CrMo steel by using the laser cladding technique, and discussed the influence of direct addition and in situ synthesizing WC on the coating performance. Through comparison, the optimal coating made by the direct addition method exhibited a higher microhardness and better wear resistance. The wear resistance and microhardness of optimal coating were 3.75 times and 1.46 times higher than that of 42CrMo steel, respectively. In another work, various WC-12Co coatings were prepared on the Inconel 718 substrate [24], and the WC-12Co fraction ranged from 0 to 30 wt%. The introduction of WC-12Co would refine the microstructure and prevent the growth of columnar grains, as shown in Figure 3a, further improving the hardness and wear resistance. Through comparison, the coating containing 30%WC-12Co had the highest microhardness of 462.63 $HV_{0.5}$ and the best wear resistance. In addition to the simple composite coating of Co/WC, Ding et al. [25] developed a Co-based alloy/WC/CaF_2 composite coating with a mass ratio of 85/10/5. During the laser cladding process, the F element was dissolved in the α-Co phase and the coating was composed of $Cr_{23}C_6$, α-Co, and WC phases. The composite coating had a high microhardness of 600 $HV_{0.2}$. Furthermore, the resulting composite coating greatly enhanced the fretting wear resistance at both room temperature and high temperature of 500 °C.

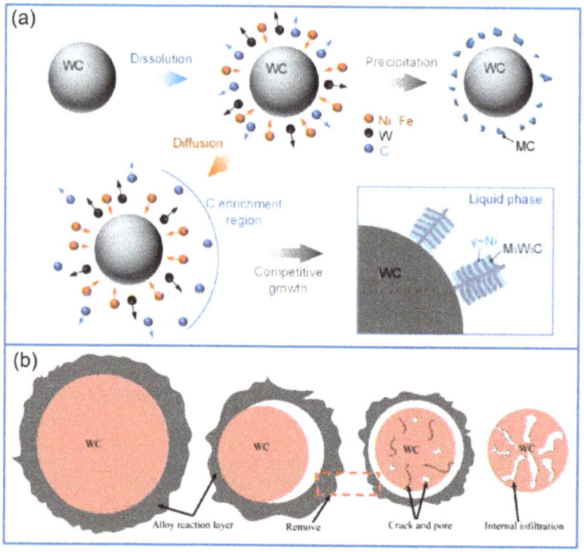

Figure 3. Scheme of carbide formation mechanism (**a**) [24]; Copyright (2021) Elsevier. The slight reduction in the size of WC particles, and defects such as ablation cracks and pores appear on WC particles (**b**) [26]. Copyright (2021) Elsevier.

Compared with Ni-based and Co-based alloy powders, Fe-based alloy powders have the advantages of high performance and low cost. There are lots of reports about WC-reinforced Fe-based coatings. In this field, Xiao et al. [26] fabricated Fe-WC coatings on 15CrNiMo steel via laser cladding and discussed the influence of WC fraction on the microhardness, structure, and wear resistance of the composite coating. The addition of WC greatly enhanced the microhardness of Fe-based coating, even increased to 1029.2 $HV_{0.2}$ from 729.9 $HV_{0.2}$. Furthermore, the wear resistance of WC-Fe composite coating was 1.3 times that of Fe coating. The transition mechanism of WC particles is presented in Figure 3b. The enhancement of wear resistance was ascribed to the presence of WC, W_2C, M_7C_3, $M_{23}C_6$, and η phases. By laser cladding the mixed powders of Fe60 and WC, Li et al. [27] in situ synthesized Fe-(Cr, W)$_{23}C_6$-WC composite coatings on the substrate of 16Mn steel. When the WC content was 20 wt%, the composite coating had the maximum hardness and the best wear resistance. The improved performance was attributed to the in situ generation of network-shaped (Cr, W)$_{23}C_6$-WC ceramic phases in α-Fe dendritic matrix, and the evolution process of phase was $WC_{primary}$ + $liquid_0$→$WC_{in\ situ}$ + $liquid_1$→$WC_{in\ situ}$ + $liquid_2$ + γ-Fe→$WC_{in\ situ}$ + $M_{23}C_6$ + α-Fe. Based on the Fe/TiC composite coating, Zeng et al. [28] prepared Fe50/TiC/WC ternary coating on the substrate of AISI 1045 steel and investigated the influence of the WC fraction on microhardness and the friction coefficient. Through optimization, the Fe50/TiC/WC ternary coating containing 10% WC presented the best performance. Compared with the pristine substrate, the microhardness was enhanced three times, the friction coefficient was 27%, and the wear volume was only 1.5%.

2.3. *WC/Other Metal Composite Coatings*

H13 is a commonly used tool steel with excellent wear resistance and high-temperature strength. Based on the H13 powder matrix, WC and Y_2O_3 powder were selected as reinforcing phases, and H13/WC + Y_2O_3 composite coating was reported to modify 8407 steel [29]. The presence of Y_2O_3 promoted the decomposition of WC, leading to the formation of uniform coating. By adjusting the content of Y_2O_3 and WC, the coating made by H13 + 1%Y_2O_3 + 10%WC exhibited the best performance. The hardness at room temperature and 500 °C was 48.4% and 32.3% higher than that of the 8407 steel, respectively. The improved hardness was ascribed to the generation of the M_7C_3 strengthening phase.

Eutectic high-entropy alloys (EHEAs) have been considered a new type of alloy with a promising application potential, due to the combination properties of high-entropy alloys and eutectic alloys [30]. To further enhance the mechanical performance of EHEAs, WC-reinforced coating has been fabricated on their surface via laser cladding technique. For example, Li et al. [31] developed $AlCoCrFeNi_{2.1}$/WC EHEAs composite coatings for modifying the 316 L stainless steel. With an increasing WC content, the $Cr_{21}W_2C_6$ and Cr_7C_3 phases were gradually generated in the composite coating. The composite coating containing 30 wt% WC possessed a maximum hardness of 572.3 $HV_{1.0}$ and the best wear resistance.

Invar alloy, also known as low expansion alloy, is a magnetic metal alloy composed of iron (Fe) and nickel (Ni). Its main components are 36% nickel and 64% iron, with a face-centered cubic structure. The outstanding feature of Invar alloy is its extremely low coefficient of thermal expansion, which makes it suitable for applications that require extremely high precision. Based on the matrix of Invar alloy powders, WC and Y_2O_3 were used as reinforcing agents for improving the surface properties of the Invar alloy substrate [32]. When Y_2O_3 powder is added, the coating thickness and melting efficiency can be improved. Moreover, a high fraction of Y_2O_3 would induce the transformation from cellular dendritic crystals to cellular crystals. When the Y_2O_3 fraction was 5%, the microhardness of the composite coating was 193.1 HV, which was much higher than that of the substrate.

To sum up, all kinds of WC-reinforced metal composite coatings have been introduced in this section. According to the types of metal matrix, several parts are classified, including WC/Ni, WC/Co, WC/Fe, and WC/other metal composite coatings. Compared with another metal matrix, many more works focus on the development of WC-reinforced Ni composite coatings. Furthermore, various alloy powders, such as H13, HEAs, and Invar alloy have been selected as metal matrices for fabricating WC-reinforced alloy coatings. Among this literature, the research direction mostly involves the effect of WC content and particle size, the optimization of process parameters, coating structure design, a combination of WC and other ceramic phases or oxides, in situ synthesis of WC, and the selection of metal matrix, and so on. Regarding the performance enhancement of WC-reinforced composite coatings, most works have discussed the influence of WC addition on hardness, tensile stress and fracture toughness, wear resistance, corrosion resistance, and so on. Considering the difference in testing conditions and metal matrix, it is very difficult to compare the coating performance in different literature.

3. TiC-Reinforced Composite Coatings

TiC has the merits of high hardness, high strength, high melting point, corrosion resistance, and wear resistance, which is widely used in the mechanical industry, aerospace, building materials, and electrical fields. TiC-reinforced metal composite coatings exhibit excellent hardness and wear resistance, good high-temperature resistance, outstanding chemical stability, and low friction coefficient. The preparation methods of TiC-reinforced coatings can be divided into direct feeding method and in situ synthesis method. In this section, the recent advances in various TiC-reinforced laser cladding coatings are introduced, including Ni-based, Fe-based, high-entropy alloys based, and Ti-based coatings.

3.1. TiC/Ni Composite Coatings

Among the existing works about TiC/Ni composite coatings, based on the feeding method, direct feeding method, and in situ generation method can be classified. Furthermore, the research topics mostly focus on the TiC content and particle size, process parameters, in situ reinforcement of TiC, and so on. Regarding the direct addition of TiC, Cai et al. [33] discussed the influence of TiC fraction on the structure and performance of TiC/Ni60 composite coating. Various precipitates were generated in the coatings with the change in TiC content. As the TiC content increased, the wear resistance and bending strength both decreased. The main reason was attributed to the stress concentration around the large-size particles, and the generated cracks were easily extended, leading to brittle fractures. Meng et al. [34] investigated the effect of the nano-TiC addition method on the performance of Ni-based self-fusing alloy coatings. Through comparison, the Ni4 sample with chemical electroless coating of Ni-B-TiC presented the optimal wear resistance and the maximum microhardness, due to the bimodal structure generated by the nonuniform dispersion of nano TiC in the composite coating. To obtain the optimum composite coating, Chen et al. [35] discussed the influence of process parameters on the performance of TiC/Ni45 composite coating. In their work, 10 wt% TiC and Ni45 self-fluxing powder were cladding on the surface of 45 steel. The best parameters were obtained: powder feed rate 243.7 mg/s, laser powder 380 W, and scanning speed 5 mm/s. The microhardness of the resulting coating was about 2.5 times that of 45 steel.

Except for adding TiC powders directly, TiC can be in situ synthesized in the Ni matrix. In this field, Chen et al. [36] fabricated TiC/Ni functionally gradient coatings (FGC) by using Ni45 powder, Ti particles, and Ni-coated graphite, and the coating microstructure of the remelting zone is given in Figure 4a. The microhardness of the TiC/Ni coating gradually increased from the bottom to the top, and the maximum value reached 1036.25 HV. The wear rate gradually decreased from the bottom to the top, and the top of the coating exhibited the smallest wear rate of 4.872×10^{-6} $mm^3 \cdot N^{-1} \cdot m^{-1}$. To prepare TiC/Ni composite coating, Ti powder, and Ni-coated graphite were mixed with In625 matrix to in situ synthesize the TiC phase. The microstructure is shown in Figure 4b,c. The robust bonding between matrix and in situ-generated TiC would remove the plastic of conventional coatings by grinding method [37]. In addition, Wang et al. [38] fabricated in situ TiC-reinforced coating by using the mixed powders of 96 wt% TC4 and 4 wt% SiC. During the thermoforming process, SiC was decomposed to produce TiC in situ. The optimized C16 coating obtained at 1600 W exhibited the highest microhardness (351.8 $HV_{0.2}$), with a 38.1% wear rate of the substrate.

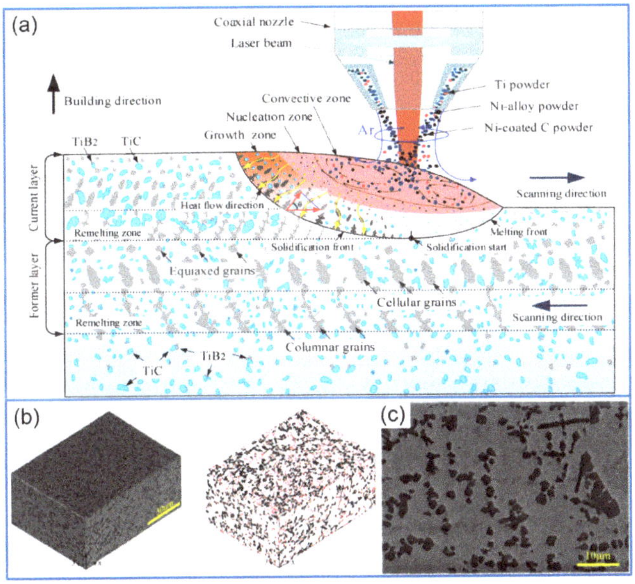

Figure 4. The effect of in situ TiC on the microstructure of the remelting zone between the layers of TiC/Ni coating (**a**) [36]; Copyright (2022) Elsevier. Microstructure of in situ TiC-reinforced coating (**b**,**c**) [37]. Copyright (2025) Elsevier.

To further enhance the mechanical property of TiC-reinforced coating, other ceramic phases can be in situ introduced into the metal matrix. For example, various fractions of TiB_2-TiC were in situ synthesized in the Ni-based alloy matrix by adjusting the dosage of Ni-coated graphite, Ti, and B_4C [39]. The in situ TiB_2-TiC coating possessed an enhanced corrosion resistance. Moreover, the wear volume was 46.4% lower than that of pure Ni-based coating. Qi et al. [40] also fabricated TiC-TiB_2 binary phases reinforced Ni-based coating by using Ni-coated graphite, TA0, and B_4C powders. A high Ni-coated graphite fraction would decrease the nucleation rate and the size of the ceramic phase. Through comparison, the reinforced coating prepared by 8 wt% Ni-coated graphite and 3 wt% B_4C presented the best wear resistance, and the wear volume had a 26.47% reduction. Moreover, the corrosion potential increased by 39%, and corrosion current density decreased significantly. Zheng et al. [41] synthesized TiB_2- and TiC-reinforced Ni50A composite coating on AISI 1045 steel by using the powders of Ni50A, Ti, and B4C. The influence of scanning rate, laser power, and Ti/B_4C powder ratio on the coating properties was investigated. By comparing the hardness and corrosion resistance of different coatings, the best process parameters were obtained. The Ti/B_4C ratio was 4:1, the scanning rate was 8 mm/s, and the laser power was 1200 W.

In addition to the Ni powers, various Ni-based alloy powders have been selected as matrices to prepare TiC/Ni-based alloy composite coating. For example, to improve the erosion resistance and hardness of AISI 420 stainless steel, NiCr-TiC composite coating was prepared by laser cladding of NiCr-TiC powders on a stainless steel surface [42]. The result demonstrated that the hardness of the coating significantly increased with the addition of TiC particles. During the experiment in the erosion test, the NiCr-TiC coating presented the lowest mass loss, while pure AISI 420 stainless steel had the maximum mass loss. In addition, TiC/Inconel 625 composite coating was fabricated by extremely high-speed laser cladding technology (EHLA) toto improve the corrosion resistance of AISI 1045 [43]. The high-speed cladding accelerated the solidification rate of the coating and the generation of carbides. The generated CrC and NbC grains effectively blocked the corrosion channels and improved anti-corrosion performance. Compared with the coating prepared by conventional laser cladding, the EHLA coating exhibited better corrosion resistance.

3.2. *TiC/Fe Composite Coatings*

TiC can be used for enhancing the mechanical performance of Fe or stainless steel matrix. In this respect, Zhu et al. [44] discussed the influence of TiC content on the reinforced 410 martensitic stainless steel (MSS) coatings. When the TiC content was up to 15%, pores and microcracks would be formed in the coating. The increasing TC content in the coating would enhance the wear resistance and microhardness. Through comparison, the coating containing 10% TiC possessed the best comprehensive performance, much better than that of pure MSS coating. Hui et al. [45] fabricated WC-10Co_4Cr-xTiC coatings on the substrate of H13 steel by adding TiC into WC/10Co_4Cr powders. The relationship between the wear resistance and TiC fraction was disclosed. The result showed that the porosity of the coating was reduced after adding TiC, and the composition of the coating was W_2C, WC, Cr_7C_3, Co_3W_3C, and TiC. The friction coefficients increased and the wear rate decreased.

Based on the matrix of ferrite stainless steel (SMC) powder, Ti powder and Cr_3C_2 powder were introduced to in situ synthesize TiC-reinforced SMC coatings [46]. The generation of TiC was related to the replacement of Cr in the reaction process, which led to the improvement of the corrosion resistance of the coating. Furthermore, with an increase in TiC content, the wear resistance and hardness of composite coating were greatly improved. In addition to generating a single TiC-reinforcing phase, Yang et al. [47] fabricated TiC and VC binary phases reinforced Fe-Cr-C alloy (2Cr13), in which, nickel-coated graphite, Ti, and V powders (C/Ti/V molar ratio of 2:1:1) were used to in situ generate composite coating with a high wear resistance. The optimized composite coating containing the 5% group exhibited better wear-resisting properties. The abrasion resistance was even enhanced to more than 21 times higher than that of semi-steel roll material (160CrNiMo) substrate. Jiang et al. [48] prepared TiC/NbC coating on 304 stainless steel by using the 304 powder, B_4C powder, TC4 powder, and Nb powder. The hard phases Cr_7C_3 and $Cr_{23}C_6$ were produced with an increasing (Ti, Nb) C-reinforcing agent, which facilitated the enhancement of hardness. In addition, the friction coefficient and wear rate of the coating were affected by the laser power. In another work [49], B4C, Ti powder, and 316 L stainless steel powder were used to in situ synthesize TiC/TiB_2-reinforced composite coatings, and the resulting composite coating contained the phase of TiB_2, TiC, Fe_2B, Cr_3C_2, and Fe_3C. The 2# coating prepared by adding 5%B_4C-15%Ti composite powder exhibited the highest microhardness of 961.94 $HV_{0.3}$, and the sample also presented the smallest volume loss and smallest friction coefficient among all samples.

3.3. TiC/High Entropy Alloys Composite Coatings

TiC-reinforced high-entropy alloys (HEAs) coatings have been well focused in recent years. In this field, Chen et al. [50] developed TiC-reinforced FeCoCrNiCu HEA composite coatings by adjusting the fraction of TiC particles, and revealed their wear resistances at RT and 600 °C, respectively, as shown in Figure 5a. The coating with 15 wt% TiC exhibited a superior wear resistance at RT, and the microhardness was 531.65 $HV_{0.5}$. When tested at 600 °C, TiO_2 and the Cr_2O_3 tribo-film generated on the coating surface effectively resisted the friction by facilitating the lubrication. In addition to directly adding TiC particles, TiC particle-reinforced AlCoCrFeNi-based coatings were in situ synthesized on the substrate of 40CrNiMo by using the Ti powder and Cr_3C_2 [51]. The in situ-generated TiC particles greatly improved the wear resistance and hardness. In addition, Gao et al. [52] fabricated in situ TiC-reinforced CoCrFeNi HEA composite coatings by using Ti powder, C powder, and CoCrFeNi HEA powder. During the laser cladding process, the Cr_7C_3 and TiC phases were formed in the coating. Different from directly adding TiC particles, in situ-generated TiC promoted the nucleation rate and facilitated the refinement of grain size (Figure 5b,d). The enhanced microhardness and wear resistance were ascribed to the fine grain, second phase strengthening, and solid solution strengthening. Li et al. [53] prepared WC/TiC composite reinforced AlCoCrFeNi coating by adding Ti powder and WC, and investigated the microstructure, wear resistance, and hardness of the resulting coating. Due to the presence of WC and in situ-generated TiC particles and solid solution strengthening, the hardness and wear resistance of AlCoCrFeNi-WC/TiC coating enhanced to 808.7 $HV_{0.3}$ and 14 times, respectively. In addition, W and in situ TiC-reinforced CoCrFeNi HEA coatings were synthesized on 45 steel [54]. The quality of composite coating and its hardness and wear resistance were significantly improved by adding W and in situ TiC, due to the synergistic effect between TiC and W. The highest hardness of the coating reached 1122.43 $HV_{0.2}$.

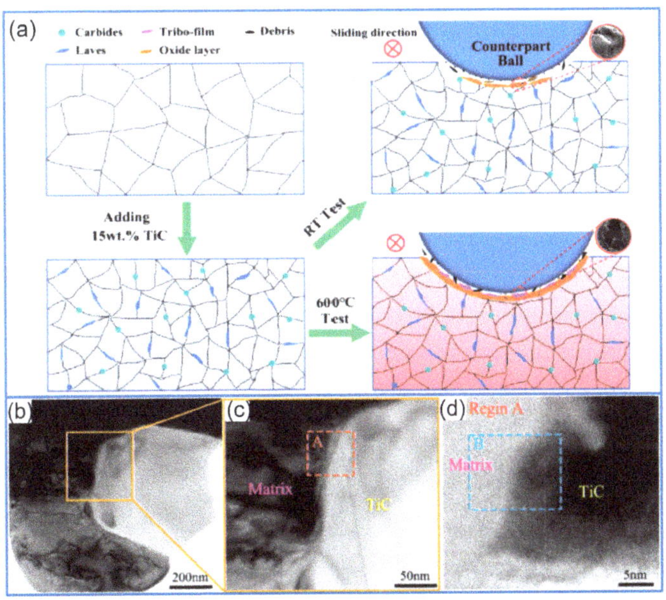

Figure 5. Microstructure evolution for TiC-free coating and 15 wt% TiC coating, and the wear behaviors at different temperatures (**a**) [50]; Copyright (2024) Elsevier. TEM images of TiC and HEA matrix (**b**), and corresponding magnified region (**c**,**d**) [52]. Copyright (2024) Elsevier.

Except for the change in TiC fraction and the type of HEAs, Zhuang et al. [55] investigated the influence of ultrasound on TiC distribution in 4 wt% TiC-reinforced CrMnFeCoNi coatings. Ultrasonic amplitude greatly affected the distribution of TiC particles. A low amplitude of 5 μm would lead to the particle aggregation. When enhanced to 10–20 μm, the fishbone-shape distribution of TiC was obtained. In addition, the increased ultrasonic amplitude would increase the hardness and reduce the corrosion resistance of the composite coating.

3.4. TiC/Ti Alloys Composite Coatings

Among the existing metal materials, Ti and Ti alloys can serve as biomedical materials and structural materials for chemical engineering, shipbuilding, aerospace, etc. Among Ti alloys, TC4 (Ti-6Al-4V) exhibits superior corrosion resistance, heat resistance, and strength. However, the poor wear resistance and low hardness further limit their wide application. Considering the good wettability of Ti alloys and TiC-reinforcing agents, Chen et al. [56] fabricated TiC/TiAl coatings on TC4 substrate and revealed the influence of TiC fraction on the microstructure and performance of coating. Four phases of TiAl, Ti_3Al, TiC, and Ti_2AlC were formed in the coating. Through comparison, the coating containing 5 vol% TiC presented the lowest friction coefficient and wear rate. While the 15 vol% coating had the highest hardness of 799 $HV_{1.0}$.

By using the Ti6Al4V alloy as the matrix, TiC/TiB composite biocompatible coatings were in situ synthesized with pure B_4C and pure Ti powder as reactants [57]. In fact, the coating was composed of the TiB, TiC, Ti_3Al, $TiVC_2$, and α-Ti phases. The introduction of TiB greatly improved the mechanical performance of the coating, including microhardness and fatigue wear resistance. In addition, Zhang et al. [58] prepared TiC_x-reinforced $CrTi_4$-based composite coatings on Ti6Al4V substrate by using the powders of NiCr-Cr_3C_2, CeO_2, and Ti6Al4V, and discussed the effect of CeO_2 fraction on the microstructure of the coating. The addition of CeO_2 did not affect the phase composition (TiC_x and $CrTi_4$), while effectively preventing the generation of pores and cracks. The decomposed CeO_2 would recrystallize to produce Ce oxides dispersed at the TiC_x/$CrTi_4$ boundary. Compared with the substrate of Ti6Al4V, the wear resistance and microhardness of TiC_x-reinforced $CrTi_4$-based coating were increased by 57% and 23% [59].

In summary, various TiC-reinforced metal composite coatings are summarized in this section, such as TiC/Ni, TiC/Fe, TiC/HEAs, and TiC/Ti alloy composite coatings. Unlike the WC-reinforced metal coatings, most of the reports have focused on the in situ synthesis of TiC by adopting different reactants, such as Ti powers, nickel-coated graphite, B_4C, Cr_3C_2 or WC, and so on. Compared with the coating prepared by directly adding TiC, in situ-generated TiC facilitated the nucleation rate and achieved the refinement of grain size. Moreover, solid solution strengthening and second-phase strengthening also contributed to the enhancement of hardness and wear resistance. Therefore, TiC has been widely adopted as the reinforcing agent for enhancing the hardness and wear resistance of laser cladding coatings. Considering the advantages of in situ-generated TiC, in addition to the TiC-reinforced phases, other reinforced agents are introduced into the metal matrix, such as WC, Cr_3C_2, TiB_2, NbC, VC, etc. Different from the WC, HEAs, and Ti alloys widely serve as the matrix for TiC, and lots of literature has reported the TiC/HEAs and TiC/Ti alloys composite coatings, further indicating the specialization and universality of TiC-reinforcing agents.

4. NbC-Reinforced Composite Coatings

As an important reinforcing agent, niobium carbide (NbC) has received widespread attention in the field of materials science, due to the advantages of high melting point

(3490 °C), high hardness (24.6 GPa), and outstanding mechanical properties. Adding NbC-reinforcing agent to metal materials can form small and uniformly distributed NbC particles. These particles effectively hinder dislocation motion, thereby significantly improving the strength and hardness of metal materials. To date, various NbC-reinforced metal coatings have been developed, including NbC/Ni composite coatings, NbC/Fe composite coatings, and NbC/HEAs composite coatings.

4.1. NbC/Ni Composite Coatings

NbC acts as a reinforcing agent for improving the mechanical performance of Ni-based coating by direct addition or in situ synthesis method. For example, Gaddam et al. [60] fabricated NbC/Ni composite coating through the in situ reaction between 90 wt% Ni and 10 wt% NbC. The authors disclosed the microstructure and the phase evolution of composite coating. By using Ni-coated graphite, Ni625, and Nb powder as raw materials, NbC-reinforced Ni-based coating was in situ synthesized on the 42CrMo steel [61]. The amount of additive greatly affected the corrosion resistance and mechanical performance of the coating. When the NbC amount was 15%, the resulting coating had the highest hardness of 441.9 HV. While the best corrosion resistance was obtained with the addition of 10%. When the addition was 20%, few NbC would be formed in the coating, which led to the degradation of wear resistance and hardness of the coating. In addition, Lian et al. [62] also prepared NbC-enhanced Ni-based coatings by using B_4C and Nb powders and discussed the influence of Nb and B_4C dosage on the coating performance. The hardness could be improved by increasing the content of B_4C, and the highest hardness reached 1794.1 $HV_{1.0}$. Different from the change of hardness, wear resistance would increase firstly with an increasing B4C content, and then reduce. when the Nb/B4C ratio was 1:1.3, the resulting coating presented the best wear resistance. The authors also focused on the research on in situ xNbC-reinforced Ni-20WC coating [63], and revealed the influence of NbC content on the structure, hardness, and corrosion resistance of the composite coating. An increased NbC fraction would lead to the agglomeration of NbC particles, surface roughening, and WCp nucleation. When x was 20 wt%, a refined structure was obtained with uniform distribution. The resulting coating had an enhanced microhardness (3.6 times of matrix) and optimal corrosion resistance.

4.2. NbC/Fe Composite Coatings

Various NbC/Fe composite coatings have been fabricated mostly by in situ synthesis methods using different reactants. For example, by adding the mixed powders of C, Nb, and Cr_3C_2, different morphologies of NbC particles were in situ-generated for enhancing the wear resistance of Fe-based coating [64]. By adjusting the ratio of raw materials, the products of $Cr_{23}C_6$ and $Cr_{0.19}Fe_{0.7}Ni_{0.11}$ compounds were formed at grain boundaries. Through comparison, cross-shaped NbC-reinforced coating exhibited a better wear resistance than that of rectangular NbC. In addition, Cao et al. [65] also prepared in situ NbC-modified Fe-based coating. In their work, various NbC phases were generated, including particles, dendrites, polyhedrons, networks, and petals (Figure 6a). The result demonstrated that the morphology of NbC had no great effect on the coating microhardness, while an increasing NbC fraction would enhance the coating microhardness. Chen et al. [66] fabricated in situ NbC-reinforced Fe-based coating by using Nb and B4C powders and disclosed the in situ synthesis mechanism by combining experiment and simulation. The coating composition was composed of [Fe-Cr] solid solution and Fe_2B, NbC, and B_4C-reinforced phases. During the cladding process, NbC particles (1.03 μm) were in situ-generated at the grain boundary and played a fine-grained strengthening effect. Compared with Fe-based coating, the reinforced coating showed a higher hardness (4.16 times) and lower volume loss (reduced

5 times). Among the literature about NbC-reinforced Fe-based coatings, several works discuss the influence of the addition method on coating performance. To solve this problem, Zhang et al. [67] fabricated NbC/Fe composite coatings on SS304 substrate by direct addition and in situ synthesis method, and the sample was denoted as A-NbC and I-NbC, respectively. Different from A-NbC, the I-NbC coating achieved a fine dispersion of the reinforcement phase, as shown in Figure 6b. The microhardness of I-NbC and A-NbC coating was 534.9 ± 4 HV and 524.6 ± 12 HV. However, the I-NbC coating had a minimum wear rate and friction coefficient, and the enhanced wear resistance was ascribed to the generation of passivation film by introducing Cr and Nb atoms.

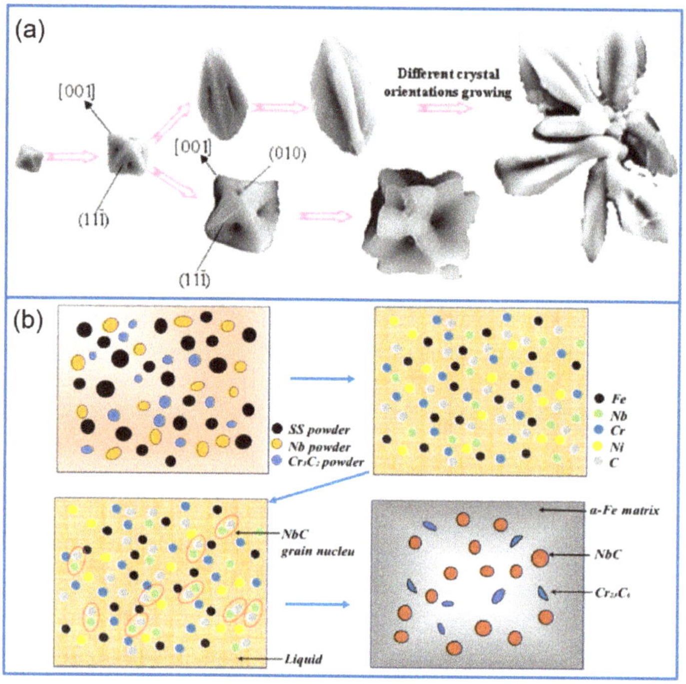

Figure 6. Formation schemes of two typical NbC petals (**a**) [65]; Copyright (2016) Elsevier. Illustration for laser cladding process of I-NbC composite coating (**b**) [67]. Copyright (2024) Elsevier.

4.3. NbC/HEAs Composite Coatings

CrMnFeCoNi high-entropy alloys (HEAs) with a face-centered cubic (FCC) feature have received widespread attention, due to their good fracture toughness and plasticity. To further enhance the mechanical properties of CrMnFeCoNi HEAs, Sun et al. [68] fabricated ceramic particles enhanced CrMnFeCoNi coating by using the powders of B4C, TiC, or NbC. During the laser cladding process, TiC and NbC were stably dispersed in the CrMnFeCoNi matrix, while the metal matrix reacted with B_4C to produce M_2B and $M_{23}C_6$ phases. The result confirmed that the coating containing NbC nanoparticles exhibited a better wear resistance than that of the coating reinforced by TiC, due to the mechanism of dislocation reinforcement and dispersion reinforcement. In this respect, Wu et al. [69] prepared NbC-reinforced FeNiCoCr HEAs coatings on the 316 SS, and discussed the influence of NbC content on the corrosive wear performance and corrosion resistance. Through comparison, the coating of FeNiCoCr + 10 wt% NbC possessed a superior corrosion resistance, and the corrosive wear rate was as low as 1.89×10^{-5} mm^3/Nm. The coating microhardness was 440.8 $HV_{0.2}$, which was greatly enhanced than that with no NbC.

To sum up, NbC-reinforced composite coatings are highlighted in this section. Among the existing metal matrices, NbC-reinforced Ni-based, Fe-based, and HEAs coatings have been widely reported in recent years. Regarding the preparation method, NbC-based coatings are usually fabricated by in situ synthesis method, and the raw materials usually include Nb, C, Nb, Cr_3C_2, B_4C, or other carbides. Due to the high melting point and high hardness of NbC, the addition of NbC in the composite coating would significantly enhance the hardness, wear resistance, and corrosion resistance. To date, the number of documents about NbC-reinforced laser cladding coatings is much less than that of WC- and TiC-based composite coating.

5. $Ti_{n+1}AlC_n$ (MAX Phase) Reinforced Composite Coatings

MAX phase ($M_{n+1}AX_n$, n = 1–3) exhibits a typical nano-laminated structure [70], in which, X is a C or N element, A is a metal element of III–IV_A, M is a Ti element. Nano-layered MAX phases integrate the typical characteristics of ceramic and metal, and present a superior wear resistance, corrosion resistance, good thermal conductivity, and oxidation resistance. When X is a C element, Ti_2AlC or Ti_3AlC_2 belonged to bimetallic carbides. The nano-layered structure of MAX enables them an excellent self-lubricating property. Thus, serving as reinforced phases, Ti_2AlC and Ti_3AlC_2 have been used for preparing laser cladding coatings to improve corrosion resistance and wear resistance. Among the existing reports, various MAX-reinforced metal coatings including Fe-based, Ni-based, Co-based, and Ti-based composite coatings have been developed. Based on the feeding method, direct addition and in situ synthesis of MAX phases are classified.

5.1. Direct Addition of MAX Phases

Adopting Ti_3AlC_2 or Ti_2AlC as a reinforced phase, Rui et al. [71] prepared three kinds of Ni60-xTi_3AlC_2 (x = 4%, 8%, and 12%) composite coatings on S355 steel and disclosed the influence of Ti_3AlC_2 content on corrosion electrochemical performance of the coating. The composition of the coating was TiC, Al_4C_3, Ti_3AlC_2, Ti_xAl_y, and Al_3Ti phases. Through comparison, the Ni60-12%Ti_3AlC_2 coating showed the optimal salt spray corrosion resistance, due to the generation of TiO_2 and α-Al_2O_3 oxide film. Zhang et al. [72] investigated the influence of different Ti_2AlC fractions on the structure and performance of Ni60/WC/xTi_2AlC composite coatings. The coating composition was TiC, Al_2O_3, WC, and W_2C phases, which exhibited irregular bulk and dendritic structures. The increased Ti_2AlC fraction would refine the grain size and enhance the hardness. The coating containing 15%Ti_2AlC presented the smallest friction coefficient and wear rate. Moreover, the addition of Ti_2AlC also improved the tribocorrosion property and abrasive wear performance. In addition, Ti_3AlC_2 enhanced Co-based coatings were cladded on the surface of H13 steel [73], and the effect of Ti_3AlC_2 fraction on the structure and friction-wear property was investigated. The result implied that the Co-5%Ti_3AlC_2 coating possessed the highest hardness and the lowest wear rate. The superior wear resistance was attributed to the enhanced oxidation resistance derived from Al-generated Al_2O_3 film, and the self-lubrication feature of Ti_3AlC_2. By using the raw materials of Ti_3AlC_2, Ti_2AlC, Co, and Ni powders, Lu et al. [74] fabricated CoNi-50Ti_3AlC_2 and CoNi-50Ti_2AlC coatings on the surface of Ti6Al4V alloy to improve the tribological property. Different compositions were formed in the two coatings by using different MAX materials, as shown in Figure 7a,b. The $TiAl_3$ and Ti_8C_5 phases were generated in the coating with Ti_2AlC. While $TiAl_3$, Ti_8C_5, and TiC phases were generated from Ti_3AlC_2. Compared with the decomposition of Ti_8C_5 in CoNi-50Ti_2AlC coating, Ti_8C_5 phases remained in CoNi-50Ti_3AlC_2 coating, further enhancing the wear resistance.

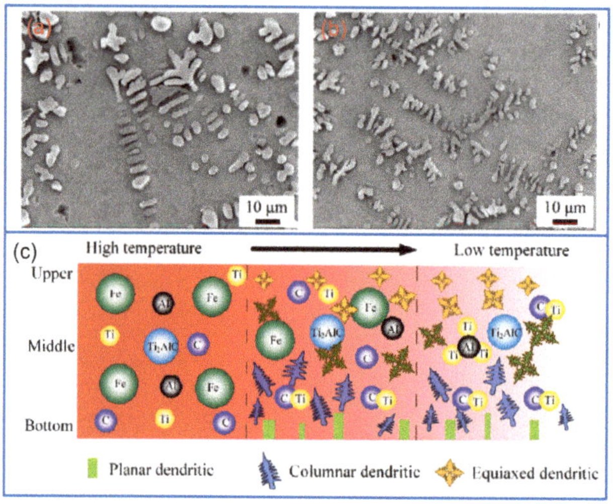

Figure 7. Morphologies of the sample of CoNi-50Ti_2AlC (**a**) and CoNI-50Ti_3AlC_2 (**b**) [74]; Copyright (2024) Elsevier. The reaction process of Ti_2AlC in the melting pool (**c**) [75]. Copyright (2024) Elsevier.

In addition to the direct addition of Ti_2AlC powder, Zhou et al. [76] prepared core-shell structured Ti_2AlC@Ni powder for fabricating Ni-based cermet coating by high-speed laser cladding method. During the preparation process, the Ti_2AlC phase decomposed into Al_2O_3, and TiC phases dispersed in the coating. Benefiting from the grain refinement, dispersion strengthening, and solid solution strengthening, the coating hardness was 2.8 times, and the wear resistance was 8.2 times (5 N load) that of the substrate. In addition, Jing et al. [75] also adopted Ni-coated Ti_2AlC powders (Ti_2AlC-Ni) for fabricating Fe-Ti_2AlC-Ni coatings on the substrate of $Cr_{12}MoV$. When the fraction of Ti_2AlC was between 5 and 10 wt%, the coatings had no cracks, with a good surface. The reaction process was provided in Figure 7c, and the coating composition was verified as Ti_3Al, TiC hard phases, and Ti_2AlC self-lubrication phase. The in situ formation of hard phases enhanced the coating hardness. Through comparison, the Fe-based coating with 10 wt% Ti_2AlC exhibited the lowest wear rate and frictional coefficient.

Different from the direct addition of Ti_2AlC and metal powders, Hua et al. [77] fabricated TC4/Ti_2AlC/B composite coating on Ti6Al4V (TC4) alloy by adding B powder. Moreover, the ultrasonic impact treatment (UIT) was conducted to further enhance the coating performance. The composition of the coating was verified as TiC, TiB, Ti3Al, Ti_2AlC phase, Fe-Ti-V, and α-Ti solid solutions, and the composition had no change with different B contents and UIT. The addition of ceramics phases greatly enhanced the microhardness of the coating. The optimized coating with 6 wt% B/Ti_2AlC/TC4 and UIT exhibited the lowest friction coefficient and the best wear resistance, due to the synergistic effect of TiO_2, Al_2O_3 oxide films, and Ti_2AlC lubricant, as well as the enhanced compressive stress by UIT.

5.2. In Situ Synthesis of MAX Phases

The in situ synthesis of Ti_2AlC (MAX) phases is different from the direct addition method, and the elements of Ti, Al, and C must be added to generate MAX phases, and the ratio of different components should be considered. For example, Richardson et al. [78] synthesized Ti-Al-C MAX phase coatings on Ti substrate by adding the Ti/Al/C powder with a stoichiometric ratio of 2/1/1. The surface composition of the coating was $TiC_{0.64}$, Ti_2AlC, Ti_3Al, and TiAl phases. The maximum hardness of the sub-surface region reached

811 ± 11 HV. In addition, Richardson et al. [79] also revealed the effect of thermal annealing on the Ti-Al-C system in laser-clad coatings. In their work, a furnace annealing at 1350 °C was conducted to induce the formation of large-scale MAX phase coatings. The result indicated that the reaction between α-Ti supersaturated with 33 at% C, Ti_yAl_z, and TiC_x generated dendritic Ti_2AlC phase grains.

By adding TiC, TiN, and TiAl mixed powders, Sun et al. [80] fabricated Ti-Al-C-N protective coatings on the surface of the Ti6Al4V alloy, and confirmed the in situ generation of the MAX phase in the coating. By adjusting the TiC/TiAl/TiN ratio, the coating with the TiC/TiAl/TiN ratio of 1/1/1 presented the best combination performance. The maximum hardness of the coating was 712–950 HV, about twice the substrate. Furthermore, the coating also exhibited the best fretting wear resistance. The enhanced wear resistance was ascribed to the strengthening role of twin Ti_2AlC/Ti_2AlN phases, with no serious stress concentration. Based on the raw materials of TiC powder, Al powder, and TiAl alloy powder, Tian et al. [81] synthesized Ti-Al-C MAX phase reinforced coatings by laser cladding and post-laser treatment. During the laser cladding process, a core-shell structure consisting of the Ti_2AlC shell and TiC core was generated. The late treatment promoted the diffusion of Al and C atoms, further inducing the nucleation-growth of the Ti_2AlC phase to form MAX phase structures. The post-treatment greatly enhanced the content of the MAX phase in the coating, about 9.03 times higher than that in the cladded coating. The post-laser treatment led to the uniform distribution of microhardness in the coating, and the average microhardness reached 467.5 HV. In another work, the TiC/TiAl mixed powders were used to prepare MAX phase coatings on the substrate of Ti6Al4V alloys [82]. The composition of the coating was Ti_2AlC, TiC, Ti-Al compounds, and a few Ti_3AlC phases. An increased TiC fraction would increase the content of the resulting Ti_2AlC phase, while Al powders improved the quality of the coating. Compared with the Ti alloy substrate, the hardness and wear resistance of the Ti-Al-C composite coating were significantly improved.

In addition to the formation of Ti-Al-C, other metal-reinforced phases can be introduced. For example, by using the mixed powders of Ti4822, Nb, and TiC, Wu et al. [83] fabricated Ti-Al-C-Nb_x composite coatings on TC4 alloy substrate, and investigated the influence of Nb fraction on the structure and tribological performance of the composite coating. Nb fraction greatly affected the microstructure of the coating, further affecting the coating performance. A high Nb fraction would induce the decomposition of Ti_3AlC_2 to produce TiC and Ti_2AlC, due to the introduction of Nb atoms into the Ti_3AlC_2 phase. Through comparison, the coating with 2 at% Nb presented the smallest wear rate and friction coefficient.

Serving as reinforced phases, bimetallic carbides Ti_2AlC and Ti_3AlC_2 belong to the typical MAX phases, which combine the merits of ceramic phases and metal, as well as the typical 2D nano-layered structure. The purpose of adding MAX phases is to improve the hardness and wear resistance of the coating. In this section, the research progress of various MAX-reinforced metal coatings has been introduced. Among the literature about the direct addition of MAX, the researchers majorly focus on the influence of MAX content, the difference between Ti_3AlC_2 and Ti_2AlC, the surface Ni wrapping of Ti_2AlC, or introduction of the B element. While for the works about in situ generation of MAX, the research hotspots mostly focus on the selection of raw materials and their ratio, such as Ti/Al/C mixed powders, TiC/Al/TiAl alloy powder, and TiC/TiN/TiAl powder. The resulting MAX products are different from the direct addition of MAX. In addition to the MAX phases, other products are introduced in the coating, which is determined by the raw materials. In order to obtain MAX phase product in the composite coatings, some researchers have adopted thermal annealing treatment to induce the generation of MAX phases. Compared with WC, TiC, and NbC, the Ti_2AlC- and Ti_3AlC_2-reinforced coatings are more complicated, and the problems involve the universality of raw materials, the uncertainty of MAX substance, and the formation of multi-component products.

6. Cr_3C_2, TaC-Reinforced Composite Coatings

6.1. Cr_3C_2-Reinforced Coatings

Cr_3C_2 ceramic powders exhibit some advantages of superior oxidation resistance, high hardness, and corrosion resistance, which have served as strengthening phases for enhancing the mechanical performance and wear resistance of different metal alloys. Cr_3C_2/Ni composite coatings have been widely developed. In this respect, Su et al. [84] deposited Ni60A/Cr_3C_2 composite coating on the substrate of $60Si_2Mn$ steel and discussed the influence of process parameters. The priority order of coating quality was scanning speed, feeding rate, and laser power. The coating composition was NiO, γ-Ni, $M_{23}C_6$ and M_7C_3. The optimized T6 coating was prepared with 5 mm/s scanning speed, 1800 W laser power, and 8 g/min feeding rate, which presented a high hardness of 902 $HV_{0.1}$, with enhanced wear resistance. In addition, the Cr_3C_2-reinforced Ni60A-Ag self-lubricating coating was cladded on a Cu alloy substrate [85]. Ag particles and Cr_7C_3 dendrites were dispersed in the γ-Ni matrix uniformly, further improving the hardness and plastic deformation resistance. Moreover, the friction coefficient of the coating was less than 0.25, much lower than that of Cu alloy. Similar to Ni60A/Cr_3C_2 composite coating, Ni625-xCr_3C_2 coatings were also cladded on the substrate of C45E4 steel [86]. The hardness of composite coating increased with an increasing Cr_3C_2 content. Moreover, the wear rate and friction coefficient were greatly reduced compared with pure Ni625 coating. The Ni625-10%Cr_3C_2 coating exhibited optimal corrosion resistance, and the corresponding mechanism was presented in Figure 8a.

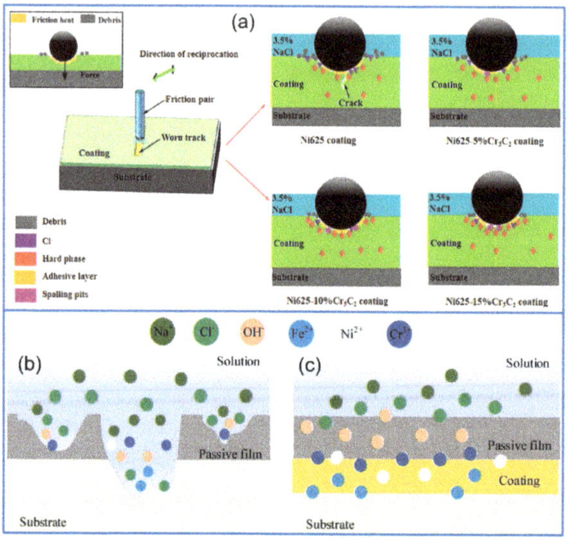

Figure 8. Wear models on Ni625−xCr_3C_2 coatings in 3.5% NaCl solution (**a**) [86]; Copyright (2024) Elsevier. The schematic diagram of the corrosion mechanism of pure alloy (**b**) and composite coating (**c**) [87]. Copyright (2024) Elsevier.

In addition to the Ni-based composite coating, Cr_3C_2-reinforced Fe-based and other alloys have been reported. For example, Mao et al. [87] developed Cr_3C_2/15-5PH stainless steel composite coatings by laser cladding. The introduction of Cr_3C_2 induced the phase transition from α-Fe to γ-Fe and produced new phases of M_7C_3 and $Cr_{23}C_6$. The composite coating containing 15 wt% Cr_3C_2 presented a superior corrosion resistance in 3.5 wt% NaCl solution, as given in Figure 8b,c. In addition, the Cr_3C_2/Fe_3Al composite coating was cladded on carbon structural steel to improve the hardness and wear resistance [88]. Fan et al. [89] prepared Cr_3C_2-reinforced Ni_3Al coatings and discussed the effect of Cr_3C_2 content on the structure, mechanical properties, and friction behaviors of the coating. The introduced Cr_3C_2 enhanced the coating toughness and hardness, as well as the wear resistance. Through comparison, the coating with 15 wt% Cr_3C_2 presented the minimum friction coefficients and wear rate. Aghili et al. [90] fabricated NiCr-Cr_3C_2 coating on the TiAl substrate by cladding the powder mixture and investigated the oxidation behavior of coating at 900 °C. The XRD result confirmed that the coating consisted of γ solid solution containing Cr_7C_3, Cr_3C_2, and $Cr_{23}C_6$ phases. Compared with the TiAl substrate, the NiCr-Cr_3C_2 coating presented a better oxidation resistance, due to the formation of an oxide surface layer. In addition, the authors also discussed the influence of processing parameters on coating properties and microhardness [91]. The microhardness of optimized NiCr-Cr_3C_2 coating reached 820 HV, much higher than that of pure TiAl substrate (320 HV). Venkatesh et al. [92] prepared Cr_xC_y-NiCrMoNb composite coatings on the surface of SA516 steel. In their experiment, Cr_7C_3 and Cr_3C_2 acted as starting materials, and only Cr_7C_3 dendrites were found in the cladding coating, indicating that Cr_7C_3 was in a stable phase during the process. The increasing scanning speed and laser power greatly decreased the fraction of carbides in the coating. The lower carbide content would induce a decrease in coating hardness, while the fine hard phase spacing and solid solution strengthening would cause an increase in hardness.

6.2. *TaC-Reinforced Coatings*

TaC has a high hardness of 1550 HV, a high melting point of 3980 °C, and low resistance, which can be used for fabricating composite coatings with a good wear resistance and corrosion resistance. Unlike those traditional metal carbides reported above, there are few reports about TaC-reinforced metal coatings. Herein, TaC-reinforced metal coatings are introduced, including the direct addition of TaC, and in situ-generated TaC.

Li et al. [93] prepared TaC/Stellite X-40 Co-based composite coatings on the nickel-aluminum bronze (NAB) substrate. The resulting coating contained Cr_3C_2, TaC, and Co_3Ta phases in the γ-Co matrix, further improving the electrochemical corrosion and wear resistance. The optimized coating containing 20 wt% TaC showed the highest hardness of 771.7 $HV_{0.2}$, the minimum friction coefficient and wear rate, as well as a high self-corrosion potential. Except for adding TaC powders, TaC can be in situ synthesized in different metal matrices. For example, Yu et al. [94] prepared TaC/Ni-Al-Cr composite coatings by using Ta, C, and Ni-Al-Cr powders, which, in situ-generated TaC and enhanced the bonding strength between the matrix and TaC. The composite coating had a better wear resistance than that of Ni-Al-Cr coating. By using the raw materials of Ni, Ta powder, TiC, and Ni-coated graphite, Hu et al. [95] developed Ni_3Ta-TaC-reinforced Ni-based coatings on $5Cr_5MoSiV$ steel. The addition of TiC powder accelerated the in situ formation of TaC and Ni_3Ta, and TaC particles were generated around TiC, which avoided the formation of cracks and Ni_3Ta intermetallic compounds. The wear resistance of the composite coating was 4 times higher than the substrate. The authors also discussed the difference between Ni-Ta coating and Ni-Ta-C coating [96]. The composition of Ni-Ta-C coating was Ni_3Ta, γ-Ni, and TaC phases. Due to the generation of TaC and Ni_3Ta reinforced phases, the hardness and wear resistance of Ni-Ta-C coating were both better than that of Ni-Ta coating. Liu et al. [97] prepared Ta-reinforced cobalt-based coatings by using Ta powder and Stellite F cobalt-based alloy powder. The in situ-generated TaC phases reduced the impact of particles in the erosion process, further enhancing the erosion resistance. The coating containing 30 wt% Ta presented the best comprehensive performance.

To sum up, Cr_3C_2, TaC-reinforced composite coatings have been introduced in this part. Various Cr_3C_2-reinforced composite coatings have been investigated, and the topics involve the effect of process parameters and Cr_3C_2 content, the synergistic effect with Ag particles, and the selection of metal matrices, such as Ni, Fe, and different alloys. Among the existing literature, most of the Cr_3C_2-based coatings are fabricated by direct addition method, instead of in situ synthesis. However, TaC is usually in situ synthesized in different metal matrices by using the raw materials of Ta, C, and metal powders. The in situ-generated TaC phases greatly improve the coating hardness, wear resistance, and erosion resistance.

7. Conclusions and Prospects

7.1. Conclusions

The use of laser cladding technique to prepare metal carbide-reinforced coatings overcomes the difficult processing problem of ceramic materials, which can obtain composite coatings with an excellent performance in working conditions. Especially after adding different carbide-reinforcing agents, the hardness, wear resistance, corrosion resistance, and high-temperature oxidation resistance of the substrate can be significantly improved. Therefore, metal carbide-reinforced laser cladding coatings have important research and practical application value. According to the type of metal carbides, five sections are classified in this review, including WC-reinforced coatings, TiC-reinforced coatings, NbC-reinforced coatings, $Ti_{n+1}AlC_n$ (MAX) reinforced coatings, and Cr_3C_2, TaC-reinforced coatings. The selection of raw materials, fabrication method, microstructure/composition, and application performance of various carbide-reinforced composite coatings are highlighted in this review. The application of various metal carbides and performance enhancement of carbide-reinforced coatings are summarized. From Table 1, the application range, performance advantages and caution of each metal carbide can be easily obtained, which will help researchers to choose a suitable reinforcing phase for fabricating high-performance laser cladding coatings. In addition to the summary in Tables, some important conclusions are also drawn and listed as follows:

(1) As a typical metal carbide reinforcing agent, WC-reinforced coatings have been extensively reported, including WC/Ni, WC/Co, WC/Fe, and WC/other metal composite coatings. The research topics mostly focus on the effect of WC content and particle size, the optimization of process parameters, coating structure design, a combination of WC and other ceramic phases or oxides, the in situ synthesis of WC, the selection of metal matrix, and so on. The main function of WC in composite coatings is to enhance the hardness, mechanical properties, wear resistance, and corrosion resistance.

(2) Various TiC-reinforced composite coatings are introduced, including TiC/Ni, TiC/Fe, TiC/HEAs, and TiC/Ti alloy composite coatings. The research topics mostly involve the in situ synthesis of TiC by selecting different raw materials, such as Ti powers, nickel-coated graphite, B_4C, Cr_3C_2, or WC. Different from the direct addition of TiC particles in a metal matrix, in situ-generated TiC would refine the grain size by adjusting the nucleation rate. Furthermore, the function of solid solution strengthening and second phase strengthening further enhances the coating hardness and wear resistance.

(3) NbC-reinforced Ni-based, Fe-based, and HEAs coatings have been developed in recent years. Compared with WC- and TiC-based composite coating, the amount of literature about NbC-reinforced composite coatings is much less. In terms of

processing method, NbC-reinforced composite coatings are usually prepared by in situ synthesis method, and the powders of Nb, C, Nb, Cr_3C_2, B_4C, or other carbides are adopted as raw materials. The high melting point and high hardness of NbC enable the NbC-reinforced composite coatings with a high hardness, wear resistance, and corrosion resistance.

(4) Bimetallic carbides Ti_2AlC and Ti_3AlC_2 belong to MAX phases with a typical 2D nano-layered structure. By using MAX as fillers, the researchers have discussed the effect of MAX dosage, the difference between Ti_3AlC_2 and Ti_2AlC, the surface wrapping, and the introduction of other nonmetal elements. While for the works about in situ generation of MAX, the researchers pay more attention to the selection of raw materials and the optimization of component ratio. The in situ-generated MAX products are different from the direct addition of MAX, and thermal annealing treatment is required to induce the formation of MAX phases. The issues about MAX reinforced coatings are related to the universality of raw materials, the uncertainty of MAX substance, and the formation of multi-component products.

(5) In respect to Cr_3C_2-reinforced composite coatings, researchers have discussed the effect of process parameters and Cr_3C_2 content, the synergistic effect with Ag particles, and the selection of metal matrices, such as Ni, Fe, and different alloys. Most Cr_3C_2-based coatings are prepared by direct addition of Cr_3C_2. TaC-reinforced coatings are mostly prepared by in situ synthesis by using the mixed powders of Ta, C, and metal powders or other metal carbides. The introduction of TaC phases significantly improves the hardness, wear resistance, and erosion resistance of coatings.

Table 1. The application of various metal carbides in laser cladding coatings.

Carbides	Metal Matrix	Performance Enhancement	Caution
WC	Reinforced Ni, Co, Fe, various alloys	Enhancing the hardness, mechanical performance, wear/corrosion resistance	WC content is <20 wt%
TiC	Reinforced Ni, Fe, Ti, HEAs	Enhancing the hardness, wear/corrosion resistance; in situ-generated TiC is better	When TiC content is >15%, pores and microcracks
NbC	Reinforced Ni, Fe, HEAs	Enhancing the hardness, wear/corrosion resistance	Compared with direct addition, in situ NbC is better
$Ti_{n+1}AlC_n$	Reinforced Ni, Co, Fe, Ti	2D nano-layered structure, enhancing the hardness, wear/corrosion resistance	Require thermal annealing to produce MAX phases
Cr_3C_2	Reinforced Ni, Fe, TiAl	Enhancing the hardness, wear/corrosion/oxidation resistance	Cr_7C_3 was stable phase, much stabler than Cr_3C_2
TaC	Reinforced Ni, Co	Enhancing the hardness, wear/corrosion resistance	In situ formation of TaC avoids the cracks

7.2. Prospects

Although significant research achievements and progress have been achieved in the field of laser cladding coatings in recent years, there are still many issues that require further in-depth study and improvement.

(1) The selection of metal carbides in practical application and optimization design. Metal carbides have diversity, and the suitable selection of one or several carbides is really important. Moreover, the molding method including direct addition or in situ synthesis determines the industrialization process of composite coatings. Therefore, researchers must be familiar with the advantages/disadvantages and functionality of each metal carbide, and choose the appropriate metal carbide for the practical requirement.

(2) Defects such as pores and cracks may occur during the actual processing of laser cladding carbide reinforced coatings. The defects generated in composite coatings

are generally related to the type, dosage of the filler, and processing parameters. Therefore, the researchers should select appropriate addition amounts and processing parameters, based on the physical feature, and particle size of various carbide fillers.

(3) The design and preparation of multi-functional ceramic coating materials, that is, how to add one or several metal carbides to prepare coatings with multiple excellent properties. This requires researchers to be familiar with the compatibility between metal substrates and carbides, and to select appropriate reinforcing agents based on the type of substrate. For example, a strong tough coating can be obtained by combining MAX-phase ceramics with metals. If the conditions permit, theory calculation results are required to guide the selection of metal carbides.

Author Contributions: Conceptualization, X.H. and C.G.; validation, G.W. and W.D.; writing—original draft preparation, D.J.; writing—review and editing, C.G. All authors have read and agreed to the published version of the manuscript.

Funding: This research was funded by the Guangdong Provincial Key Laboratory of Battery Recycling and Reuse (2021B1212050002).

Institutional Review Board Statement: Not applicable.

Informed Consent Statement: Not applicable.

Data Availability Statement: Not applicable.

Conflicts of Interest: The authors declare no conflicts of interest.

References

1. Wang, K.; Zhang, Z.; Xiang, D.; Ju, J. Research and Progress of Laser Cladding: Process, Materials and Applications. *Coatings* **2022**, *12*, 1382. [CrossRef]
2. Liu, Y.; Ding, Y.; Yang, L.; Sun, R.; Zhang, T.; Yang, X. Research and progress of laser cladding on engineering alloys: A review. *J. Manuf. Process.* **2021**, *66*, 341–363. [CrossRef]
3. Wu, Q.; Long, W.; Zhang, L.; Zhao, H. A review on ceramic coatings prepared by laser cladding technology. *Opt. Laser Technol.* **2024**, *176*, 110993. [CrossRef]
4. Das, A.K. Effect of rare earth oxide additive in coating deposited by laser cladding: A review. *Mater. Today Proc.* **2022**, *52*, 1558–1564. [CrossRef]
5. Zhu, L.; Xue, P.; Lan, Q.; Meng, G.; Ren, Y.; Yang, Z.; Xu, P.; Liu, Z. Recent research and development status of laser cladding: A review. *Opt. Laser Technol.* **2021**, *138*, 106915. [CrossRef]
6. Sun, S.; Wang, J.; Xu, J.; Cheng, X.; Jing, C.; Chen, Z.; Ru, H.; Liu, Y.; Jiao, J. Preparing WC-Ni coatings with laser cladding technology: A review. *Mater. Today Commun.* **2023**, *37*, 106939. [CrossRef]
7. Liu, Y.; Wang, K.; Fu, H. Improvement of the High Temperature Wear Resistance of Laser Cladding Nickel-Based Coating: A Review. *Metals* **2023**, *13*, 840. [CrossRef]
8. Liu, D.; Yang, X.; Zhao, A.; Cheng, X.; Zhang, Q. Preparation of nickel-based composite coatings by laser cladding technology: A review. *Int. J. Adv. Manuf. Technol.* **2024**, *134*, 3107–3137. [CrossRef]
9. Wang, H.; Liu, K.; Li, J.; Geng, S.; Jing, L.; Skuratov, V. Reinforcements/matrix micro-interface evolution and properties of in-situ Ni60A/WC coatings prepared by laser cladding. *Surf. Coat. Technol.* **2024**, *484*, 130834. [CrossRef]
10. Hu, Z.; Li, Y.; Lu, B.; Tan, N.; Cai, L.; Yong, Q. Effect of WC content on microstructure and properties of high-speed laser cladding Ni-based coating. *Opt. Laser Technol.* **2022**, *155*, 108449. [CrossRef]
11. Liu, Y.; Gu, X.; Lou, C.; Kang, L.; Hou, Q.; Ma, C. Influence of WC ceramic particles on structures and properties of laser cladding Ni50-WC coatings. *J. Mater. Res. Technol.* **2023**, *26*, 14–21. [CrossRef]
12. Li, Y.; Shi, Y.; Jiang, G.; Tang, S.; Wu, J.; Wang, J. Effect of WC on microstructure, molten pool and properties of Ni based coatings by multi-layer laser cladding. *Opt. Laser Technol.* **2024**, *174*, 110612. [CrossRef]
13. Shen, X.; He, X.; Gao, L.; Su, G.; Xu, C.; Xu, N. Study on crack behavior of laser cladding ceramic-metal composite coating with high content of WC. *Ceram. Int.* **2022**, *48*, 17460–17470. [CrossRef]
14. Tan, N.; Hu, Z.; Zhou, Y.; Li, Y.; Lu, B.; Hu, D.; Liu, Y.; Li, Q. Effect of WC particle size on the microstructure and tribological properties of high-speed laser cladding Ni/WC composite coatings. *Mater. Today Commun.* **2024**, *39*, 109066. [CrossRef]

15. Wang, Q.; Li, Q.; Zhang, L.; Chen, D.X.; Jin, H.; Li, J.D.; Zhang, J.W.; Ban, C.Y. Microstructure and properties of Ni-WC gradient composite coating prepared by laser cladding. *Ceram. Int.* **2022**, *48*, 7905–7917. [CrossRef]
16. Bian, Y.; Li, S.; Tian, C.; Chen, B.; Dong, B.; Shu, Z.; Guo, J.; He, X.; Yu, G. Effect of powder feeding rate on mass fraction of WC particles, microstructure evolution and micro-hardness of WC-Ni composite coating by laser cladding. *J. Alloys Compd.* **2025**, *1013*, 178432. [CrossRef]
17. Li, S.; Huang, K.; Zhang, Z.; Zheng, C.; Li, M.; Wang, L.; Wu, K.; Tan, H.; Yi, X. Wear mechanisms and micro-evaluation of WC + TiC particle-reinforced Ni-based composite coatings fabricated by laser cladding. *Mater. Charact.* **2023**, *197*, 112699. [CrossRef]
18. Li, M.; Yang, J.; Han, B.; Song, L.; Li, P.; Dong, W.; Xue, X. Comparative investigation on microstructures and properties of WC/Cr_3C_2 reinforced laser cladding Ni-based composite coatings subjected to ultrasonic impact treatment. *Mater. Today Commun.* **2023**, *34*, 105219. [CrossRef]
19. Chen, J.; Lian, G.; Lin, T.; Lu, H.; Wang, Y. Effects of the proportions of carbon on the microstructure and properties of NbC-reinforced Ni-WC composite coatings by laser cladding in-situ synthesis. *Mater. Today Commun.* **2024**, *38*, 107896. [CrossRef]
20. Chen, J.; Lian, G.; Feng, M.; Zhang, W.; Chen, R. Microstructure evolution and properties of (Nb,M)C (M=Ti,V and Zr) reinforced Ni-WC coatings by laser cladding. *J. Alloys Compd.* **2025**, *1010*, 177539. [CrossRef]
21. Zhao, J.; Li, R.; Feng, A.; Feng, H. Effect of rare earth La_2O_3 particles on structure and properties of laser cladding WC-Ni60 composite coatings. *Surf. Coat. Technol.* **2024**, *479*, 130569. [CrossRef]
22. Jiang, C.; Zhang, J.; Chen, Y.; Hou, Z.; Zhao, Q.; Li, Y.; Zhu, L.; Zhang, F.; Zhao, Y. On enhancing wear resistance of titanium alloys by laser cladded WC-Co composite coatings. *Int. J. Refract. Met. Hard Mater.* **2022**, *107*, 105902. [CrossRef]
23. Lv, N.; Yue, H.; Guo, C.; Dai, W.; Zhang, J.; Li, Q.; Zhao, G.; Hao, G. A comparative investigation on the effects of reinforcement phase addition methods on laser melting deposited WC/Co coatings. *J. Manuf. Process.* **2024**, *129*, 134–146. [CrossRef]
24. Xu, P.; Zhu, L.; Xue, P.; Yang, Z.; Wang, S.; Ning, J.; Meng, G.; Lan, Q.; Qin, S. Microstructure and properties of IN718/WC-12Co composite coating by laser cladding. *Ceram. Int.* **2022**, *48*, 9218–9228. [CrossRef]
25. Ding, H.; Cao, Y.; Hua, K.; Tong, Y.; Li, N.; Sun, L.; Li, X.; Wu, H.; Wang, H. Fretting wear resistance at ambient and elevated temperatures of 316 stainless steel improved by laser cladding with Co-based alloy/WC/CaF_2 composite coating. *Opt. Laser Technol.* **2023**, *163*, 109428. [CrossRef]
26. Xiao, Q.; Sun, W.L.; Yang, K.X.; Xing, X.F.; Chen, Z.H.; Zhou, H.N.; Lu, J. Wear mechanisms and micro-evaluation on WC particles investigation of WC-Fe composite coatings fabricated by laser cladding. *Surf. Coat. Technol.* **2021**, *420*, 127341. [CrossRef]
27. Li, J.; Zhu, Z.; Peng, Y.; Shen, G. Phase evolution and wear resistance of in-situ synthesized $(Cr, W)_{23}C_6$-WC composite ceramics reinforced Fe-based composite coatings produced by laser cladding. *Vacuum* **2021**, *190*, 110242. [CrossRef]
28. Zeng, X.; Wang, Q.; Chen, C.; Lian, G.; Huang, X. Effects of WC addition on the morphology, microstructure and mechanical properties of Fe50/TiC/WC laser claddings on AISI 1045 steel. *Surf. Coat. Technol.* **2021**, *427*, 127781. [CrossRef]
29. Yang, Z.; Hao, H.; Gao, Q.; Cao, Y.; Han, R.; Qi, H. Strengthening mechanism and high-temperature properties of H13 + WC/Y_2O_3 laser-cladding coatings. *Surf. Coat. Technol.* **2021**, *405*, 126544. [CrossRef]
30. Liu, J.; Li, Z.; Lin, D.; Tang, Z.; Song, X.; He, P.; Zhang, S.; Bian, H.; Fu, W.; Song, Y. Eutectic high-entropy alloys and their applications in materials processing engineering: A review. *J. Mater. Sci. Technol.* **2024**, *189*, 211–246. [CrossRef]
31. Li, Z.; Xie, D.; Liu, Y.; Lv, F.; Zhou, K.; Jiao, C.; Gao, X.; Wang, D.; Liu, Y.; Zu, H.; et al. Effect of WC on the microstructure and mechanical properties of laser-clad $AlCoCrFeNi_{2.1}$ eutectic high-entropy alloy composite coatings. *J. Alloys Compd.* **2024**, *976*, 173219. [CrossRef]
32. Du, M.; Wang, L.; Gao, Z.; Yang, X.; Liu, T.; Zhan, X. Microstructure and element distribution characteristics of Y_2O_3 modulated WC reinforced coating on Invar alloys by laser cladding. *Opt. Laser Technol.* **2022**, *153*, 108205. [CrossRef]
33. Cai, Q.; Li, G.; Wu, B.; Xu, S.; Wang, L.; Guo, Y. Effect of TiC content on microstructure and properties of TiC/Ni60 coatings on Ti6Al4V alloy deposited by laser cladding. *Opt. Laser Technol.* **2024**, *168*, 109854. [CrossRef]
34. Meng, Q.; Wang, C.; Liu, T.; Song, Q.; Xue, B.; Cui, H. Microstructure and performance optimization of laser cladding nano-TiC modified nickel-based alloy coatings. *Surf. Coat. Technol.* **2024**, *479*, 130583. [CrossRef]
35. Chen, L.; Yu, T.; Chen, X.; Zhao, Y.; Guan, C. Process optimization, microstructure and microhardness of coaxial laser cladding TiC reinforced Ni-based composite coatings. *Opt. Laser Technol.* **2022**, *152*, 108129. [CrossRef]
36. Chen, L.; Chen, Y.; Chen, X.; Yu, T.; Wang, Z. Microstructure and properties of in situ TiC/Ni functionally gradient coatings by powder-fed laser cladding. *Ceram. Int.* **2022**, *48*, 36789–36801. [CrossRef]
37. Chen, L.; Guan, C.; Ma, Z.; Cui, Z.; Zhang, Z.; Yu, T.; Gu, R. Modeling and simulation of grinding surface morphology for laser cladding in situ TiC reinforced Ni-based composite coatings. *Surf. Coat. Technol.* **2025**, *498*, 131831. [CrossRef]
38. Wang, H.; Wu, M.; Miao, X.; Jin, X.; Cui, C. TiC in-situ strengthening mechanism and crack initiation mechanism of SiC/TC4 composite coatings by laser cladding. *Ceram. Int.* **2024**, *50 Pt A*, 52558–52571. [CrossRef]
39. Qi, X.; Li, Y.; Cui, W.; Du, J.; Zhao, Y.; Li, F. Improving the wear and corrosion properties of laser cladded Ni-based composite coatings via regulating in-situ TiB_2-TiC. *Ceram. Int.* **2025**, *51*, 9442–9454. [CrossRef]

40. Cui, W.; Li, Y.; Li, F.; Qi, X.; Sun, X.; Pan, Z.; Niu, J. Wear and corrosion properties of in-situ TiC–TiB_2 modified Ni-based composite coatings with different B/C ratios prepared by laser cladding. *Ceram. Int.* **2024**, *50 Pt B*, 2424–2435. [CrossRef]
41. Zheng, Y.; Lian, G.; Lu, H.; Chen, C.; Huang, X. Defects, organization, and properties of TiB_2–TiC Bi-ceramic phase by laser cladding in situ synthesis. *Ceram. Int.* **2024**, *50 Pt A*, 41097–41116. [CrossRef]
42. Saeedi, R.; Shoja Razavi, R.; Bakhshi, S.R.; Erfanmanesh, M.; Ahmadi Bani, A. Optimization and characterization of laser cladding of NiCr and NiCr–TiC composite coatings on AISI 420 stainless steel. *Ceram. Int.* **2021**, *47*, 4097–4110. [CrossRef]
43. Ge, T.; Chen, L.; Gu, P.; Ren, X.; Chen, X. Microstructure and corrosion resistance of TiC/Inconel 625 composite coatings by extreme high speed laser cladding. *Opt. Laser Technol.* **2022**, *150*, 107919. [CrossRef]
44. Zhu, H.; Ouyang, M.; Hu, J.; Zhang, J.; Qiu, C. Design and development of TiC-reinforced 410 martensitic stainless steel coatings fabricated by laser cladding. *Ceram. Int.* **2021**, *47*, 12505–12513. [CrossRef]
45. Hui, X.; Dejun, K. Microstructure and tribological properties of laser cladded TiC reinforced WC-10Co4Cr coatings at 500 °C. *Mater. Today Commun.* **2024**, *40*, 109548. [CrossRef]
46. Zhang, H.F.; Wang, L.; Zhang, S.; Wu, C.L.; Zhang, C.H.; Sun, X.Y. Design, fabrication, microstructure and properties of in-situ synthesized TiC reinforced stainless steel matrix composite coating by laser cladding. *Mater. Charact.* **2023**, *204*, 113177. [CrossRef]
47. Yang, X.; Chen, Y.; Zhang, Z.; Liu, Y.; Tao, S.; Li, G.; Chen, H. Study on microstructure and properties of laser-clad Fe-based (Ti, V)C composite coatings. *Surf. Coat. Technol.* **2023**, *464*, 129552. [CrossRef]
48. Jiang, G.Y.; Zhang, J.W. The influence of laser power on the microstructure and friction performance of laser-prepared TiC-NbC composite coatings on stainless steel surfaces. *Mater. Today Commun.* **2024**, *41*, 110812. [CrossRef]
49. Chen, L.; Yu, T.; Guan, C.; Zhao, Y. Microstructure and properties of metal parts remanufactured by laser cladding TiC and TiB_2 reinforced Fe-based coatings. *Ceram. Int.* **2022**, *48*, 14127–14140. [CrossRef]
50. Chen, G.D.; Liu, X.B.; Yang, C.M.; Zhang, F.Z.; Li, X.G.; Zheng, J.; Liu, J. Strengthening mechanisms of laser cladding TiC/FeCoCrNiCu high-entropy composite coatings: Microstructure evolution and wear behaviors. *Tribol. Int.* **2024**, *199*, 109979. [CrossRef]
51. Li, Y.; Wang, K.; Fu, H.; Guo, X.; Lin, J. Microstructure and wear resistance of in-situ TiC reinforced AlCoCrFeNi-based coatings by laser cladding. *Appl. Surf. Sci.* **2022**, *585*, 152703. [CrossRef]
52. Gao, Z.; Niu, Z.; Gao, Z.; Li, J.; Bai, G.; Ke, L.; Yu, Y.; Zhang, C. Microstructure and wear behavior of in-situ synthesized TiC-reinforced CoCrFeNi high entropy alloy prepared by laser cladding. *Appl. Surf. Sci.* **2024**, *670*, 160720. [CrossRef]
53. Li, Y.; Fu, H.; Ma, T.; Wang, K.; Yang, X.; Lin, J. Microstructure and wear resistance of AlCoCrFeNi-WC/TiC composite coating by laser cladding. *Mater. Charact.* **2022**, *194*, 112479. [CrossRef]
54. Li, S.; Niu, W.; Lei, Y.W.; Zheng, Y. Microstructure and wear performance of ex/in-situ TiC reinforced $CoCrFeNiW_{0.4}Si_{0.2}$ high-entropy alloy coatings by laser cladding. *J. Alloys Compd.* **2024**, *1002*, 175458. [CrossRef]
55. Zhuang, D.D.; Tao, W.W.; Ni, H.M.; Wang, A.Z.; Du, B.; Zhang, S.H.; Lian, X.L.; Wang, D.; Feng, Y.J. TiC distribution and properties of TiC-CrMnFeCoNi coating fabricated by laser cladding with ultrasound. *Surf. Coat. Technol.* **2023**, *468*, 129744. [CrossRef]
56. Chen, N.; Xiao, H.; Ren, L.; Huang, F.; Chen, Y.; Cao, S.; Wu, H.; Zhu, L. Microstructure and tribological properties of laser-cladded TiCx/TiAl composite coatings on TC4 alloy. *Tribol. Int.* **2024**, *192*, 109236. [CrossRef]
57. Chen, T.; Li, W.; Liu, D.; Xiong, Y.; Zhu, X. Effects of heat treatment on microstructure and mechanical properties of TiC/TiB composite bioinert ceramic coatings in-situ synthesized by laser cladding on Ti6Al4V. *Ceram. Int.* **2021**, *47*, 755–768. [CrossRef]
58. Zhang, Z.; Yang, F.; Zhang, H.; Zhang, T.; Wang, H.; Xu, Y.; Ma, Q. Influence of CeO_2 addition on forming quality and microstructure of TiCx-reinforced CrTi4-based laser cladding composite coating. *Mater. Charact.* **2021**, *171*, 110732. [CrossRef]
59. Zhang, Z.; Yang, F.; Zhang, H.; Zhang, T.; Wang, H. Microstructure and element distribution of laser cladding TiCx-reinforced CrTi4-based composite coating with CeO_2/Ce_2O_3. *Mater. Lett.* **2021**, *283*, 128772. [CrossRef]
60. Gaddam, S.; Nartu, M.S.K.K.Y.; Chesetti, A.; Mantri, S.A.; Mishra, R.S.; Dahotre, N.B.; Banerjee, R. Hierarchical phase evolution during direct laser deposition of an in-situ Ni-NbC composite. *Scripta Mater.* **2023**, *226*, 115225. [CrossRef]
61. Zhang, Y.; Yu, T.; Sun, J.; Sun, Z.; Wang, Y. Effect of in-situ NbC content on the microstructure and mechanical properties of Ni625 composite coating by laser cladding. *Ceram. Int.* **2024**, *50 Pt C*, 48074–48083. [CrossRef]
62. Lian, G.; Yue, K.; Chen, C.; Huang, L.; Zheng, M. Influences of powder ratios on the mechanical properties of in-situ synthetic NbC coatings. Mater. *Today Commun.* **2023**, *35*, 106318.
63. Lian, G.; Chen, J.; Lu, H.; Chen, C.; Huang, X. The microstructure and properties of reinforced Ni-20WC coatings by the laser-cladding in-situ synthesis of xNbC. *Ceram. Int.* **2025**, *51*, 13135–13151. [CrossRef]
64. Xi, W.; Song, B.; Sun, Z.; Yu, T.; Wang, J.; Sun, Q. Effect of various morphology of in situ generated NbC particles on the wear resistance of Fe-based cladding. *Ceram. Int.* **2023**, *49*, 10265–10272. [CrossRef]
65. Cao, Y.; Zhi, S.; Gao, Q.; Tian, X.; Geng, T.; Guan, X.; Qin, C. Formation behavior of in-situ NbC in Fe-based laser cladding coatings. *Mater. Charact.* **2016**, *119*, 159–165. [CrossRef]

66. Chen, L.; Yu, T.; Xu, P.; Zhang, B. In-situ NbC reinforced Fe-based coating by laser cladding: Simulation and experiment. *Surf. Coat. Technol.* **2021**, *412*, 127027. [CrossRef]
67. Zhang, H.F.; Zhang, S.; Wu, H.; Wang, R.; Zhang, C.H.; Wu, C.L.; Chen, J.; Chen, H.T. Mechanical properties and corrosion resistance of laser cladding iron-based coatings with two types of NbC reinforcement. *Surf. Coat. Technol.* **2024**, *479*, 130558. [CrossRef]
68. Sun, D.; Zhu, L.; Cai, Y.; Yan, Y.; Ge, F.; Shan, M.; Tian, Y.; Han, J.; Jiang, Z. Tribology comparison of laser-cladded CrMnFeCoNi coatings reinforced by three types of ceramic (TiC/NbC/B_4C). *Surf. Coat. Technol.* **2022**, *450*, 129013. [CrossRef]
69. Wu, H.; Wang, Z.Y.; Wang, M.S.; Wang, R.; Zhang, S.; Zhang, C.H.; Wu, C.L.; Chen, H.T.; Chen, J. Microstructure evolution, corrosion and corrosive wear properties of NbC-reinforced FeNiCoCr-based high entropy alloys coatings fabricated by laser cladding. *Eng. Fail. Anal.* **2025**, *171*, 109352. [CrossRef]
70. Zhang, Z.; Duan, X.; Jia, D.; Zhou, Y.; van der Zwaag, S. On the formation mechanisms and properties of MAX phases: A review. *J. Eur. Ceram. Soc.* **2021**, *41*, 3851–3878. [CrossRef]
71. Rui, S.; Chao, Z.; Dejun, K. Salt spray corrosion behavior and electrochemical performance of laser cladded Ni60–Ti_3AlC_2 composite coatings in 3.5% NaCl solution. *Mater. Today Commun.* **2023**, *36*, 106550. [CrossRef]
72. Zhang, K.; Kong, D. Microstructure, tribocorrosion and electrochemical properties of Ti_2AlC reinforced Ni60WC coatings by high–speed laser cladding. *Mater. Charact.* **2025**, *221*, 114723. [CrossRef]
73. Zhou, J.; Kong, D. Friction–wear performances and oxidation behaviors of Ti_3AlC_2 reinforced Co–based alloy coatings by laser cladding. *Surf. Coat. Technol.* **2021**, *408*, 126816. [CrossRef]
74. Lu, Y.; Peng, Y.; Chang, X.; Kong, D. Ti_2AlC and Ti_3AlC_2 reinforced CoNi coatings by laser cladding: Nanostructures, tribological properties and density functional theory calculations. *J. Manuf. Process.* **2024**, *131*, 736–749. [CrossRef]
75. Jing, S.; Li, Y.; Cai, Y.; Tan, N.; Li, Q.; Zhang, G.; Li, G. Investigation on microstructure and tribological properties of Ti_2AlC-Ni reinforced Fe-based prepared by high-speed laser cladding. *Surf. Coat. Technol.* **2024**, *489*, 131143. [CrossRef]
76. Zhou, L.; Ma, G.; Wang, H.; Wang, W.; Mou, H.; Zhu, X.; Zhao, H.; Li, Y.; Tan, N. High-speed laser cladded Ni-based cermet coating with high ceramic phase content derived from core-shell structured powder. *Surf. Coat. Technol.* **2024**, *489*, 131110. [CrossRef]
77. Hua, S.W.; Pang, M.; Ji, F.Q.; Chen, J.; Liu, G. Microstructure and tribological properties of Ti_2AlC-B particle-enhanced self-lubricating coatings on Ti6Al4V by ultrasonic impact treatment and laser cladding. *Mater. Today Commun.* **2023**, *34*, 105165. [CrossRef]
78. Richardson, P.; Cuskelly, D.; Brandt, M.; Kisi, E. Microstructural analysis of in-situ reacted Ti_2AlC MAX phase composite coating by laser cladding. *Surf. Coat. Technol.* **2020**, *385*, 125360. [CrossRef]
79. Richardson, P.; Cuskelly, D.; Brandt, M.; Kisi, E. Effects of furnace annealing on in situ reacted Ti_2AlC MAX phase composite coatings deposited by laser cladding. *Surf. Coat. Technol.* **2021**, *405*, 126597. [CrossRef]
80. Sun, L.; Cao, Y.; Ding, H.; Wang, Y.; Ma, Q.; Hua, K.; Wang, H. Fabrication of composite coatings containing in-situ reacted Ti_2AlC/Ti_2AlN MAX phases by laser cladding and investigation of fretting wear mechanism. *Ceram. Int.* **2024**, *50 Pt B*, 39138–39149. [CrossRef]
81. Tian, Y.; Xiao, H.; Ren, L.; Feng, J.; Xiao, Y.; Chen, N.; Zhou, X. A new strategy to fabricate Ti_2AlC MAX coatings by the two-step laser method. *Surf. Coat. Technol.* **2022**, *448*, 128944. [CrossRef]
82. Chu, M.; Xiao, H.; Ren, L.; Mo, T.; Lin, B. Microstructure and properties of Ti–Al–C composite coatings prepared by laser cladding. *Ceram. Int.* **2024**, *50 Pt B*, 12498–12509. [CrossRef]
83. Wu, H.; Xiao, H.; Chen, N.; Chu, M.; Lin, B.; Zhang, Z.; Fu, G.; Mo, T. Effect of Nb content on the tribological properties of laser-cladded Ti-Al-C MAX phase composite coatings. *Tribol. Int.* **2025**, *204*, 110426. [CrossRef]
84. Su, Z.; Li, J.; Shi, Y.; Ren, S.; Zhang, Z.; Wang, X. Effect of process parameters on microstructure and tribological properties of Ni60A/Cr_3C_2 laser cladding on 60Si2Mn steel. *Surf. Coat. Technol.* **2023**, *473*, 130005. [CrossRef]
85. Cao, S.; Liang, J.; Wang, L.; Zhou, J. Effects of NiCr intermediate layer on microstructure and tribological property of laser cladding Cr_3C_2 reinforced Ni60A-Ag composite coating on copper alloy. *Opt. Laser Technol.* **2021**, *142*, 106963. [CrossRef]
86. Yang, C.; Lu, Y.; Kong, D. Laser cladded Ni625–xCr_3C_2 coatings: Microstructure, tribocorrosion and electrochemical properties. *Surf. Coat. Technol.* **2024**, *478*, 130487.
87. Mao, K.; Du, Y.; Cao, H.; Peng, Y.; He, G.; Liang, Q.; Tu, J. Effects of Cr_3C_2 addition on microstructure and corrosion properties of Cr_3C_2/15–5PH composite coatings on 12Cr13 by laser cladding. *Mater. Today Commun.* **2024**, *41*, 110375. [CrossRef]
88. Wang, Y.; Feng, Y.; Sun, X.; Liu, S.; Chen, G. Effect of Process Parameters on the Microstructure and Wear Resistance of Fe3Al/Cr_3C_2 Composites. *Coatings* **2024**, *14*, 384. [CrossRef]
89. Fan, X.; Li, W.; Yang, J.; Zhu, S.; Cui, S.; Li, X.; Shao, L.; Zhai, H. Effect of Cr_3C_2 content on microstructure, mechanical and tribological properties of Ni3Al-based coatings. *Surf. Coat. Technol.* **2023**, *466*, 129597. [CrossRef]

90. Aghili, S.E.; Shamanian, M.; Amini Najafabadi, R.; Keshavarzkermani, A.; Esmaeilizadeh, R.; Ali, U.; Marzbanrad, E.; Toyserkani, E. Microstructure and oxidation behavior of NiCr-chromium carbides coating prepared by powder-fed laser cladding on titanium aluminide substrate. *Ceram. Int.* **2020**, *46*, 1668–1679. [CrossRef]
91. Aghili, S.E.; Shamanian, M. Investigation of powder fed laser cladding of NiCr-chromium carbides single-tracks on titanium aluminide substrate. *Opt. Laser Technol.* **2019**, *119*, 105652. [CrossRef]
92. Venkatesh, L.; Samajdar, I.; Tak, M.; Doherty, R.D.; Gundakaram, R.C.; Prasad, K.S.; Joshi, S.V. Microstructure and phase evolution in laser clad chromium carbide-NiCrMoNb. *Appl. Surf. Sci.* **2015**, *357*, 2391–2401. [CrossRef]
93. Li, Z.; Yan, H.; Zhang, P.; Guo, J.; Yu, Z.; Ringsberg, J.W. Improving surface resistance to wear and corrosion of nickel-aluminum bronze by laser-clad TaC/Co-based alloy composite coatings. *Surf. Coat. Technol.* **2021**, *405*, 126592. [CrossRef]
94. Yu, T.; Tang, H. Microstructure and high-temperature wear behavior of laser clad TaC-reinforced Ni-Al-Cr coating. *Appl. Surf. Sci.* **2022**, *592*, 153263. [CrossRef]
95. Hu, D.; Liu, Y.; Chen, H.; Liu, J.; Wang, M. Effect of TiC addition on the microstructure and properties of Ni3Ta–TaC reinforced Ni-based wear-resistant coating. *Ceram. Int.* **2021**, *47*, 23194–23202. [CrossRef]
96. Hu, D.; Liu, Y.; Chen, H.; Wang, M.; Liu, J. Microstructure and properties of in-situ synthesized Ni3Ta-TaC reinforced Ni-based coatings by laser cladding. *Surf. Coat. Technol.* **2021**, *405*, 126599. [CrossRef]
97. Liu, Y.; Ding, T.; Lv, H.; Hu, D.; Zhang, Y.; Chen, H.; Chen, Y.; She, J. Microstructure and properties of Ta-reinforced cobalt based composite coatings processed by direct laser deposition. *Surf. Coat. Technol.* **2022**, *447*, 128874. [CrossRef]

molecules

Article

The Influence of Y_2O_3 Dosage on the Performance of Fe60/WC Laser Cladding Coating

Haiyan Jiang [1], Dazhi Jiang [2], Chenguang Guo [2] and Xiaodong Hong [3,*]

[1] Basic Department, Liaoning Institute of Science and Technology, Benxi 117004, China
[2] School of Mechanical Engineering, Liaoning Technical University, Fuxin 123000, China
[3] School of Materials and Energy, Foshan University, Foshan 528000, China
* Correspondence: hongxiaodong@lntu.edu.cn

Abstract: To prepare a high-performance Fe-based laser cladding coating, herein, various Fe60/WC/
Y_2O_3 coatings are deposited on the surface of 42CrMo steel plate via a laser cladding technique. The WC dosage is fixed as 10 wt%, while the dosage of Y_2O_3 ranges from 0 to 7.5 wt%. The influences of Y_2O_3 dosage on the coating hardness, wear resistance, and corrosion resistance are investigated. With the addition of Y_2O_3, the feature peak of WC disappears, and the peaks of $M_{23}C_6$ gradually weaken, indicating that Y_2O_3 promotes the decomposition of WC and suppresses the formation of new metal carbides. When the dosage of Y_2O_3 is 2.5 wt%, a grid-like structure is formed on the coating surface, suggesting uniform distribution of decomposed W within the Fe matrix. When the Y_2O_3 dosage exceeds 5 wt%, a large amount of CO_2 gas is released, leading to an increase in surface pores. Through a comparison, the optimal dosage of Y_2O_3 is 2.5 wt%, and the resulting 3# coating has the highest hardness of 861.97 HV. Moreover, the 3# coating also shows the minimum friction coefficient and the minimum wear volume, reflecting its superior wear resistance. The polished coating serves as a working electrode, and the corrosion resistance is tested in 3.5% NaCl solution. The sample containing 2.5 wt% Y_2O_3 has the highest corrosion potential and the lowest corrosion current density, indicating excellent corrosion resistance. The enhanced performance is ascribed to the improved surface quality and the formation of a W-reinforced grid structure. The high-performance coating has promising application potential in material and component repair.

Keywords: laser cladding coating; WC; Y_2O_3; wear resistance; corrosion resistance

1. Introduction

Surface modification technology plays a crucial role in improving materials' properties and extending their service lives [1]. Among the existing surface modification techniques, laser cladding coating technology is a process method in which a high-energy laser beam is engaged to rapidly melt cladding materials, such as alloy powder, ceramics, etc., onto the substrate surface [2]. This forms a metallurgical bonded surface coating, thereby significantly improving the wear resistance, corrosion resistance, and other properties of the substrate materials [3]. In terms of specific applications, laser cladding technology is a new technology with many economic benefits. It can prepare high-performance alloy surfaces on inexpensive metal substrates without affecting the properties of the substrate, reduce costs, and save precious and rare metal materials [4]. Therefore, researchers have attached great importance to research on and applications of laser cladding technology. In addition to studies on metal or alloy cladding coatings, various metal carbides have been widely

 https://doi.org/10.3390/molecules30234598

used as reinforcing phases for fabricating carbide-reinforced metal-based cladding coatings, further enhancing the overall properties [5,6]. For example, WC is often used to enhance the performance of alloy coatings [7] owing to its high hardness and excellent wear resistance. Introducing WC into Fe-based coatings promotes the formation of a dense and uniform microstructure, effectively improving both the wear and corrosion resistance of the coating. However, an excessive amount of WC will compromise the surface quality of the sample and generate a large number of pores, thereby decreasing the coating performance [8]. To reduce the WC dosage and avoid performance degradation, appropriate amounts of rare earth oxides have been proven effective in significantly improving the wear and corrosion resistance of coatings [9,10]. In recent years, rare earth oxide Y_2O_3 has exhibited great potential for material modification owing to its high melting point and superior corrosion resistance [11]. When applied in Fe-based laser cladding coating, Y_2O_3 acts as a refiner, leading to enhanced hardness and wear resistance by grain refinement and microstructure optimization. For example, Li et al. [12] introduced TiB/TiC/Y_2O_3 reinforcement phases into a Ti-6Al-4V matrix using laser cladding technology and discussed the impact of Y_2O_3 on the coating microstructure and mechanical properties. The results demonstrated that the addition of Y_2O_3 refined the primary phase structure, resulting in the uniform distribution of the internal constituents. Furthermore, the coating containing Y_2O_3 exhibited improved resistance to micro-cutting and cracking. Liang et al. [13] fabricated Ni/WC/Y_2O_3 coatings on 316L stainless steel with different dosages of Y_2O_3. The results indicated that an appropriate amount of Y_2O_3 effectively refined the microstructure and suppressed the precipitation of hard carbide phases. Moreover, the incorporation of rare earth elements enhanced the solid solution's strengthening effect on the cladding coating, leading to improved wear resistance and electrochemical corrosion resistance of composite coatings. Tao et al. [14] developed a Ni-based alloy coating reinforced by WC micro-particles and nano-Y_2O_3 on ultra-high-strength steel. The results confirmed that the addition of nano Y_2O_3 inhibited the dissolution of WC during the cladding process. The 2 μm WC-Y_2O_3/Ni composite exhibited the lowest friction coefficient. The existing literature indicates that the synergistic effect between WC and Y_2O_3 greatly enhances the performance of Ti- and Ni-based laser cladding coatings. However, works have seldom focused on the coating performance of WC/Y_2O_3-reinforced Fe-based cladding coatings. Herein, this work investigates the effect of Y_2O_3 content on the microstructure, hardness, wear resistance, and corrosion resistance of Fe60/WC cladding coating containing 10 wt% WC. The research result offers significant guidance for fabricating high-performance Fe-based coatings, paving the way for their application in advanced material and component repair.

2. Results and Discussion

Figure 1a shows spherical Fe60 particles with an average particle size of 50–100 μm. Following the method in Table 1, five types of mixed powders were prepared by ball milling for 4h, and their surface morphologies are illustrated in Figure 1b–e. After adding WC particles, the 1# powder in Figure 1b consists of spherical Fe60 particles mixed with irregular WC particles. Figure 1c–e correspond to the 2#–5# powders containing different fractions of Y_2O_3. As the Y_2O_3 fraction rises, the proportion of spherical particles on the sample surface obviously decreases, while the fine debris increases, indicating the presence of a large number of fine Y_2O_3 particles in the mixed powder. Especially for the 5# powder, there are a few big particles and a large number of fine particles, reflecting the effective integration of Fe60, WC, and Y_2O_3 particles.

Five types of mixed powders were used for laser cladding onto the surface of 42CrMo steel plate, and the cladding coating is shown in Figure 2. From the 1# sample without Y_2O_3, a metallic luster is observed on coating surface. As the amount of Y_2O_3 increases, the cladding coating color gradually darkens. When the amount of Y_2O_3 is more than 5 wt%, a significant number of pores appear on the surfaces of the 4# coating and 5# coating. The 5# coating shows notably poorer surface quality than the 4# coating. During the laser cladding process, the amount of gas generated in the laser cladding equipment significantly increases with an increase in Y_2O_3 dosage due to the decomposition of metal oxides at a high temperature. The decomposed Y_2O_3 releases a large amount of oxygen, which reacts with the carbon in Fe60 matrix to form CO_2 gas. This gas evolution results in the formation of many pores on the 4# and 5# coatings. The presence of extensive pores deteriorates surface quality and decays mechanical properties of the cladding coatings.

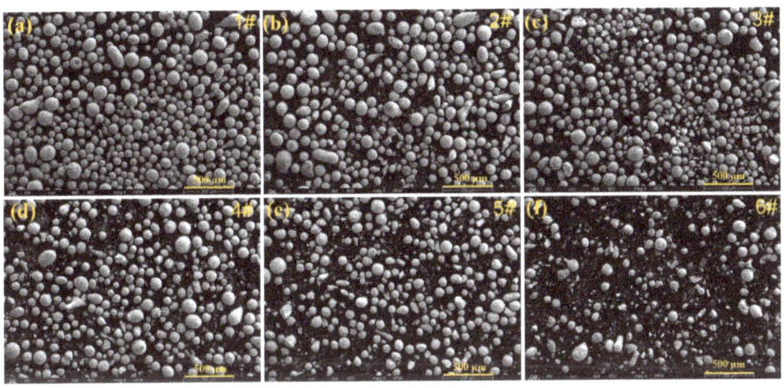

Figure 1. Morphology of various powders: (**a**) Fe60 powders, (**b**) Fe60 + WC, (**c**–**f**) Fe60 + WC + xY_2O_3.

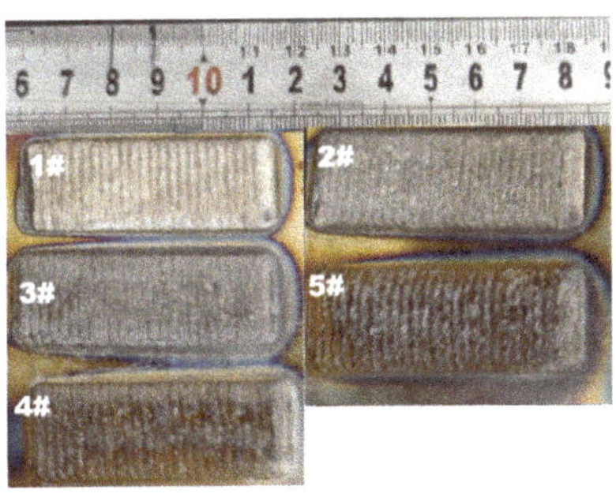

Figure 2. Surface photos of five kinds of laser cladding coatings.

X-ray diffraction (XRD) analysis was performed to reveal the influence of Y_2O_3 dosage on the phase composition and structure of F60/WC cladding coating. In Figure 3, the XRD pattern of the 1# Fe60/WC coating exhibits a strong peak of WC along with weaker peaks of α-Fe and $M_{23}C_6$ carbides [15], reflecting the presence of residual WC and the generation of new metal carbides. With an increase in Y_2O_3 addition, the peak of WC disappears, and the peak intensity of $M_{23}C_6$ gradually weakens, suggesting that Y_2O_3 promotes the decomposition of WC. Furthermore, the addition of Y_2O_3 facilitates the phase transition from α-Fe to γ-Fe. When the content of Y_2O_3 reaches 7.5 wt%, the peaks of carbides are nearly absent in the 5# sample. The results indicate that a higher Y_2O_3 content accelerates the decomposition of WC. During the laser cladding process, the amount of gas generated gradually increases with increasing Y_2O_3 content, and the resulting CO_2 gas release leads to a decrease in carbon content within the molten pool. Consequently, the excessive consumption of carbon reduces the formation of $M_{23}C_6$ carbides [16].

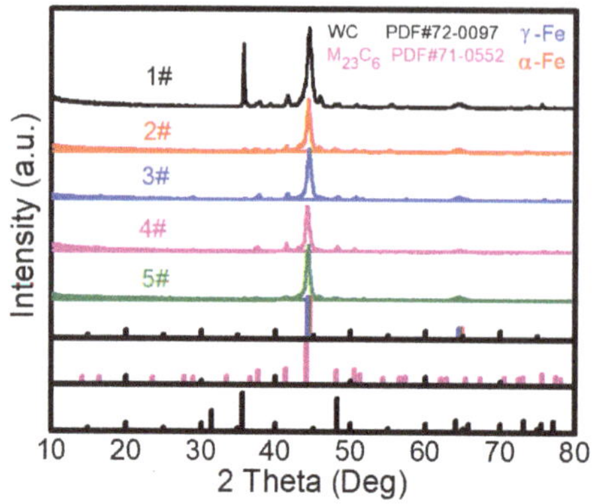

Figure 3. XRD patterns of five kinds of laser cladding coatings.

The metallographic morphologies of the interfacial region and corresponding cladding coating are shown in Figure 4. The lower part of the fusion zone exhibits a Weibull structure, which is a characteristic of the welding overheating zone. The upper fusion layer consists mainly of a eutectic structure. The presence of planar and cellular crystals in the fusion zone of all samples indicates sound metallurgical bonding between the coating and the substrate. The addition of WC in the 1# coating leads to a hypoeutectic structure, which is composed of primary γ-Fe and a eutectic structure. With the introduction of Y_2O_3, the eutectic content in the hypoeutectic structure initially increases and then decreases. A more distinct eutectic structure generally corresponds to a higher carbide content. The presence of carbide phases within the γ-Fe contributes to enhanced wear resistance. These results imply that a moderate amount of Y_2O_3 effectively promotes eutectic formation, whereas excessive Y_2O_3 is detrimental to its development [17]. In Figure 4e, the amount of eutectic structure does not increase further, which is related to the excessive consumption of carbon in the molten pool due to the reaction with Y_2O_3, resulting in a reduced carbon content available for carbide formation.

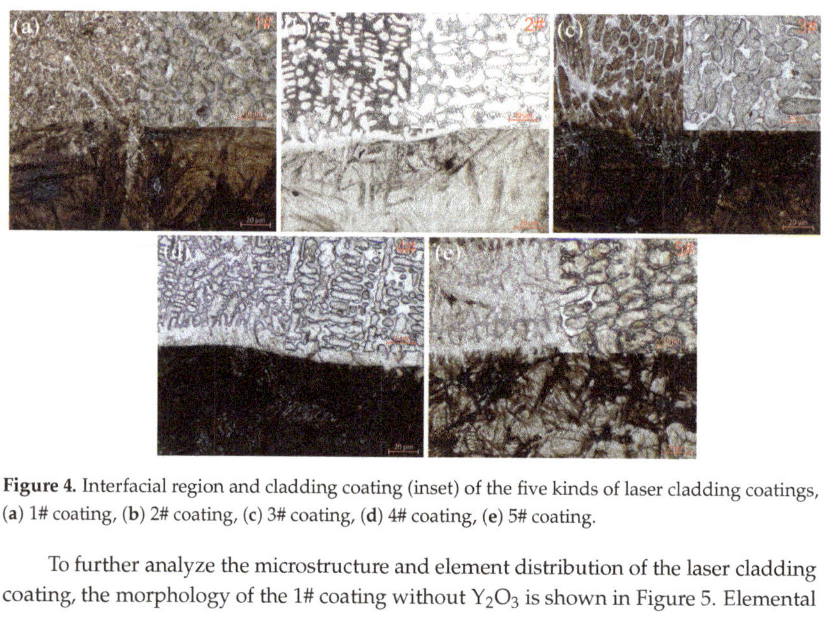

Figure 4. Interfacial region and cladding coating (inset) of the five kinds of laser cladding coatings, (**a**) 1# coating, (**b**) 2# coating, (**c**) 3# coating, (**d**) 4# coating, (**e**) 5# coating.

To further analyze the microstructure and element distribution of the laser cladding coating, the morphology of the 1# coating without Y_2O_3 is shown in Figure 5. Elemental mapping was conducted in the region containing residual WC particle. It was observed that the white particle was enriched in W and C, indicating the particle identity as residual WC-reinforcing phase. The signals of Fe and Cr in the adjacent matrix region are complementary to these W signals, indicating that WC is embedded within the Fe matrix. The diffusion of W from the bright particle into the surrounding area can be seen clearly, indicating that the WC particle undergoes decomposition during the laser cladding process, and the dissolved W atoms diffuse into the Fe matrix.

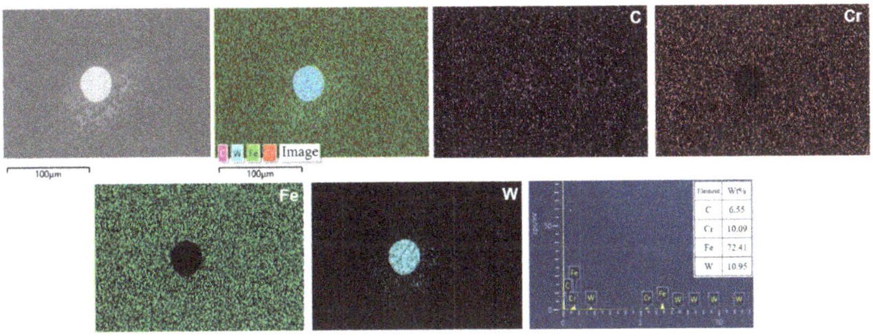

Figure 5. Microstructure and elemental mapping of 1# cladding coating without Y_2O_3.

To reveal the effect of Y_2O_3 on the microstructure of Fe60/WC coating, Figure 6 provides the morphology and element distribution of the 3# coating with 2.5% Y_2O_3. It is difficult to find areas of residual WC on the coating surface, and a grid-like structure can be observed clearly. The distribution map of W reveals that W is mainly concentrated within these grid regions, whereas the depressed areas correspond to the ferrite matrix. This microstructural evolution is attributed to the role of Y_2O_3 in promoting the decomposition of WC during laser cladding. The resulting W atoms are uniformly distributed within the newly generated grid framework, thereby reinforcing the Fe matrix. In addition, the C fraction in 3# coating is only 1.33 wt%, much lower than the 6.55 wt% in the 1# coating (Figure 5). This drastic reduction verifies that the addition of Y_2O_3 particles greatly decreases the C fraction in the Fe matrix. Furthermore, the overall decrease in the W fraction suggests its successful dissolution and homogeneous dispersion within the Fe matrix.

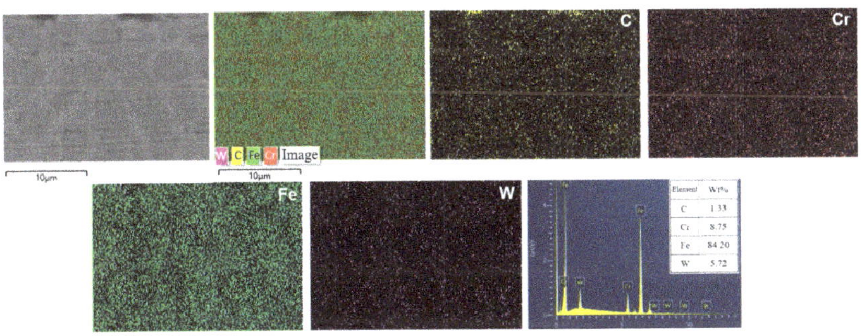

Figure 6. Microstructure and elemental mapping of 3# coating containing 2.5% Y_2O_3.

Figure 7a shows the microhardness curves of different cladding coatings. The hardness of these coatings shows a gradual increasing trend from the steel plate substrate towards the cladding coating, indicating that the cladding coating region possesses superior hardness compared to the substrate. While the substrate hardness remains relatively consistent, the coatings themselves exhibit significant variation. As the amount of Y_2O_3 increases, the hardness of the cladding coating shows a general decreasing trend. When the addition amount of Y_2O_3 is 2.5 wt%, the 3# coating has the highest average hardness of 861.97 HV. When the addition amount exceeds 2.5 wt%, the hardness of the cladding coating tends to decrease, which is attributed to the increased surface porosity caused by the gas evolution during cladding. In addition to hardness, friction tests were conducted on these cladding coatings by using Si_3N_4 balls under a 10 N load, a speed of 120 mm min^{-1}, and a stroke of 7 mm. The friction coefficients of different cladding coatings are shown in Figure 7b. During the initial 5 min of the test, all friction coefficient curves fluctuate significantly. During the subsequent wear process, the fluctuation of the curve is alleviated. After 20 min, these fluctuations are mitigated in the subsequent phase and gradually stabilize. The average friction coefficients for coatings 1#–5# are 0.711, 0.713, 0.675, 0.766, and 0.727, respectively. Among them, the 3# coating has a minimum friction coefficient of 0.675, which is in accordance with the highest hardness. Therefore, the 3# coating (Fe60/WC with 2.5 wt% Y_2O_3) exhibits the best wear resistance.

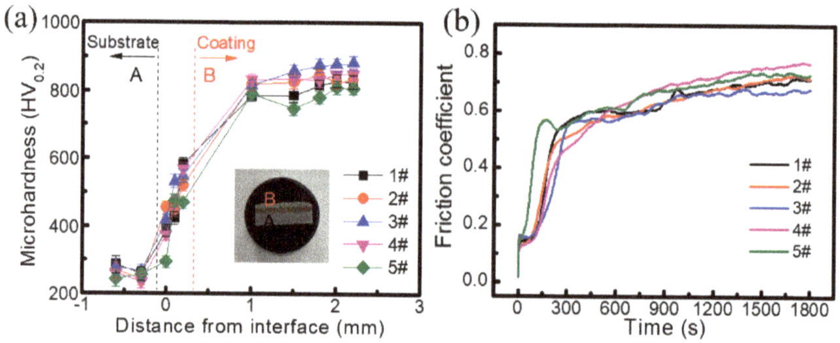

Figure 7. (**a**) Microhardness curves of different coatings upward along the fusion line and (**b**) friction coefficient curves of different coatings.

The wear volumes of the cladding coating after friction testing were measured by using a laser surface profilometer, as shown in Figure 8. The measured values for coatings 1#–5# are 2.7×10^{-3}, 2.5×10^{-3}, 1.8×10^{-3}, 2.8×10^{-3}, and 3.4×10^{-3} mm^3, respectively. Among them, the 3# coating presents the smallest wear volume. This result is completely consistent with its minimum friction coefficient, reflecting a positive correlation between wear volume and friction coefficient, further verifying that the Fe60/WC coating with 2.5 wt% Y_2O_3 possesses the optimal wear resistance.

Figure 9 shows the worn surface morphologies of the five cladding coatings. The 1# coating exhibits severe abrasive wear, characterized by the presence of large particles and debris particles. This can be attributed to its low hardness, which facilitates material detachment during sliding, thereby accelerating abrasive wear. In contrast, no such particles are observed on the wear region of the other four coatings containing Y_2O_3. Among them, the 3# coating shows a relatively smooth surface with only slight plastic deformation, which is consistent with its highest hardness. When the addition amount of Y_2O_3 reaches 7.5 wt%, it is difficult to observe wear marks on the worn surface of the 5# coating, which is likely due to the surface pores generated during sample fabricating. These surface pores increase the roughness and promote material loss, which increase its wear amount. The result is consistent with the largest wear volume shown in Figure 8f.

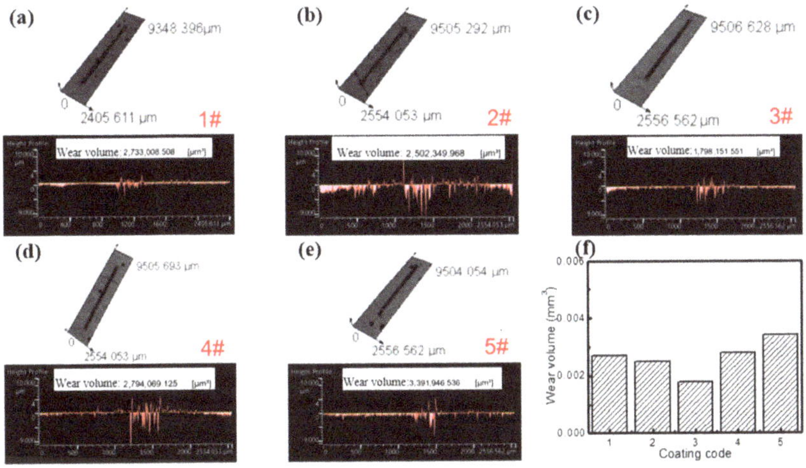

Figure 8. Wear volumes of different coatings and performance comparison, (**a**) 1# coating, (**b**) 2# coating, (**c**) 3# coating, (**d**) 4# coating, (**e**) 5# coating, (**f**) the comparison of different coatings.

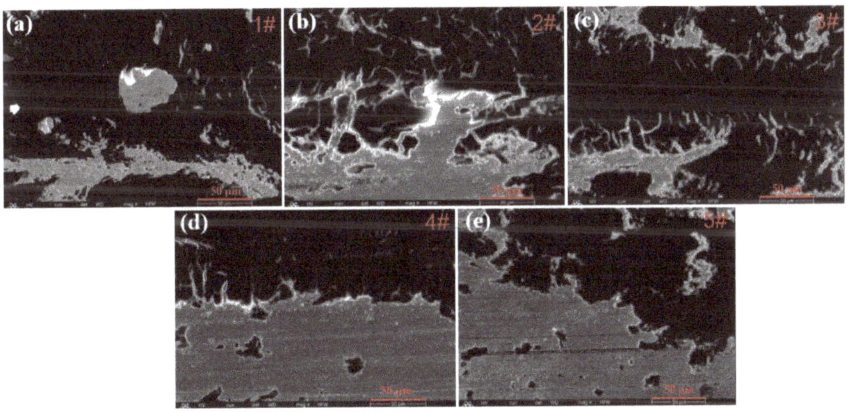

Figure 9. Morphologies of the five kinds of coatings after the friction test, (**a**) 1# coating, (**b**) 2# coating, (**c**) 3# coating, (**d**) 4# coating, (**e**) 5# coating.

The polished coating was encapsulated with epoxy resin and served as a working electrode. The dynamic potential polarization (Tafel) curves of different samples were measured in 3.5% NaCl solution [18,19] and are presented in Figure 10a. The corrosion current in the anode region increases with an increase in corrosion potential. There was no obvious anodic passivation zone observed in the five samples, which would have resulted in the direct contact of the electrolyte with the material surface and the formation of pitting corrosion. These Tafel curves were linearly fitted by using Origin software [20,21]. The intersection of two fitted lines was determined by extrapolation and is denoted as the corrosion potential and corrosion current density of the sample. The corrosion potential and corrosion current density of five cladding coatings were obtained and are illustrated in Table 1 and Figure 10b. The corrosion potentials (E_{corr}s) of the five coatings are −0.726 V, −0.837 V, −0.704 V, −0.812 V, and −0.838 V, respectively. The corrosion current densities (i_{corr}s) of the five coatings are 2.06×10^{-5}, 3.01×10^{-5}, 1.30×10^{-5}, 3.56×10^{-5}, and 3.91×10^{-5} A cm^{-2}, respectively. In corrosion analysis, E_{corr} and i_{corr} reflect the corrosion kinetics and thermodynamic properties of materials, respectively. A more negative E_{corr}

indicates a higher thermodynamic tendency for corrosion, whereas a larger i_{corr} corresponds to a faster kinetic rate of corrosion. Generally speaking, a higher corrosion potential and a lower corrosion current reflect the superior corrosion resistance of the coating material. Among these coatings, the 3# coating presents the most positive corrosion potential (−0.704 V) and the lowest corrosion current density (1.30×10^{-5} A cm^{-2}). Therefore, the 3# sample (Fe60/WC with 2.5% Y_2O_3) possesses the best corrosion resistance. The optimal corrosion resistance of 3# sample is attributed to the synergistic effect of its improved surface quality and the refined internal microstructure.

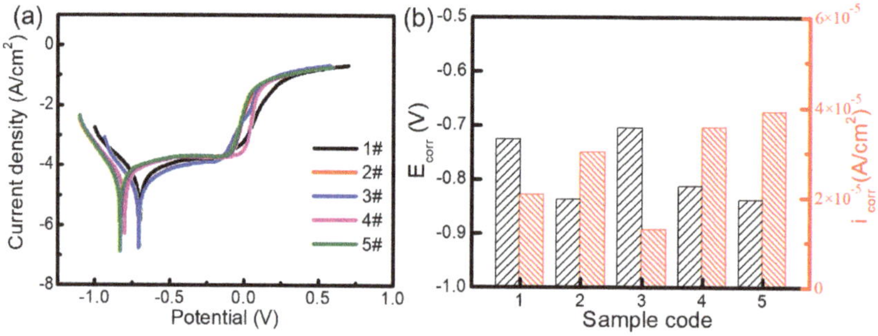

Figure 10. Electrochemical corrosion in 3.5% NaCl solution: (**a**) potentiodynamic polarization curves; (**b**) corrosion potential and corrosion current density.

Table 1. Electrochemical parameters of five coatings obtained from Figure 10a.

Sample	Solution	E_{corr}/V_{SCE}	i_{corr}/A cm^2
1#	3.5% NaCl	−0.726	2.06×10^{-5}
2#	3.5% NaCl	−0.837	3.01×10^{-5}
3#	3.5% NaCl	−0.704	1.30×10^{-5}
4#	3.5% NaCl	−0.812	3.56×10^{-5}
5#	3.5% NaCl	−0.838	3.91×10^{-5}

After electrochemical ×corrosion testing in 3.5% NaCl solution, the surface morphology of different cladding coatings was observed using SEM. In Figure 11a, the surface of the 1# sample (with only WC) displays a limited number of pits alongside moderate corrosion. With an addition of Y_2O_3, the surface morphology of the coating undergoes a significant change. Among these samples, the 5# sample shows the most extensive pitting area, mainly due to intergranular corrosion of Fe grains in the pits and producing a large amount of $Fe(OH)_3$ corrosion products. Therefore, an excessive amount of Y_2O_3 filler ultimately degrades the coating's corrosion resistance. Through a comparison, the 3# sample containing 2.5% Y_2O_3 and 10% WC presents the smallest pitting area, further proving the outstanding corrosion resistance. This observation is in excellent agreement with the electrochemical corrosion data presented in Figure 10b. To further verify the negative impact of a high Y_2O_3 fraction on the corrosion resistance of the Fe60/WC coating, Figure 11f provides a high-magnification SEM image of the corroded 5# coating. Abundant layered corrosion products are generated on the sample surface. Both the protruding and recessed areas form a porous structure stacked by $Fe(OH)_3$ nanosheets, with a size of several micrometers. The formation o a nanosheet morphology is associated with the excessive consumption of carbon in the molten pool, causing a great decrease in metal carbide faction in the eutectic structure. Consequently, the γ-Fe within the eutectic structure is severely corroded to produce metal hydroxides. In contrast, the 3# sample containing 10% WC and 2.5% Y_2O_3

exhibits the smallest pitting area, further demonstrating its superior corrosion resistance in the 3.5% NaCl solution.

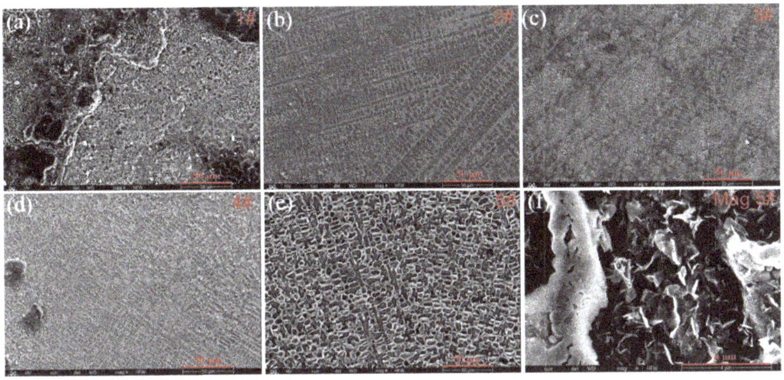

Figure 11. Microstructures of different coatings after the electrochemical test: (**a**) 1# coating, (**b**) 2# coating, (**c**) 3# coating, (**d**) 4# coating, (**e**) 5# coating, and (**f**) magnification of 5# coating.

3. Materials and Methods

3.1. Fabrication of Laser Cladding Coatings

Polished 42CrMo steel plate (Xingwang Metal, Dongguan, China) was selected as the substrate material. Commercialized Fe60 powders (Yaoyi Alloys, Shanghai, China) were used as metal matrix, and the components are provided in Table 2. WC powders (~50 μm) and Y_2O_3 powders (10~20 μm) were selected as ceramic fillers and modifiers. Mixed powders of Fe60, WC, and Y_2O_3 were ball-milled for 4 h by using zirconia spheres. An MFSC 2000W fiber laser device (Chuangxin Laser Co., Ltd., Shanghai, China) was used to prepare laser cladding coating with a power of 1000 W and 2 mm laser spot. The laser cladding process was performed under Ar gas. The scanning speed was 6 mm s^{-1}, and the overlap rate was 50%. Synchronous powder feeding was conducted with a gas-powder speed of 10 L·min^{-1}. Five kinds of cladding coatings were prepared on 42CrMo steel plate, as illustrated in Table 3.

Table 2. The components of Fe60 powders.

Fe	C	Cr	Si	Ni	B
60.2 wt%	2.8 wt%	20.5 wt%	3.0 wt%	9.5 wt%	4.0 wt%

Table 3. Material contents of different cladding coatings.

Sample Code	Fe60 (g)	WC (g)	Y_2O_3 (g)	Y_2O_3 Fraction
1#	36	4	0	0%
2#	35	4	0.5	1.3%
3#	34	4	1	2.5%
4#	33	4	2	5.0%
5#	33	4	3	7.5%

3.2. Material Characterization and Measurement

Scanning electron microscope by Ultra 60 (Zeiss, Shanghai, China) was used for observing the microstructure of the powder and cladding coating, coupled with an energy-dispersive spectrometer (EDS) for measuring the elemental contents. X-ray diffractometer coded as XRD-6100 (Shimazu, Guangzhou, China) was utilized for characterizing the crystal structure and composition, and the scanning rate was $5^{\circ}\cdot min^{-1}$. Prior to metallographic structure, the cladding coating needed to be etched by using a dilute $HCl/FeCl_3$ solution with a mass ratio of 1/10. Then, anhydrous ethanol was used to rinse the etched surface.

Vickers hardness tester (FM-810, Future-Tech, Tokyo, Japan) was used to measure the coating hardness, in which a 1000 g load was applied on the coating surface for 12 s to obtain the hardness value. Rtec rotary friction testing machine (MFT-5000, Rtec-Instruments, New York, NY, USA) was used to test the friction coefficient of different coatings, and the wear volume was obtained via laser surface profilometer (OLS5100-SAF, Olympus, Tianjin, China). During the friction process, Si_3N_4 milling ball was used by applying a 10 N load. The sliding speed was fixed as 120 $mm\cdot min^{-1}$. In corrosion testing, the polished cladding coating was used as a working electrode, and 3.5 wt% NaCl solution was used as an electrolyte [22]. The counter electrode was platinum sheet (1.5×1.5 cm^2), and reference electrode was Calomel electrode. A CHI660E electrochemical workstation (Chenhua 660E, Chenhua, Shanghai, China) was used to measure the potentiodynamic polarization (Tafel) curve of different cladding coatings.

4. Conclusions

To fabricate high-performance Fe-based laser cladding coatings, various $Fe60/WC/Y_2O_3$ coatings were deposited on the surface of 42CrMo steel substrates via a laser cladding technique. The influences of the Y_2O_3 content on the microstructure, hardness, wear resistance, and corrosion resistance were investigated systematically. The main findings are summarized as follows:

(1) With the addition of Y_2O_3, the characteristic diffraction peaks of WC disappear, while the peaks of $M_{23}C_6$ gradually weaken. This indicates that the introduction of Y_2O_3 promotes the decomposition of WC and inhibits the formation of metal carbides. SEM result verifies the formation of abundant grid-like structures in the 3# sample with 2.5 wt% Y_2O_3, further demonstrating the uniform distribution of decomposed W within the Fe matrix. When the Y_2O_3 content is more than 5 wt%, excessive Y_2O_3 introduces a large amount of O, which reacts with C to form CO_2 gas. This reaction reduces the content of newly generated carbides and increases the number of surface pores.

(2) The dosage of Y_2O_3 has a significant impact on the hardness of the cladding coating. The coating hardness initially increases and then decreases with increasing Y_2O_3 addition. When the content of Y_2O_3 is 2.5 wt%, the 3# sample presents the maximum average hardness of 861.97 HV, which is 3.3 times that of the substrate. The friction coefficient of different cladding coatings was tested using Si_3N_4 grinding balls. The 3# sample also exhibits the lowest friction coefficient (0.675) and the smallest wear volume of 1.8×10^{-3} mm^3. Therefore, the $Fe60/WC/Y_2O_3$ cladding coating with 2.5 wt% Y_2O_3 demonstrates the optimal wear resistance.

(3) Electrochemical measurement was conducted using polished cladding coating encapsulated with epoxy resin as the working electrode. Tafel curves of different samples were tested. The 3# sample (with 2.5% Y_2O_3) presents the most positive corrosion potential (−0.704 V) and the lowest corrosion current density (1.30×10^{-5} A cm^{-2}). This indicates its superior corrosion resistance in 3.5% NaCl solution, which is attributed to the improved surface quality and the formation of a W-reinforced grid structure.

(4) The incorporation of Y_2O_3 additive enhances the interfacial bonding between the metal matrix and WC reinforcement by promoting the decomposition of WC. This microstructural refinement results in a significant improvement in the coating's hardness, wear resistance, and corrosion resistance. These findings provide a viable strategy for fabricating high-performance composite coatings on steel substrates, which can

substantially enhance the durability and reliability of critical equipment components. Consequently, this approach contributes to extending the service life of repaired parts, improving operational reliability, and reducing maintenance costs.

Author Contributions: Conceptualization, H.J. and X.H.; methodology, H.J.; investigation, D.J.; writing—review and editing, X.H. and C.G. All authors have read and agreed to the published version of the manuscript.

Funding: This research was funded by Guangdong Basic and Applied Basic Research Foundation, grant number 2022A1515140159.

Data Availability Statement: The original contributions presented in this study are included in the article. Further inquiries can be directed to the corresponding author.

Conflicts of Interest: The authors declare no conflict of interest.

References

1. Zhao, P.; Shi, Z.; Wang, X.; Li, Y.; Cao, Z.; Zhao, M.; Liang, J. A Review of the Laser Cladding of Metal-Based Alloys, Ceramic-Reinforced Composites, Amorphous Alloys, and High-Entropy Alloys on Aluminum Alloys. *Lubricants* **2023**, *11*, 482. [CrossRef]
2. Zhu, L.; Xue, P.; Lan, Q.; Meng, G.; Ren, Y.; Yang, Z.; Xu, P.; Liu, Z. Recent research and development status of laser cladding: A review. *Opt. Laser Technol.* **2021**, *138*, 106915. [CrossRef]
3. Liu, Y.; Ding, Y.; Yang, L.; Sun, R.; Zhang, T.; Yang, X. Research and progress of laser cladding on engineering alloys: A review. *J. Manuf. Process.* **2021**, *66*, 341–363. [CrossRef]
4. Wang, K.; Zhang, Z.; Xiang, D.; Ju, J. Research and Progress of Laser Cladding: Process, Materials and Applications. *Coatings* **2022**, *12*, 1382. [CrossRef]
5. Jiang, D.; Wang, G.; Dong, W.; Hong, X.; Guo, C. Recent Advance on Metal Carbides Reinforced Laser Cladding Coatings. *Molecules* **2025**, *30*, 1820. [CrossRef]
6. Sun, S.; Wang, J.; Xu, J.; Cheng, X.; Jing, C.; Chen, Z.; Ru, H.; Liu, Y.; Jiao, J. Preparing WC-Ni coatings with laser cladding technology: A review. *Mater. Today Commun.* **2023**, *37*, 106939. [CrossRef]
7. Liu, L.; Lin, Y.; Peng, L.; Kang, X.; Wang, X. Progress in Microstructure Design and Control of High-Hardness Fe-Based Alloy Coatings via Laser Cladding. *Coatings* **2024**, *14*, 1351. [CrossRef]
8. Wu, T.; Shi, W.; Xie, L.; Gong, M.; Huang, J.; Xie, Y.; He, K. Study on the effect of Ni60 transition coating on microstructure and mechanical properties of Fe/WC composite coating by laser cladding. *Opt. Laser Technol.* **2023**, *163*, 109387. [CrossRef]
9. Yang, Z.Z.; Hao, H.; Gao, Q.; Cao, Y.B.; Han, R.H.; Qi, H.B. Strengthening mechanism and high-temperature properties of H13 + WC/Y_2O_3 laser-cladding coatings. *Surf. Coat. Technol.* **2021**, *405*, 126544. [CrossRef]
10. Zhang, Z.; Yang, F.; Zhang, H.; Zhang, T.; Wang, H.; Xu, Y.; Ma, Q. Influence of CeO_2 addition on forming quality and microstructure of TiC_x-reinforced $CrTi_4$-based laser cladding composite coating. *Mater. Charact.* **2021**, *171*, 110732. [CrossRef]
11. Zhang, Z.; Yang, Q.; Yu, Z.; Wang, H.; Zhang, T. Influence of Y_2O_3 addition on the microstructure of TiC reinforced Ti-based composite coating prepared by laser cladding. *Mater. Charact.* **2022**, *189*, 111962. [CrossRef]
12. Li, J.; Luo, X.; Li, G.J. Effect of Y_2O_3 on the sliding wear resistance of TiB/TiC-reinforced composite coatings fabricated by laser cladding. *Wear* **2014**, *310*, 72–82. [CrossRef]
13. Liang, F.; Li, K.; Shi, W.; Zhu, Z. Effect of Y_2O_3 Content on Microstructure and Corrosion Properties of Laser Cladding Ni-Based/WC Composite Coated on 316L Substrate. *Coatings* **2023**, *13*, 1532. [CrossRef]
14. Tao, L.; Yang, Y.; Zhu, W.; Sun, J.; Wu, J.; Xu, H.; Yan, L.; Yang, A.; Xu, Z. Stress Distribution in Wear Analysis of Nano-Y_2O_3 Dispersion Strengthened Ni-Based μm-WC Composite Material Laser Coating. *Materials* **2024**, *17*, 121. [CrossRef]
15. Li, J.; Zhu, Z.; Peng, Y.; Shen, G. Phase evolution and wear resistance of in-situ synthesized $(Cr, W)_{23}C_6$-WC composite ceramics reinforced Fe-based composite coatings produced by laser cladding. *Vacuum* **2021**, *190*, 110242. [CrossRef]
16. Du, M.; Wang, L.; Gao, Z.; Yang, X.; Liu, T.; Zhan, X. Microstructure and element distribution characteristics of Y_2O_3 modulated WC reinforced coating on Invar alloys by laser cladding. *Opt. Laser Technol.* **2022**, *153*, 108205. [CrossRef]
17. Xu, Y.; Fu, S.; Lu, H.; Li, W. Process optimization, microstructure characterization, and tribological performance of Y_2O_3 modified Ti_6Al_4V-WC gradient coating produced by laser cladding. *Surf. Coat. Technol.* **2024**, *478*, 130496. [CrossRef]
18. Li, Y.Z.; Wang, X.; Zhang, G.A. Corrosion behaviour of 13Cr stainless steel under stress and crevice in 3.5 wt.% NaCl solution. *Corros. Sci.* **2020**, *163*, 108290. [CrossRef]
19. Feng, L.; Yang, Y.; Zhao, Y.; Ma, K.; Cui, J. Corrosion behaviors and mechanism of $Al_xCrFeMnCu$ high-entropy alloys in a 3.5 wt% NaCl solution. *Corros. Sci.* **2024**, *233*, 112087. [CrossRef]

20. Wu, L.; Ma, A.; Zhang, L.; Zheng, Y. Intergranular erosion corrosion of pure copper tube in flowing NaCl solution. *Corros. Sci.* **2022**, *201*, 110304. [CrossRef]
21. Kim, S.H.; Kang, C.Y.; Bang, K.S. Weld metal impact toughness of electron beam welded 9% Ni steel. *J. Mater. Sci.* **2001**, *36*, 1197–1200. [CrossRef]
22. Wu, H.; Wang, Z.Y.; Wang, M.S.; Wang, R.; Zhang, S.; Zhang, C.H.; Wu, C.L.; Chen, H.T.; Chen, J. Microstructure evolution, corrosion and corrosive wear properties of NbC-reinforced FeNiCoCr-based high entropy alloys coatings fabricated by laser cladding. *Eng. Fail. Anal.* **2025**, *171*, 109352. [CrossRef]

molecules

MDPI

Article

Enhanced Coercivity and Tb Distribution Optimization of Sintered Nd-Fe-B Magnets by TbF_3 Grain Boundary Diffusion Facilitated by Ga

Ling Wang [1,*], Wenjiao Li [1], Xiaopeng Wang [1], Zejun Deng [2] and Shujuan Gao [1,*]

[1] Department of Chemical and Material Engineering, Lyuliang University, Lvliang 033001, China; 20151031@llu.edu.cn (W.L.); wangxpfoxwish@163.com (X.W.)
[2] School of Materials Science and Engineering, Central South University, Changsha 410083, China; zejun.deng@csu.edu.cn
* Correspondence: 20121013@llu.edu.cn (L.W.); shujuangao@llu.edu.cn (S.G.)

Abstract: The grain boundary diffusion process employing a mixed diffusion source, comprising heavy rare-earth elements and low-melting metals, significantly enhances the coercivity (H_{cj}) of sintered Nd-Fe-B magnets. In the present study, Tb and Ga were deposited onto the surface of Nd-Fe-B magnets to serve as a diffusion source for improving hard magnetic properties. The effects of varying deposition sequences of Tb and Ga on the magnetic properties and microstructure of the magnets were analyzed. The findings demonstrate that TbF_3 grain boundary diffusion facilitated by Ga effectively increases the efficiency of Tb substitution, leading to enhanced coercivity. When Tb and Ga are deposited simultaneously, coercivity shows a notable improvement of 53.15% compared to the untreated magnet, with no reduction in remanence. Additionally, thermal stability is enhanced, resulting in superior overall magnetic properties. Microstructural analysis reveals that Ga promotes the diffusion of Tb into the magnet. In the magnet where Tb and Ga are co-deposited, the formation of a thinner and more uniform $(Nd,Tb)_2Fe_{14}B$ shell–core structure, along with the greater infiltration depth of Tb, leads to a broader distribution of core–shell structures within the magnet. This effectively increases the anisotropy fields (H_A) of the main phase grains, preventing the nucleation of antiferromagnetic domains at the edges of main-phase grains, thereby enhancing coercivity. Furthermore, the corrosion resistance of the magnet subjected to mixed diffusion is improved. This study provides a foundation for producing highly efficient magnets with a lower content of heavy rare-earth elements. The simplicity and flexibility of the process make it highly suitable for industrial applications.

Keywords: Nd-Fe-B; grain boundary diffusion; electrophoretic deposition; magnetic properties

1. Introduction

Permanent magnet materials are crucial in the realms of modern science, technology, and industry. From everyday items like smartphones, headphones, and hard disk drives, to core components in new energy vehicles, 5G communications, high-speed rail transportation, and advanced manufacturing equipment like MRI machines and precision instruments, permanent magnet materials are indispensable. These applications not only rely on the high energy products and stable magnetic performance of permanent magnetic materials, but also on their lightweight, high efficiency, and environmentally friendly and energy-saving characteristics. Among the various permanent magnet materials, Nd-Fe-B stands out due to its exceptionally high magnetic energy products and coercivity, making it the ideal choice for achieving high performance, miniaturization, and lightweight in electronic and electromechanical products.

As the development of electric vehicles and robotics technology continues, the status of Nd-Fe-B permanent magnets will become increasingly prominent. However, their low Curie temperature and poor stability at elevated temperatures restrict their applicability [1,2]. To enhance high-temperature coercivity, heavy rare earth (HRE) elements that can form a (HRE, $Nd)_2Fe_{14}B$ phase with higher magnetic anisotropy fields, such as Dy or Tb, are commonly introduced. However, the addition of Dy or Tb decreases the maximum energy product ($(BH)_{max}$) and remanence (B_r) due to the antiferromagnetic coupling between Dy or Tb atoms and Fe atoms within the (Nd, $HRE)_2Fe_{14}B$ lattice, leading to a reduction in saturation magnetization (M_s) [3,4]. Furthermore, given the limited availability of heavy rare earths, reliance solely on Dy or Tb for improving magnet performance has become increasingly impractical. Research has also demonstrated [5–7] that antiferromagnetic domains often nucleate near grain boundaries—specifically, in the peripheral regions of the main phase grains—before expanding and growing. Consequently, confining heavy rare earth elements to the outer regions of the main phase grains can achieve enhanced coercivity while minimizing the loss of remanence caused by HRE addition. This realization led to the development of the grain boundary diffusion process (GBDP) [8,9]. During this process, heavy rare earth elements such as Dy or Tb infiltrate the magnets predominantly through grain boundaries, forming an HRE-enriched shell with higher anisotropic fields and continuous grain boundary phases that improve magnetic isolation between main phases. This method has successfully reduced heavy rare earth usage, achieving both resource conservation and cost efficiency.

GBD technology has been extensively utilized for producing magnets with low heavy rare earth (HRE) content and high coercivity [10–13]. Among the critical processes in the GBD technique, coating preparation plays a pivotal role, directly affecting the productivity, coating quality, production costs, and ultimately, the diffusion efficiency of HRE elements into the magnet, influencing the cost-effectiveness of the final GBD product. Various methods have been employed for coating magnets, including dip coating, chemical bath deposition, chemical vapor deposition, vacuum evaporation plating, and sputtering of alloys or pure metals [14–17]. However, these techniques face challenges in large-scale industrial application, such as complex procedures, high costs, uneven coating, or poor interfacial adhesion. Recently, electrophoretic deposition (EPD) technology has garnered significant attention [18–21]. This method involves depositing DyF_3 or TbF_3 powders dispersed in a medium onto the magnet surface under an external electric field. Compared to other techniques, EPD offers advantages such as high deposition efficiency, uniform and dense film formation, lower deposition costs, controllable coating thickness, and simpler operations, making it more suitable for industrial-scale production.

Heavy rare earth elements, such as Dy, Tb, and their corresponding oxides, fluorides, and hydrides, are frequently employed as diffusion sources. In 2012, Soderznik et al. [22] first applied DyF_3 powder onto the surface of sintered magnets using EPD, resulting in a significant improvement in H_{cj} from 1200 to 1620 kA/m. In 2015, Cao et al. [23] used the EPD technique to deposit DyF_3 powder on the magnet surface, leading to a 50% increase in H_{cj} from 1274 to 1911 kA/m. In 2016, Cao et al. [24] continued their work by preparing TbF_3 diffusion magnets with varying RE content via the EPD method, suggesting that a higher RE content enhanced the Tb diffusion process. In 2023, Wang et al. [25] established that increasing the Tb content significantly boosted the coercivity of the magnet during the EPD process of TbF_3 powder. Nevertheless, substantial increases in the usage of HRE can result in considerable cost increases. Several recent studies have investigated the use of mixed diffusion sources, combining HRE with low melting point metals (such as Zn, Cu, Al) [26–30]. When compared with HRE diffusion magnets, mixed diffusion magnets show a more pronounced enhancement in H_{cj}. Moreover, the use of this mixed diffusion source further improves the utilization efficiency of HRE, while also lowering production costs.

Recent studies have demonstrated that adding small amounts of Ga can significantly enhance the magnetic properties of Nd-Fe-B sintered magnets [31,32]. The observed increase in coercivity has been associated with improved wettability of the liquid phase during sintering and the formation of additional Nd-rich phases at the grain boundaries. Current research on non-rare earth Ga mainly focuses on traditional doping methods, with limited reports on its use as a diffusion source for grain boundary diffusion.

In this study, TbF_3 and $Ga(NO_3)_3$ powders were deposited onto the surface of sintered Nd-Fe-B magnets via EPD to create a diffusion source. The GBDP was then completed through heat treatment. The impact of Ga-assisted Tb grain boundary diffusion on the microstructure, magnetic properties, and corrosion resistance of sintered Nd-Fe-B magnets was evaluated.

2. Results and Discussion

Figure 1a presents the demagnetization curves of both the original and diffused magnets at 298 K. As shown in Figure 1b, the coercivity (H_{cj}) of the original sample is 1527.87 kA/m, while the remanence (B_r) is 1.28 T. After the diffusion of Tb at the grain boundary, the coercivity of the "Tb magnet" increased to 1603.50 kA/m, representing a 4.95% improvement, while the remanence slightly decreased to 1.27 T. These findings suggest that Tb grain boundary diffusion effectively enhances coercivity, although a minor reduction in remanence occurs. It is known that Tb predominantly concentrates around the primary phase of the magnet following GBDP, aiding in the formation of the $Tb_2Fe_{14}B$ phase. Given that the anisotropy field of this phase is considerably higher than that of the $Nd_2Fe_{14}B$ phase and is positioned within the structure where demagnetization initiates [33], the demagnetization process is significantly reduced. In contrast to the "Tb magnet", the coercivity of the Tb and Ga mixed diffusion magnet was substantially increased, whether the deposition occurred layer by layer or simultaneously. The coercivity of the "Tb/Ga magnet", "Ga/Tb magnet", and "Tb-Ga magnet" reached 1682.32 kA/m (an increase of 10.11%), 1910.03 kA/m (25.01%), and 2339.97 kA/m (53.15%), respectively. As shown in prior studies [31,32,34], Ga elements lower the melting point of the Nd-Fe alloy, enhance the wettability of the grain boundary phase, and create effective diffusion paths. This results in improved GBDP efficiency, leading to an increase in the intrinsic coercivity of sintered magnets. It is worth noting that in the case of the "Tb/Ga magnet", after Tb deposition, the Tb film tends to peel off considerably in $Ga(NO_3)_3$ solution during the subsequent deposition stage, which causes a significant reduction in the Tb content in the diffusion source and, as a result, a less pronounced increase in coercivity.

The H_{cj} of the "Tb-Ga magnet" is the highest among the four, with an increase of up to 53.15%. This improvement is significantly greater than that observed in magnets diffused solely with Tb or those with layer-by-layer diffusion of Ga and Tb. This suggests that the simultaneous deposition of Tb and Ga provides a more substantial enhancement of coercivity.

In addition, unlike the "Tb magnet", the magnetic remanence (B_r) of the three magnets with mixed diffusion sources remained unchanged or slightly increased. Both the "Tb/Ga magnet" and "Ga/Tb magnet" exhibited minor enhancements in B_r. This implies that incorporating a low-melting-point element assisted in the diffusion of Tb, resulting in a marked improvement in coercivity while keeping B_r nearly constant. Consequently, this led to a noticeable increase in $(BH)_{max}$. Overall, introducing Ga into the diffusion source significantly improves the overall magnetic properties of the magnet and optimizes Tb utilization.

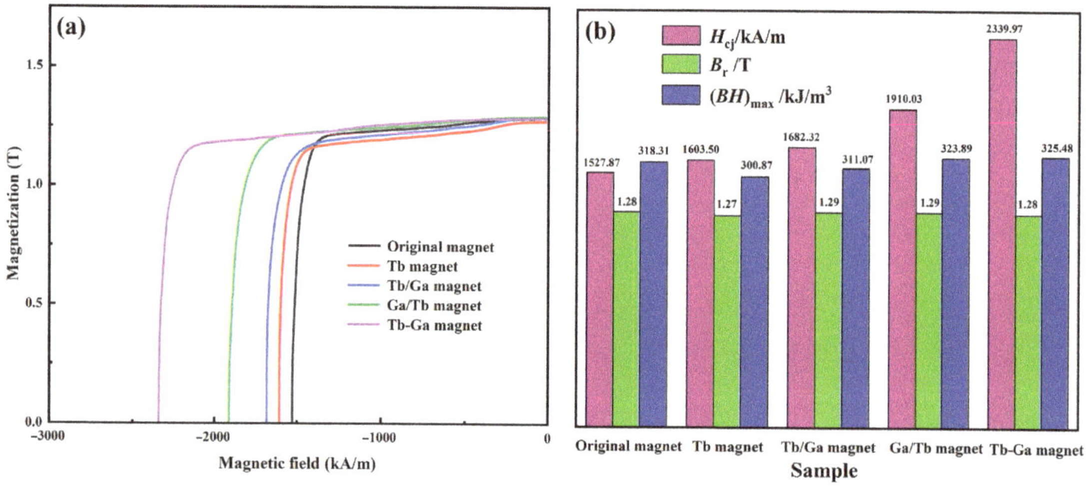

Figure 1. (**a**) Demagnetization curves of the original magnet, Tb magnet, Tb/Ga magnet, Ga/Tb magnet and Tb-Ga magnet at 298 K. (**b**) The histogram of the magnetic properties.

The temperature stability of a magnet indicates its capacity to preserve its magnetic properties under different external temperatures. The remanence temperature coefficient (α) and coercivity temperature coefficient (β) are commonly used to describe the temperature stability of magnets. They are defined as follows:

$$\alpha = \frac{B_r(T) - B_r(T_0)}{B_r(T_0)(T - T_0)} \quad (1)$$

$$\beta = \frac{H_{cj}(T) - H_{cj}(T_0)}{H_{cj}(T_0)(T - T_0)} \quad (2)$$

B_r (T_0) and B_r (T) represent the remanence of the samples at room temperature (T_0) and high temperature (T), respectively. H_{cj} (T_0) and H_{cj} (T) refer to the coercivity of the samples at room temperature (T_0) and high temperature (T). Figure 2a presents the remanence and B_r of the original magnet and the "Tb-Ga magnet" over the temperature range of 20 °C to 140 °C. Between 20 °C and 140 °C, the α value for the original magnet is −0.130%/°C, and for the "Tb-Ga magnet", it is −0.127%/°C, indicating that the addition of the Tb and Ga composite diffusion source improves the temperature stability of the remanence. Moreover, Figure 2a shows that the B_r of the diffused magnets undergoes only a slight reduction across various temperature conditions, suggesting that the co-diffusion of Tb and Ga has a minimal effect on the magnet's remanence. In comparison to the original magnet, the β value for the "Tb-Ga magnet" improves from −0.569%/°C to −0.542%/°C (as shown in Figure 2b), demonstrating a greater reduction in the absolute value of β. This is beneficial for maintaining better magnetic properties at higher temperatures. The coercivity temperature coefficient (β) is not only associated with the magnetic anisotropy field (H_A) of the main phase grains but also with the internal microstructure of the magnet. In the "Tb-Ga magnet", due to the diffusion of Tb elements, a $(Tb, Nd)_2Fe_{14}B$ shell structure with a larger H_A forms around the NdFeB grains (as supported by the microstructural characterization discussed later in Section 2), which aids in reducing the β absolute value of the magnet. Additionally, the Ga element, with its lower melting point, may assist in optimizing the magnet's microstructure, which also contributes to the reduction in the β absolute value. Consequently, β is effectively diminished. As a result, both the H_{cj} and thermal stability of the "Tb-Ga magnet" are improved at elevated temperatures.

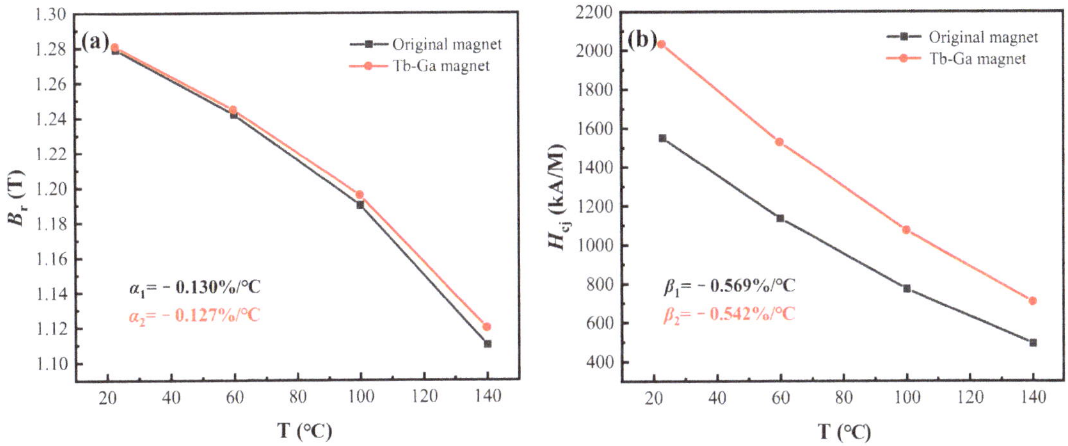

Figure 2. The temperature dependence for the B_r (**a**) and H_{cj} (**b**) of the original magnet and Tb-Ga magnet at the temperature range of 20–140 °C.

To examine the coercivity mechanism of magnets before and after diffusion, the nucleation field model [35,36] can be applied. The H_{cj} is described by the following equation:

$$H_{cj}(T) = aH_A(T) - N_{eff}M_S(T) \quad (3)$$

where $H_A(T)$ represents the anisotropy field, $M_s(T)$ is the saturation magnetization, parameter a reflects the reduction in the anisotropy field due to microstructural defects or c-axis misalignment, and N_{eff} represents the microstructural factors that influence the local effective demagnetization field, which are mainly affected by grain size, grain shape, the presence of RE-rich triple-junction phases, and intergranular exchange [37,38]. By performing a linear fit of H_{cj}/M_s versus H_A/M_s at different temperatures, the parameters a and N_{eff} for the original magnet and "Tb-Ga magnet" are determined, as shown in Figure 3. From the results, it can be observed that the sensitive factor a increased from 4.70 to 5.04, indicating that the interfacial defects between the main phase and the Nd-rich phase in the magnet were optimized after diffusion. Simultaneously, the demagnetization factor N_{eff} increased from 1011.59 to 1018.53, suggesting that the stray field within the magnet increased post-diffusion, which may be related to grain growth during the diffusion process. Additionally, the incorporation of Tb-Ga may lead to misalignment of grains near the additive. These factors likely contribute to the increase in N_{eff}, which significantly raises the local demagnetizing field [39].

Considering the increased intrinsic coercivity and improved thermal stability, it can be concluded that the Tb substitution efficiency in the "Tb-Ga magnet" is much better than in the other three diffusion magnets. The efficiency of Tb substitution in Nd-Fe-B sintered magnets is well known to be closely related to the distribution of the Tb element. To further explore the mechanism by which Ga enhances the efficiency of Tb substitution in sintered Nd-Fe-B magnets, the distribution of Tb from the diffusion interface to a depth of 80 μm was examined using EPMA, as shown in Figure 4. The results clearly show that the combined diffusion of Tb and Ga significantly reduces the accumulation of Tb at the diffusion surface of the magnets.

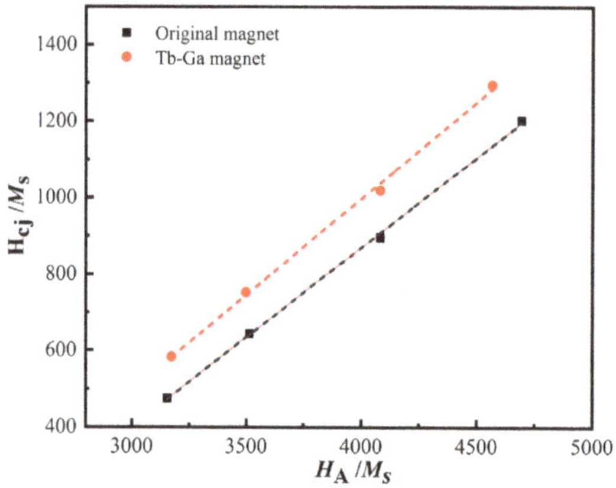

Figure 3. The dependence of H_{cj}/M_s vs. H_A/M_s of the original magnet and Tb-Ga magnet.

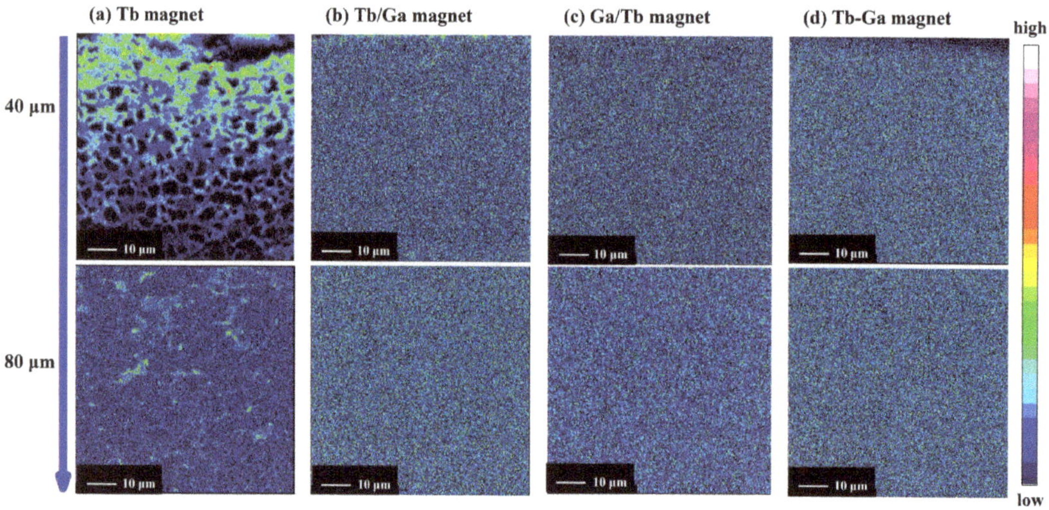

Figure 4. EPMA mappings of Tb in the 0–80 μm diffusion region for the Tb magnet (**a**), Tb/Ga magnet (**b**), Ga/Tb magnet (**c**), and Tb-Ga magnet (**d**).

Analysis of Figure 4 shows that in the "Tb magnet", a significant concentration of Tb atoms accumulates within a 20 μm region. Between 20 μm and 40 μm, the Tb atoms are distributed in a network-like pattern, and beyond 40 μm, the Tb content is very low. In contrast, the distribution of Tb atoms is much more uniform in the "Tb/Ga magnet", "Ga/Tb magnet", and "Tb-Ga magnet". At a diffusion depth of 80 μm, all three mixed diffusion magnets show relatively high and more even concentrations of Tb, indicating that Tb atoms are able to spread into GB phases and diffuse into the hard magnetic matrix phases. The absence of effective diffusion channels in the "Tb magnet" results in significant lattice diffusion of Tb in the shallow surface layers of the diffusion process. As a result, a larger quantity of Tb is unable to diffuse deeper into the layers along the grain boundaries, thereby reducing the efficiency of Tb diffusion. In contrast, the addition of low-melting-point elements in the "Tb/Ga magnet", "Ga/Tb magnet", and "Tb-Ga magnet" enhances the wettability of the Nd-rich grain boundary phases, and increases the efficiency of Tb diffusion along the grain boundaries.

Figure 5 shows the BSE-SEM images of the magnets at a diffusion depth of 100 μm from the diffusion surface, with the corresponding line scans provided in Figure 5e,f. As depicted in Figure 5(a1,a2), in the "Tb magnet", a large portion of the $Nd_2Fe_{14}B$ main phase grains (black, indicated by yellow arrow) are in direct contact. As shown in Figure 4a, the Tb concentration sharply decreases beyond 40 μm, making the formation of a core–shell structure difficult. This direct contact between the matrix grains results in exchange coupling among the ferromagnetic grains, which reduces the intrinsic coercivity of the sample.

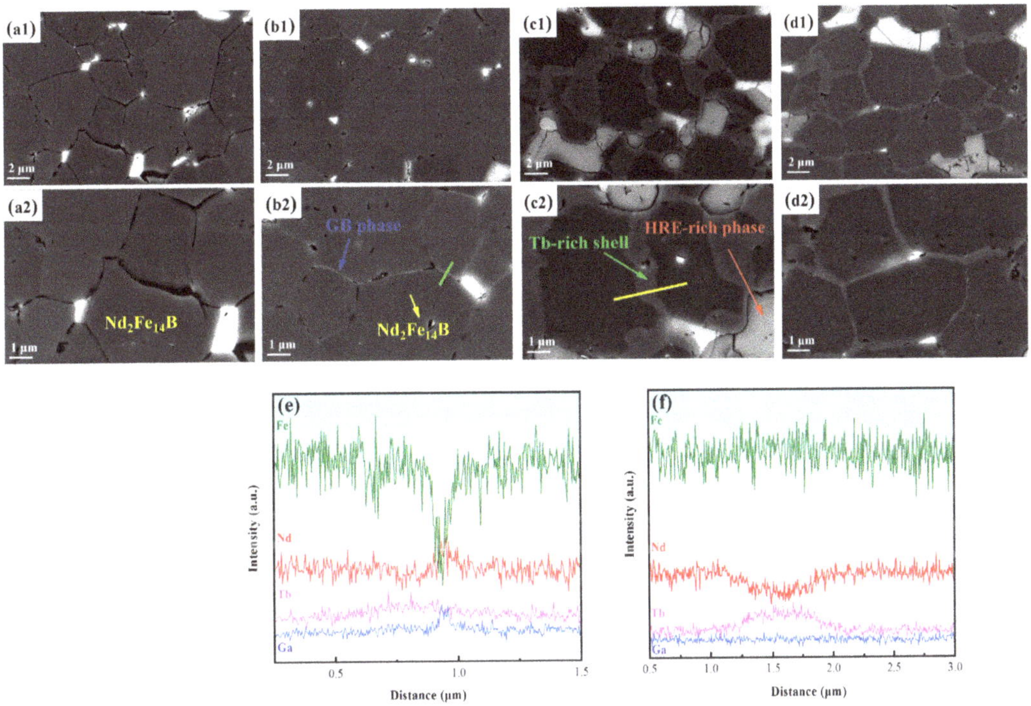

Figure 5. Cross-sectional BSE-SEM images of the Tb magnet (**a1,a2**), Tb/Ga magnet (**b1,b2**), Ga/Tb magnet (**c1,c2**), and Tb-Ga magnet (**d1,d2**). (**e**) Concentration profiles of line scanning in (**b**). (**f**) Concentration profiles of line scanning in (**c**).

The "Tb/Ga magnet" shows a predominantly smooth, thin, flaky grain boundary (GB) phase (white, indicated by blue arrow). Figure 5e indicates that the continuous GB phase is rich in Nd and Ga but lacks Fe. This supports the idea that the low-melting-point Ga element aids in forming thin grain boundaries. Ga's low melting point further reduces the melting point of the GB phase, improving its fluidity and wettability. This promotes the formation of thin grain boundary phases that surround the $Nd_2Fe_{14}B$ main phase, reducing the exchange coupling between adjacent ferromagnetic grains. As a result, the intrinsic coercivity (H_{cj}) of the magnets improves [40,41]. However, in the "Tb/Ga magnet", the core–shell structure is also not as distinct, though EDS results show that Tb slightly increases between the grain boundaries and grains. This is because, as mentioned earlier, a significant amount of Tb is lost during the Ga film deposition process, leading to a reduction in the Tb diffusion source. Consequently, only a small amount of Tb enters the main phase grains, which is insufficient to form a clear core–shell structure at the 100 μm depth observed.

In comparison to the "Tb magnet", the "Ga/Tb magnet" shows a thicker Tb-rich shell structure (gray, indicated by green arrow). Figure 5f illustrates that the shell structure is rich in Tb but lacks Nd, indicating that a considerable amount of Tb has infiltrated into the main phase grains, leading to the formation of the (Tb, Nd)$_2$Fe$_{14}$B phase. This is attributed to the low melting point element Ga, which enhances the fluidity and wettability of the GB phase, thereby creating more effective diffusion pathways. Since Tb mainly diffuses along the GB phase, the addition of Ga significantly promotes the movement of Tb into deeper parts of the magnet, causing a higher accumulation of Tb around the main phase. Consequently, Tb easily diffuses into the main phase, forming the $Tb_2Fe_{14}B$ phase, and resulting in the development of a thicker shell structure. This increases the nucleation field of reverse magnetic domains at the edge of main-phase grains, thereby enhancing coercivity [42,43]. However, it is known that overly thick Tb-rich shells can reduce the efficiency of Tb utilization [1]. Given the fixed amount of the Tb diffusion source, the formation of a thick shell reduces the amount of Tb that can penetrate deeper into the magnet. Moreover, many HRE-rich phases appear in bright gray (indicated by the red arrow), suggesting that a large portion of the Tb has replaced Nd atoms in the Nd-rich phase rather than entering the main phase to form $Tb_2Fe_{14}B$, leading to further Tb waste.

In the "Tb-Ga magnet", a thinner and more uniform shell–core structure is readily observable. Unlike the layer-by-layer deposition of Tb and Ga, the simultaneous deposition of both elements allows Tb to diffuse deeper into the magnet instead of concentrating at the main phase grains, resulting in a thinner shell structure, which contributes to the magnet's improved magnetic properties. The deeper diffusion of Tb leads to the formation of more shell–core structures. In conclusion, the structure of the "Tb-Ga magnet" positively influences coercivity and exhibits the best overall magnetic properties among the four diffusion magnets.

Figure 6 shows the XRD patterns of the original magnet, the "Tb magnet", "Tb/Ga magnet", "Ga/Tb magnet", and "Tb-Ga magnet". All XRD patterns exhibit peaks corresponding to $Nd_2Fe_{14}B$ [39,44]. When compared to the original magnet, the main phase (006) peak of the diffused magnets shifts towards higher angles. According to the Bragg equation, this shift in peak angle corresponds to a reduction in the crystal lattice parameters. Since the radius of Tb^{3+} is smaller than that of Nd^{3+}, the matrix phase unit cell contracts as Tb atoms diffuse into it, causing the peak to shift to higher angles. The more pronounced the shift, the greater the amount of Tb diffused into the main phase $Nd_2Fe_{14}B$. As shown in the figure, the "Ga/Tb magnet" and "Tb-Ga magnet" exhibit higher Tb contents, substituting Nd in the main phase compared to the other diffused magnets, which results in the formation of more (Tb, Nd)$_2$Fe$_{14}$B phases. Furthermore, the orientation degree of sintered $Nd_2Fe_{14}B$ magnets can be estimated by the intensity ratio of (00l)/(105) peaks. The intensity ratios for the three mixed diffusion magnets are lower than those of the original magnet and the "Tb magnet", suggesting that the c-axis crystallographic texture of the magnet has been weakened.

In this experiment, the original magnet and the Tb-Ga magnet were immersed in a 3.5 wt.% NaCl aqueous solution for 1 h to allow the open circuit potential on the surface to stabilize. Following this, the potentiodynamic polarization curves for both samples were recorded. The polarization curves shown in Figure 7a reveal that, in comparison to the original magnet, the curve for the Tb-Ga magnet shifts upward and to the left. This indicates that the Tb-Ga magnet has a more positive corrosion voltage (E_{corr}) and a lower corrosion current density (I_{corr}). The fitting parameters for the polarization curves, presented in Table 1, clearly indicate that the Tb-Ga magnet exhibits better corrosion resistance than the original magnet. The corrosion voltage (E_{corr}) for the Tb-Ga magnet slightly increased from −0.923 V to −0.869 V, while the corrosion current density (I_{corr}) grew by an order of magnitude, from 2.87×10^{-6} A·cm^{-2} to 1.89×10^{-6} A·cm^{-2}. The Nyquist and Bode plots for the original magnet and the Tb-Ga magnet are presented in Figure 7b,c. As illustrated in Figure 7b,c, and Table 1, compared to the original magnet, the polarization resistance (Rp) of the Tb-Ga magnet increased from 4.92 Ω·cm^{-2} to 32.09 Ω·cm^{-2}, and the impedance modulus values ($|Z|_{0.01Hz}$) showed a significant increase, further confirming the positive effect of adding Tb-Ga to the NdFeB magnet [45,46].

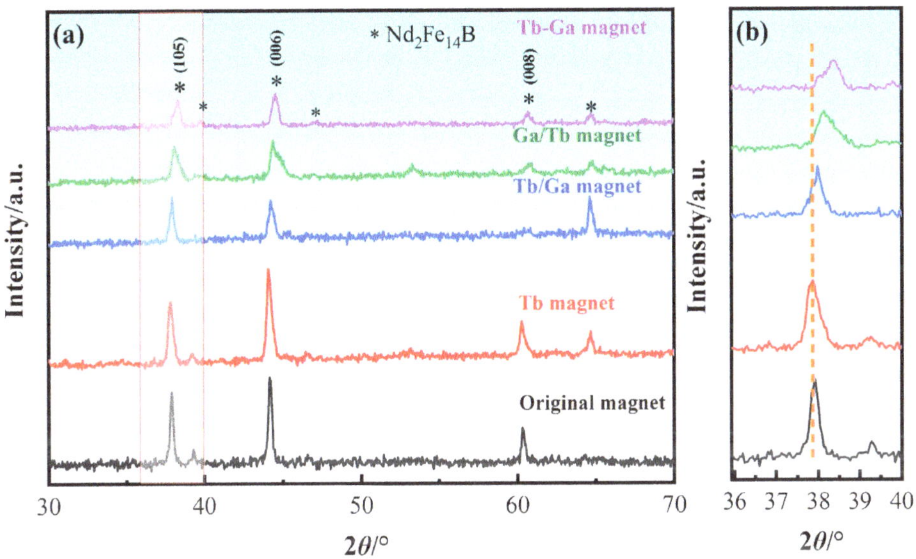

Figure 6. (**a**) XRD patterns of the different magnets. (**b**) The enlarged patterns in the range of 36~40°.

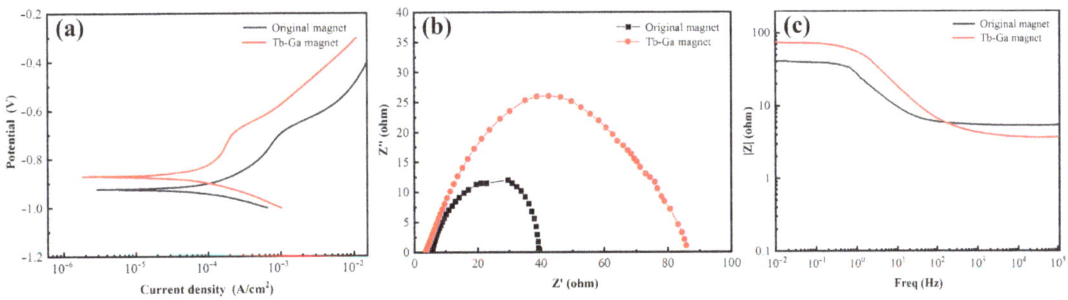

Figure 7. (**a**) Potentiodynamic polarization curves of the original magnet and Tb-Ga magnet, (**b**) Nyquist and (**c**) Bode plots of the original magnet and Tb-Ga magnet in a 3.5 wt.% NaCl aqueous solution.

Table 1. The fitting parameters of potentiodynamic polarization curves.

Sample	E_{corr} (V)	I_{corr} (A·cm^{-2})	R_p (Ω·cm^{-2})
Original magnet	−0.923	2.87×10^{-6}	14.92
Tb-Ga magnet	−0.869	1.89×10^{-6}	32.09

Overall, the electrophoretic deposition of Tb-Ga not only improves the magnet's magnetic properties but also enhances its corrosion resistance, making it a coating with excellent overall performance.

3. Materials and Methods

Commercial sintered Nd-Fe-B magnets were selected as the starting material for the EPD-based GBDP study. The original magnet has an H_{cj} of 1527.87 kA/m, a $(BH)_{max}$ of 318.31 kJ/m^3, and a remanence (B_r) of 1.28 T. The magnets were cut into thin-sheet samples with dimensions of (10 ± 0.1)mm × (10 ± 0.1)mm × (3 ± 0.1) mm (c-axis) using wire cutting. The samples were then polished with 800, 1000, and 2000 mesh sandpaper, followed by immersion in a 3 vol% HCl solution for 15 s. Afterward, they were ultrasonically cleaned with water, anhydrous ethanol, and then dried to prepare for electrophoretic deposition.

For the EPD, TbF_3 powder with an average particle size of 0.3 μm and $Ga(NO_3)_3$ (Macklin, 99.9%) powder with an average size of 50 nm were used. The TbF_3 and $Ga(NO_3)_3$ powders were mixed in an ethanol suspension. Platinum plates were employed as the anode, and the magnet was used as the cathode during the electrophoresis process. Four deposition processes were chosen to evaluate the impact of the electrophoretic deposition sequence on magnet performance. The deposition sequence and the corresponding deposited solutions are outlined in Table 2. To ensure uniformity across the experiments, the Tb concentration in the magnet was maintained at a mass fraction of 0.5 ± 0.01%. After deposition, the samples underwent heat treatment at 900 ± 5 °C for 7 h and at 500 ± 5 °C for 2 h under a vacuum of 6 ± 0.2 × 10^{-4} Pa to complete the GBDP. Figure 8 shows the schematic diagrams of the four GBDP diffusion processes.

Table 2. Electrophoretic deposition setup and process parameters.

Magnet Name	Deposition Setup	Deposited Solution	Deposition Parameters
Tb magnet	Only Tb coatings were deposited on the surface of the magnet.	10 g/L TbF_3 ethanol suspension.	U = 120 ± 0.1 V; t = 1 min
Tb/Ga magnet	Tb and Ga coatings were deposited layer by layer on the surface of the magnet.	10 g/L TbF_3 ethanol suspension;1.5 g/L $Ga(NO_3)_3$ ethanol suspension.	
Ga/Tb magnet	Ga and Tb coatings were deposited layer by layer on the surface of the magnet.	10 g/L TbF_3 ethanol suspension;1.5 g/L $Ga(NO_3)_3$ ethanol suspension.	
Tb-Ga magnet	Both Tb and Ga coatings were deposited simultaneously on the surface of the magnet.	the 10 g/L TbF_3 and 1.5 g/L $Ga(NO_3)_3$ mixed ethanol suspension.	

In the following section, the magnets subjected to diffusion are referred to as the Tb magnet, Tb/Ga magnet, Ga/Tb magnet, and Tb-Ga magnet, based on the different sequences of electrophoretic deposition used.

The corrosion resistance of sintered Nd-Fe-B magnets, both before and after electrophoretic deposition and grain boundary diffusion, was tested using a CHI660D electrochemical workstation from Chenhua Instrument Co., Ltd, Shanghai, China. The experiment employed a conventional three-electrode setup, where the sample acted as the working electrode, a saturated calomel electrode was used as the reference electrode, and a platinum electrode served as the auxiliary electrode. The exposed area of the sample was 2.68 cm^2, with a 3.5 wt.% NaCl solution as the corrosion medium. The samples were immersed for 1 h, with a scanning rate of 0.02 mV/s, and the tests were conducted at room temperature. Impedance spectroscopy was performed, scanning from a high frequency of 10^5 Hz to a low frequency of 10 mHz, with an amplitude of 10 mV.

Magnetic properties of the magnets were measured using a pulsed field magnetometer (PFM Hirst). The morphology and elemental analysis of the samples were performed using scanning electron microscopy (SEM ZEISS Sigma 360) with a backscattered electron (BSE) detector and energy dispersive spectroscopy (EDS). The phase structures of the magnets were analyzed by X-ray diffraction (XRD XD-2). Additionally, an electron probe microanalyzer (EPMA JXA-8100) was utilized to study the elemental distribution within the samples.

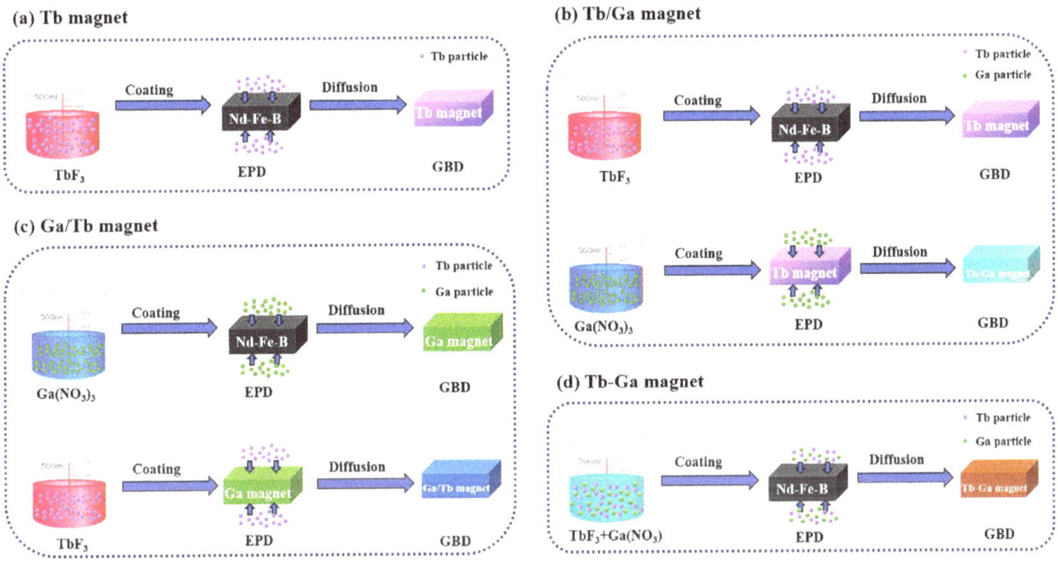

Figure 8. Schematic diagrams of GBD process of (**a**) Tb magnet, (**b**) Tb/Ga magnet, (**c**) Ga/Tb magnet and (**d**) Tb-Ga magnet.

4. Conclusions

To summarize, a Tb and Ga mixed diffusion source was prepared via the EPD technique. The magnetic properties, microstructure, and elemental distribution of the magnets were evaluated, leading to the following conclusions:

(1) The sequence of Tb and Ga deposition greatly affects the coercivity of the magnet. Among the deposition methods, the co-deposition of Tb and Ga resulted in the most significant enhancement in magnetic performance. The Tb substitution efficiency for Nd and the thermal stability of the Tb and Ga mixed diffusion magnets are notably superior to those without Ga.

(2) In comparison to layered deposition of the mixed diffusion sources, the simultaneous deposition method enhances the diffusion depth, optimizing the Tb distribution and improving the utilization of HRE resources by preventing the formation of overly thick Tb-rich grain boundaries.

(3) Electrochemical testing shows that the mixed diffusion magnets exhibit higher corrosion voltage (E_{corr}), lower corrosion current density (I_{corr}), and increased polarization resistance (R_p), significantly improving the corrosion resistance of Nd-Fe-B magnets.

Continued research may facilitate the large-scale industrial production of magnets with both excellent magnetic properties and enhanced corrosion resistance. This advancement could be realized through the co-diffusion of low melting point elements with heavy rare earth metals in Nd-Fe-B sintered magnets.

Author Contributions: Conceptualization, L.W. and S.G.; methodology, L.W.; validation, L.W.; formal analysis, W.L.; investigation, S.G.; resources, Z.D.; data curation, X.W.; writing—original draft preparation, L.W.; writing—review and editing, L.W.; visualization, X.W.; supervision, L.W.; funding acquisition, L.W. All authors have read and agreed to the published version of the manuscript.

Funding: This research was funded by the Natural Science Foundation of Shanxi (No. 202103021224318 and No. 202303021212289), the Science and Technology Plan Project of Lvliang (No. 2024GX02), and the Innovative Teaching Reform Projects in Higher Education Institutions of Shanxi Province (No. J20231328).

Institutional Review Board Statement: Not applicable.

Informed Consent Statement: Not applicable.

Data Availability Statement: Data will be made available on request.

Acknowledgments: The authors would like to thank Zhang Feiyan from Shiyanjia Lab (www.shiyanjia.com, accessed on 30th November 2024) for the magnetic property test.

Conflicts of Interest: The authors declare no conflicts of interest.

Abbreviations

The following abbreviations are used in this manuscript:

HRE	Heavy rare earths
GBDP	Grain boundary diffusion process
EPD	Electrophoretic deposition
PFM	Pulsed field magnetometer
SEM	Scanning electron microscopy
BSE	Backscattered electron
EDS	Energy dispersive spectroscopy
EPMA	Electron probe microanalyzer
XRD	X-ray diffraction

References

1. Pan, J.; Cao, S.; Li, Y.H.; Li, Q.; Ding, G.F.; Zheng, B.; Guo, S.; Fan, X.D.; Chen, R.J.; Yan, A.R. Enhancing the grain boundary diffusion efficiency of Tb for Nd-Fe-B magnets using multi-element composite diffusion source. *J. Alloys Compd.* **2024**, *984*, 174003. [CrossRef]
2. Liao, X.F.; Zhao, L.Z.; Zhang, J.S.; Xu, K.; Zhou, B.; Yu, H.Y.; Zhang, X.F.; Greneche, J.M.; Aubert, A.; Skokov, K.; et al. (Ce,La,Y)-Fe-B permanent magnets by hot deformation. *J. Mater. Res. Technol.* **2022**, *17*, 145968. [CrossRef]
3. Wang, Z.J.; Li, Y.Q.; Wu, H.H.; Ji, M.; Zhang, H.G.; Liu, W.Q.; Yue, M. New grain boundary diffusion strategy of Nd-Fe-B sintered magnets to achieve breakthroughs in Tb-availability and thick limitation. *J. Alloys Compd.* **2023**, *943*, 169174. [CrossRef]
4. Chen, F.G.; Zhang, T.Q.; Zhao, Y.; Wang, X.L.; Jiang, C.P.; Chen, J.L.; Zhao, W.Q. A novel strategy to design and fabricate Nd-Fe-B magnets. *J. Alloys Compd.* **2021**, *267*, 159102. [CrossRef]
5. Li, W.; Zhou, Q.; Zhao, L.Z.; Wang, Q.X.; Zhong, X.C.; Liu, Z.W. Micromagnetic simulation of anisotropic grain boundary diffusion for sintered Nd-FeB magnets. *J. Magn. Magn. Mater.* **2018**, *451*, 704–709. [CrossRef]
6. Chen, W.; Huang, Y.L.; Luo, J.M.; Hou, Y.H.; Ge, X.J.; Guan, Y.W.; Liu, Z.W.; Zhong, Z.C.; Wang, G.P. Microstructure and improved properties of sintered Nd-Fe-B magnets by grain boundary diffusion of non-rare earth. *J. Magn. Magn. Mater.* **2019**, *474*, 134–141. [CrossRef]
7. Sepehri-Amin, H.; Ohkubo, T.; Shima, T.; Hono, K. Grain boundary and interface chemistry of an Nd-Fe-B-based sintered magnet. *Acta Mater.* **2012**, *60*, 819–830. [CrossRef]
8. Li, W.D.; Zhang, Q.K.; Zhu, Q.H.; Xiao, S.; Xu, C.; Yang, L.J.; Zheng, B.Z.; Mao, S.D.; Song, Z.L. Formation of anti-shell/core structure of heavy rare earth elements (Tb, Dy) in sintered Nd-Fe-B magnet after grain boundary diffusion process. *Scr. Mater.* **2019**, *163*, 40–43. [CrossRef]
9. Li, W.D.; Yang, L.J.; Zhang, Q.K.; Xu, C.; Zhu, Q.H.; Song, Z.L.; Song, Z.L.; Zheng, B.Z.; Hu, F.Q.; Jiang, J.J. Effect of the grain boundary Tb/Dy diffused microstructure on the magnetic properties of sintered Nd-Fe-B magnets. *J. Magn. Magn. Mater.* **2020**, *502*, 166491. [CrossRef]

10. Huang, Q.L.; Jiang, Q.Z.; Shi, Y.; Rehman, S.U.; Wei, X.; Wang, R.S.; Zhang, Y.J.; Xu, D.Q.; Chen, D.K.; Zhong, Z.C. Enhancement of terbium efficiency by gallium and copper co-doping in (Pr, Nd)-Fe-B sintered magnets. *J. Rare Earths* **2023**, *41*, 572–577. [CrossRef]
11. Liu, Z.W.; He, J.Y.; Ramanujan, R.V. Significant progress of grain boundary diffusion process for cost-effective rare earth permanent magnets: A review. *Mater. Des.* **2021**, *209*, 110004. [CrossRef]
12. Liu Wong, Y.J.; Lin, B.S.; Chang, H.W.; Chang, W.C.; Chiu, C.H.; Mo, C.C. Enhancement of coercivity and thermal stability of $Pr_{85}Al_{15}$ doped NdFeB sintered magnets followed with grain boundary diffusion. *J. Magn. Magn. Mater.* **2023**, *571*, 170562. [CrossRef]
13. Kim, S.; Ko, D.S.; Lee, H.S.; Kim, D.; Roh, J.W.; Lee, W. Enhancing the coercivity of Nd-Fe-B sintered magnets by consecutive heat treatment–induced formation of Tb-diffused microstructures. *J. Alloys Compd.* **2019**, *780*, 574–580. [CrossRef]
14. Löewe, K.; Brombacher, C.; Katter, M.; Gutfleish, O. Temperature-dependent Dy diffusion processes in Nd-Fe-B permanent magnets. *Acta Mater.* **2015**, *83*, 248–255. [CrossRef]
15. Zhou, Q.; Liu, Z.W.; Zhong, X.C.; Zhang, G.Q. Properties improvement and structural optimization of sintered NdFeB magnets by non-rare earth compound grain boundary diffusion. *Mater. Des.* **2015**, *86*, 114–120. [CrossRef]
16. Cao, X.J.; Chen, L.; Guo, S.; Fan, F.C.; Chen, R.J.; Yan, A.R. Effect of rare earth content on TbF3 diffusion in sintered Nd-Fe-B magnets by electrophoretic deposition. *Scr. Mater.* **2017**, *131*, 24–28. [CrossRef]
17. Hirota, K.; Nakamura, H.; Inowa, T.; Honshima, M. Coercivity enhancement by thegrain boundary diffusion process to Nd-Fe-B sintered magnets. *IEEE Trans. Magn.* **2006**, *42*, 2909–2911. [CrossRef]
18. Luo, J.M.; Guan, Y.W.; Huang, Y.L.; Hou, Y.H.; Mao, H.Y. Microstructure and magnetic properties of pintered NdFeB magnets optimizedby electrophoretic deposition of Al and its grain boundary diffusion. *Rare Met. Mater. Eng.* **2020**, *49*, 2125–2131.
19. Hu, J.K.; Peng, C.X.; Zhan, H.; Bai, X.Y.; Wang, L. Effect of grain boundary diffusion in TbF3-Al mixed electrophoresis onproperties of sintered Nd-Fe-B magnets. *J. Magn. Magn. Mater.* **2023**, *566*, 170324. [CrossRef]
20. Guan, Y.W.; Huang, Y.L.; Rao, Q.; Li, W.; Hou, Y.H.; Luo, J.M.; Pang, Z.S.; Mao, H.Y. Investigation on the grain boundary diffusion of Dy2O3 film prepared by electrophoretic deposition for sintered Nd-Fe-B magnets. *J. Alloys Compd.* **2021**, *857*, 157606. [CrossRef]
21. Cao, X.J.; Chen, L.; Guo, S.; Chen, R.J.; Yan, G.L.; Yan, A.R. Magnetic and microstructural properties of DyF3-coated sintered Nd-Fe-B magnets by electrophoretic deposition. *IEEE Trans. Magn.* **2015**, *51*, 2101804. [CrossRef]
22. Soderznik, M.; Rozman, K.Z.; Kobe, S.; McGuiness, P. The grain-boundary diffusion process in Nd-Fe-B sintered magnets based on the electrophoretic deposition of DyF3. *Intermetallics* **2012**, *23*, 158–162. [CrossRef]
23. Cao, X.J.; Chen, L.; Guo, S.; Li, X.B.; Yi, P.P.; Yan, A.R.; Yan, G.L. Coercivity enhancement of sintered Nd-Fe-B magnets by efficiently diffusing DyF3 based on electrophoretic deposition. *J. Alloys Compd.* **2015**, *631*, 315–320. [CrossRef]
24. Cao, X.J.; Chen, L.; Guo, S.; Chen, R.J.; Yan, G.L.; Yan, A.R. Impact of TbF3 diffusion on coercivity and microstructure in sintered Nd-Fe-B magnets by electrophoretic deposition. *Scr. Mater.* **2016**, *116*, 40–43. [CrossRef]
25. Wang, C.; Luo, Y.; Wang, Z.L.; Yan, W.L.; Zhao, Y.Y.; Quan, N.T.; Peng, H.J.; Wu, K.W.; Ma, Y.H.; Zhao, C.L.; et al. Effect of MgCl2 on electrophoretic deposition of TbF3 powders on Nd-Fe-B sintered magnet. *J. Rare Earths* **2023**, *41*, 721–727. [CrossRef]
26. Li, J.J.; Guo, C.J.; Zhou, T.G.; Qi, Z.Q.; Yu, X.; Yang, B.; Zhu, M.G. Effects of diffusing DyZn film on magnetic properties and thermal stability of sintered NdFeB magnets. *J. Magn. Magn. Mater.* **2018**, *454*, 215–220. [CrossRef]
27. Cao, J.L.; He, J.Y.; Yu, Z.G.; Song, W.Y.; Yu, H.Y.; Fan, W.B.; Zhou, B.; Xu, Z.H.; Liu, Z.W. Alloying Pr-Tb-Cu diffusion source with Ni for enhancing both coercivity and corrosion resistance of Nd-Fe-B magnets. *J. Alloys Compd.* **2022**, *911*, 165049. [CrossRef]
28. Lu, K.C.; Bao, X.Q.; Song, X.F.; Wang, Y.X.; Tang, M.H.; Li, J.H.; Gao, X.X. Temperature dependences of interface reactions and Tb diffusion behavior of Pr-Tb-Cu-Al alloys/Nd-Fe-B magnet. *Scr. Mater.* **2021**, *191*, 90–95. [CrossRef]
29. Zhou, Z.G.; Li, J.H.; Bao, X.Q.; Zhou, Y.Y.; Gao, X.X. Improvement of bending strength via introduced (Dy, Tb)Cu phase at grain boundary on giant magnetostrictive Tb-Dy-Fe alloy by diffusing Dy–Cu alloys. *J. Alloys Compd.* **2020**, *826*, 153959. [CrossRef]
30. Chen, G.X.; Bao, X.Q.; Lu, K.C.; Lv, X.K.; Ding, Y.; Zhang, M.; Wang, C.G.; Gao, X.X. Microstructure and magnetic properties of Nd-Fe-B sintered magnet by diffusing Pr-Cu-Al and Pr-Tb-Cu-Al alloys. *J. Magn. Magn. Mater.* **2019**, *477*, 17–21. [CrossRef]
31. Mendes, T.; Silva, S.C.; Perigo, E.A.; Faria, R.N.; Takiishi, H. The effect of Cu, P, Ga and Gd on microstructure and magnetic properties in the PrFeCoBNb HD sintered magnets. *Mater. Sci. Forum* **2010**, *273*, 660661. [CrossRef]
32. Huang, Q.F.; Jiang, Q.Z.; Shi, Y.; Rehman, S.U.; Wei, X.; Li, Z.X.; Shi, D.W.; Xu, D.Q.; Zhong, Z.C. Enormous improvement of the coercivity of Ga and Cu co-doped Nd-Fe-B sintered magnet by post-sinter annealing. *J Alloys Compd.* **2021**, *894*, 162418. [CrossRef]
33. Wu, B.H.; Ding, X.F.; Zhang, Q.K.; Yang, L.J.; Zheng, B.Z.; Hu, F.Q.; Song, Z.L. The dual trend of diffusion of heavy rare earth elements during the grain boundary diffusion process for sintered Nd-Fe-B magnets. *Scr. Mater.* **2018**, *148*, 29–32. [CrossRef]
34. Pandian, S.; Chandrasekaran, V.; Markandeyulu, G.; Iyer, K.J.L.; Rama, K.V.S. Effect of Al, Cu, Ga, Nb additions on the magnetic properties and microstructural features of sintered NdFeB. *J. Appl. Phys.* **2002**, *92*, 6082–6086. [CrossRef]
35. Kronmüller, H.; Durst, K.D.; Sagawa, M. Analysis of the magnetic hardening mechanism in RE-Fe-B permanent magnets. *J. Magn. Magn. Mater.* **1988**, *74*, 291–302. [CrossRef]
36. Hirosawa, S.; Matsuura, Y.; Yamamoto, H.; Fujimura, S.; Sagawa, M.; Yamauchi, H. Magnetization and magnetic anisotropy of $R_2Fe_{14}B$ measured on single crystals. *J. Appl. Phys.* **1986**, *59*, 873–879. [CrossRef]

37. Hono, K.; Sepehri-Amin, H. Prospect for HRE-free high coercivity Nd-Fe-B permanent magnets. *Scr. Mater.* **2018**, *151*, 6–13. [CrossRef]
38. Cui, W.B.; Takahashi, Y.K.; Hono, K. Microstructure optimization to achieve high coercivity in anisotropic Nd-Fe-B thin films. *Acta Mater.* **2011**, *59*, 7768–7775. [CrossRef]
39. Liu, Y.K.; Liao, X.F.; He, J.Y.; Yu, H.Y.; Zhong, X.C.; Zhou, Q.; Liu, Z.W. Magnetic properties and microstructure evolution of in situ Tb-Cu diffusion treated hot-deformed Nd-Fe-B magnets. *J. Magn. Magn. Mater.* **2020**, *504*, 166685. [CrossRef]
40. Liu, J.; Sepehri-Amin, H.; Ohkubo, T.; Hioki, K.; Hattori, A.; Schrefl, T.; Hono, K. Effect of Nd content on the microstructure and coercivity of hot-deformed Nd-Fe-B permanent magnets. *Acta Mater.* **2013**, *61*, 5387–5399. [CrossRef]
41. Xu, X.D.; Sasaki, T.T.; Li, J.N.; Dong, Z.J.; Amin, H.S.; Kim, T.H.; Ohkubo, T.; Schrefl, T.; Hono, K. Microstructure of a Dy-free Nd-Fe-B sintered magnet with 2T coercivity. *Acta Mater.* **2018**, *156*, 146–157. [CrossRef]
42. Wu, T.F.; Cao, S.; Kou, M.P.; Xie, Y.H.; Ding, G.F.; Guo, S.; Zheng, B.; Chen, R.J.; Zhong, M.L.; Yan, A.R. Magnetic performance and microstructure of NdFeB sintered magnet by diffusing $Tb_{10}Pr_{90-x}(Cu,Al,Ga)_x$ alloys. *J Alloys Compd.* **2023**, *934*, 167888. [CrossRef]
43. Liu, Z.W.; He, J.Y. Several issues on the development of grain boundary diffusion process for Nd-Fe-B permanent magnets. *Acta Metall. Sin.* **2021**, *57*, 1155–1170.
44. Chang, R.; Bai, G.Z.; Li, Y.P.; Yan, Z.; Qin, X.F.; Guo, J.D.; Wang, F.; Xu, X.H. Synergistic regulation and optimization of microstructure, coercivity, and thermal stability in sintered NdFeB magnets through grain boundary diffusion of Dy and Al elements. *J. Mater. Res. Technol.* **2025**, *34*, 2183–2192. [CrossRef]
45. Gurrappa, I. Suitability of Nd-Fe-B permanenet magnets for biomedical applications—A corrosion study. *J. Alloys Compd.* **2002**, *339*, 241–247. [CrossRef]
46. Ramezanzadeh, B.; Niroumandrad, S.; Ahmadi, A.; Mahdavian, M.; Moghadam, M.H.M. Enhancement of barrier and corrosion protection performance of an epoxy coating through wet transfer of amino functionalized graphene oxide. *Corros. Sci.* **2016**, *103*, 283–304. [CrossRef]

molecules

Article

Effect of Pack Chromizing on Microstructure and Tribological Properties of GCr15 Bearing Steel

Dejun Yan [1], Chunbei Wei [2,*], Peng Tang [2], Shuqi Huang [2], Songsheng Lin [2], Qian Shi [2] and Xiaodong Hong [1,*]

1 Guangdong Provincial Key Laboratory of Advanced Welding Technology for Ships, School of Materials and Energy, Foshan University, Foshan 528000, China; yandejun_2003@163.com
2 State Key Laboratory of Special Materials Surface Engineering, National Engineering Laboratory of Modern Materials Surface Engineering Technology, Guangdong Provincial Key Laboratory of Modern Surface Engineering Technology, Institute of New Materials, Guangdong Academy of Sciences, Guangzhou 510651, China; tangpeng@gdinm.com (P.T.); huangshuqi@gdinm.com (S.H.); linsongsheng@gdinm.com (S.L.); shiqian@gdinm.com (Q.S.)
* Correspondence: weichunb@163.com (C.W.); hxd9917@163.com (X.H.); Tel.: +86-20-37238071 (C.W.)

Abstract: Chromizing layers are widely employed in industrial applications due to their superior wear resistance and corrosion resistance. In this study, GCr15 bearing steel was chromized by a solid powder pack chromizing method, and the influence of chromizing time on the microstructure and mechanical properties of the chromized layers was systematically investigated. The results reveal the presence of fine pores dispersed both on the surface and at the chromized layers/substrate interface. The concentration of the Cr and Fe elements displays a gradient distribution throughout the layers. The chromized layers are primarily composed of $(Cr,Fe)_{23}C_6$ and $(Cr,Fe)_7C_3$ phases. With an increase in the chromizing time, the thickness and hardness of the chromized layers are gradually increased. A large number of radial and circumferential cracks are observed both within and around the indentation regions, accompanied by spalling at the edge. The brittleness of the chromized layer is increased, and the spalling phenomenon becomes more pronounced with prolonged chromizing time. The chromizing treatment significantly improves the tribological performance of GCr15 steel, reducing its wear rate to approximately one fifth of that of the untreated substrate.

Keywords: pack chromizing; chromizing time; hardness; indentation; tribological properties

1. Introduction

GCr15 steel, one of the most representative high-carbon and low-alloy steel types, accounts for 80% of bearing steel production due to its low cost, excellent wear resistance, high strength and hardness. It has been widely used in critical mechanical components such as bearing rings, ball screws, and steel balls across marine ships and the aviation, railway and automotive industries [1–3]. With the manufacturing techniques and processing methods improving, the bearing life has been enhanced more than 400 times compared with 80 years ago [4]. However, under increasingly demanding service conditions involving heavy loads, elevated temperatures and high-speed cyclic stresses, bearing components remain vulnerable to premature failure through wear, fatigue and corrosion. Especially in the marine environment, GCr15 is subjected to severe electrochemical corrosion and wear, which significantly compromises its applicability in such settings. Therefore, it is of great importance to improve the wear resistance, corrosion resistance and high fatigue strength of the GCr15 bearing steel.

https://doi.org/10.3390/molecules30183690

To enhance surface properties, various approaches have been developed, including surface coatings [5,6], surface alloys [7,8], surface texturing [9], strengthen grinding [1,10], and heat treatment [4,11]. Chromium and its ceramic coatings made from carbides or nitrides are particularly promising due to their high hardness, wear resistance, electrochemical corrosion resistance and oxidation stability, even under harsh environmental conditions. Our previous research (not given in this paper) and a large number of studies show that the chromium and its ceramic coatings have excellent electrochemical corrosion resistance [12–14]. However, its wear resistance is highly dependent on the coating structure and fabrication process. Thus, an in-depth investigation into the factors influencing the wear resistance of the chromium-based coatings is of significant importance.

Chromium-based coatings can be produced by various techniques, such as physical vapor deposition (PVD) [15], chemical vapor deposition (CVD) [16], thermal spraying (TS) [17] and thermos-reactive diffusion (TRD) [18]. Among these, the TRD method offers distinct advantages for great adhesion strength of coatings to the substrate, the formation of uniform, continuous and dense coating with better hardness, and wear resistance, as well as a low friction coefficient. Consequently, TRD has been widely used to produce hard carbide and nitride coatings with excellent physical and mechanical properties on tool steels, such as forging, extrusion, plastics, glass working, powder metallurgy and cutting tools molds working [19–21].

Substantially, the main factors affecting the microstructure and properties of TRD coatings are as follows: process time, treatment temperature, chemical potential of the carbon interstitial element, and chemical composition of the substrate. Considering the fact that the diffusion of carbon atoms in the austenitic substrate is much faster and easier than the diffusion of other elements in the carbide coating layer, the growth rate and pattern of the chromium carbide layer mainly rest on the outward diffusion of C atoms in the steel substrate. Therefore, the carbon content of the steel substrate significantly affects the formation mechanism, microstructure and properties of the chromized layers [20,22,23]. According to this, chromization is usually divided into two categories: soft chromization and hard chromization. The former is employed on the alloys with less than 0.1 wt.% carbon to improve their corrosion and high-temperature oxidation properties. Layers of chromium-rich phases were observed, which contributes to long-term stability of corrosion resistance [24–26], whereas a hard chromization process is applied to the alloys with carbon contents exceeding 0.3 wt.% to achieve a high-hardness surface layer. This process generates hard chromium carbide phases, resulting in superior mechanical properties, including increased hardness, strong adhesion and enhanced wear resistance [21–23,27]. Lee J.W. et al. [28] studied the influence of carbon contents on the mechanical properties of steels after they were treated by the hard chromization process. It indicated that the thickness of the chromized layer obeys the parabolic rate law and increases with the chromizing time and carbon contents of the substrate. The primary phase on the chromized surface is $(Cr,Fe)_2N_{1-x}$ with small amount of $(Cr,Fe)_{23}C_6$. The thickness of the surface nitrogen-rich layer, i.e., the $(Cr,Fe)_2N_{1-x}$ phase, increases with carbon concentration and chromizing time; this can contribute to the higher critical loads and sufficient adhesion properties in the high C content substrate.

GCr15 bearing steel is a high-carbon steel with a carbon content as high as 1 wt.%. Several studies have investigated chromizing treatment for GCr15 substrate. Tan C et al. [29] fabricated chromized coatings on GCr15 bearing steel with a gas-phase chromizing technique. The coatings exhibited superior wear resistance due to their high hardness (16.6 GPa) and interfacial adhesion (~50 N). The wear rate of the chromized layer decreased dramatically from 2.03×10^{-5} $mm^3/(N.m)$ to 1.8×10^{-6} $mm^3/(N.m)$ as the load increased from 10 N to 25 N, which attributes to the formation of chromium oxide tribolayer via shear-induced oxidation. Wang H. et al. [30] prepared chromized coatings on GCr15 steel surface with a solid powder pack chromizing method (TRD) and found a dual-layer chromized structure: a porous outer layer composed mainly of $(Cr,Fe)_{23}C_6$ phase and a dense inner layer dominated by $(Cr,Fe)_7C_3$ phase. The high C content in GCr15 substrate promotes the formation of the porous $(Cr,Fe)_{23}C_6$ layer and reduces the diffusion rate of Cr. Chromizing treatment significantly improves the hardness of the GCr15 substrate, which exceeds 1700 HV. Zong X. et al. [31] fabricated chromium carbide coatings (Cr_7C_3 and $(Cr,Fe)_7C_3$) on the surface of AISI 52100 steel (equivalent to GCr15) using the TRD. They found that the hardness of the surface increases to 1730–1920 HV and the adhesion-strength quality of the coating is correlated to HF2~HF3 according to the VDI 3198 standards [32]. The chromium carbide coating can significantly improve the wear resistance of the substrate with the friction coefficient decreased from 0.46 to 0.37 and wear-weight loss decreased by 89.3%.

Despite the potential advantages mentioned above, the application of TRD technology in GCr15 steel is not extensive compared with other deposition methods. Furthermore, limited research has focused on the adhesion strength and tribological properties, particularly regarding the wear-failure mechanism of chromized layers on bearing steels. Therefore, a comprehensive investigation of TRD's detailed effects on the microstructure, adhesion, and tribological properties of GCr15 steel is warranted. Among the TRD methods, solid powder pack chromizing is a relatively mature process for preparing the chromized layer with a simple method and low cost [33]. In this study, solid powder pack chromizing treatment was applied to GCr15 bearing steel to improve its wear resistance. The adhesion strength and the tribological behavior of the chromized layers were systematically investigated to comprehend the relationships between composition, microstructure and wear properties.

2. Results and Discussion

2.1. Microstructure and Compositions

Figure 1 depicts the XRD patterns of the chromized layers treated under different chromizing time by both grazing-incidence XRD (GIXRD) and conventional XRD measurements. The chromized layer primarily consists of $Cr_{23}C_6$ (PDF#35-0783) and $(Cr,Fe)_7C_3$ (PDF# 05-0720) phases, accompanied by a minor $(Cr,Fe)_2N_{1-x}$ (PDF#19-0330) phase. The enhanced intensity of $(Cr,Fe)_2N_{1-x}$ diffraction peaks in GIXRD indicates that its preferential disperses on the surface of the chromized layer. The nitrogen in the chromized layer is primarily derived from the decomposition of NH_4Cl in the chromizing agent. The decomposed nitrogen atoms react with the surface-adsorbed Cr to form $(Cr,Fe)_2N_{1-x}$.

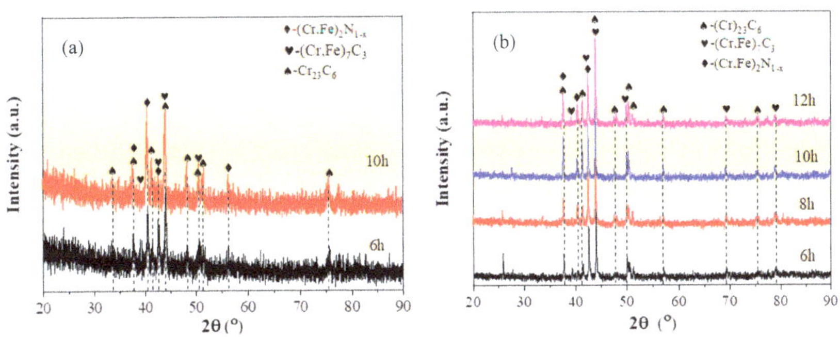

Figure 1. XRD patterns of the chromized layers with different chromizing time (**a**) grazing incidence; (**b**) normal incidence.

The formation process of the chromized layer primarily involves the inward diffusion of Cr, N atoms from the chromizing agent and the outward diffusion of substrate elements such as Fe and C. In the initial stage, the following reactions (as shown in Table 1 Equations (1)–(8)) occur in the chromizing agent at high temperatures [34,35].

Table 1. The reactions occur during the chromizing process.

Formulas	Gibbs Free Energy	Number
$NH_4Cl(s) \rightarrow NH_3 + HCl(g)$		(1)
$2NH_3 \rightarrow N_2 + 3H_2(g)$		(2)
$2HCl(g) + Cr(s) \rightarrow CrCl_2(g) + H_2(g)$		(3)
$CrCl_2(g) + Fe \rightarrow FeCl_2 + [Cr]$ (Active Cr atoms)		(4)
$CrCl_2(g) + H_2 \rightarrow 2HCl(g) + [Cr]$		(5)
$CrCl_2 \rightarrow Cl_2 + [Cr]$		(6)
$2Cr + N_2 \rightarrow 2CrN$		(7)
$4Cr + N_2 \rightarrow 2Cr_2N$		(8)
$23Cr + 6C \rightarrow Cr_{23}C_6$	ΔG_0 (KJ mol^{-1}) = $-309.616 - 0.0774$ T (298 K < T < 1773 K)	(9)
$7Cr + 3C \rightarrow Cr_7C_3$	ΔG_0 (KJ mol^{-1}) = $-174.401 - 0.0259$ T (298 K < T < 2171 K)	(10)
$3Cr + 2C \rightarrow Cr_3C_2$	ΔG_0 (KJ mol^{-1}) = $-79.078 - 0.01766$ T (298 K < T < 2130 K)	(11)

It is known that the growth of the chromized layer involves thermodynamics and kinetics issues. At a low temperature (<1000 °C), the resistance encountered during the coating growth is high and the kinetics dominate the growth process of coating. The Cr atoms/ions spill from the packed powders and diffuse into the substrate at a slow rate [36]. Cr atoms diffuse inwards and C atoms in the substrate diffuse outwards, meeting at the interface and reacting to form the chromium carbide phase. Theoretically, there would be many different types of chromium carbides including $Cr_{23}C_6$, Cr_7C_3 and Cr_3C_2, formed by the reaction. The following equations (as shown in Table 1 Equations (9)–(11)) show the reaction and the associated Gibbs free energy [21].

Previous studies indicate that the Gibbs free energy of $Cr_{23}C_6$, Cr_7C_3 and Cr_3C_2 remains nearly constant within a certain temperature range. According to the thermodynamic theory, it can be concluded that the formation order of the three kinds of chromium carbides should by $Cr_{23}C_6$ first, then Cr_7C_3 and lastly Cr_3C_2 [37]. Thus, when the ratio between Cr and C is large enough, $Cr_{23}C_6$ phase can easily form [38]. The phase structure of the chromized layer is also closely related to the C content in the substrate. For high-carbon steel, the active chromium atoms first react with the carbon atoms on the surface of the steel substrate to form $Cr_{23}C_6$. After the formation of the $Cr_{23}C_6$ layer on the surface, surface-adsorbed Cr atoms penetrate through the carbide layer to reach the substrate, which increases the inward diffusion resistance of Cr atoms, while the Cr atoms in the $Cr_{23}C_6$ phase near the substrate interface constantly diffuse inward. This results in a continuous decrease in Cr content in the $Cr_{23}C_6$ phase at the interface, coupled with the continuous outward diffusion of C atoms from the substrate, which promote the phase transformation from $Cr_{23}C_6$ to Cr_7C_3 near the interface [27]. Additionally, the outward diffusion of Fe from the substrate to replace part of the Cr atoms is more conducive to thermodynamic stability [39]. So, the $(Cr,Fe)_{23}C_6$ phase and the $(Cr,Fe)_7C_3$ phase are formed in the chromized layer. In Figure 1b, with the increase in chromizing time, the diffraction peak intensity enhances, indicating that the crystal structure of the chromized layer phase is strengthened, which may be attributed to the grain growth caused by the prolonged chromizing time.

According to the Scherrer equation,

$$D_{hkl} = \frac{K\lambda}{\beta_{hkl}\cos\theta} \tag{12}$$

In the formula, K is a constant, β is a half-height width of the diffraction peak, θ is an angle corresponding to the diffraction peak, and λ is a wavelength of the x-ray. Bringing the angle of a single diffraction peak into Equation (12), the grain thickness of the corresponding crystal plane normal is obtained. After taking multiple angles into the calculation, the grain size D can be obtained by averaging. The calculation results of grain size under different chromizing time are summarized in Table 2. It can be observed that as the chromizing time increases, the grain size of the $Cr_{23}C_6$ phase increases gradually, while the grain size of the $(Cr,Fe)_7C_3$ phase does not exhibit a significant change. In summary, the grain size of the chromized layer gradually coarsens with extended chromizing time.

Table 2. Grain size of carbide under different chromizing time.

Sample	$Cr_{23}C_6$ (nm)	$(Cr,Fe)_7C_3$ (nm)
CT6	32.2	33.6
CT8	34.0	32.5
CT10	37.4	31.4
CT12	37.2	35.8

Figure 2 shows the surface morphologies of the chromized layers under the different chromizing time. The surface of the chromized layer exhibits relatively smooth topography, with scattered protruding rhombic grains embedded on the surface. As the chromizing time increases, the number and size of rhombic grains gradually increase. A large number of fine pores are distributed on the chromized layer surface. The formation of these pores may be attributed to the migration of vacancies in the subsurface of substrate, inward diffusion of Cr atoms into the substrate and decomposition of gas atoms from the NH_4Cl activator in the pack mixture [22,40]. EDS analysis was performed on different regions of the chromized surface, as shown in Table 3. It can be found that the protruding rhombic grains are primarily enriched with Cr and N, while the remaining areas consist mainly of Cr and C. This indicates that the chromized layer predominantly comprises chromium carbide phases, while the scattered rhombic grains represent nitrogen-rich compounds. The presence of a small number of nitrogen-rich compounds in the chromized layer is related to active nitrogen atoms decomposed from the activator NH_4Cl in the chromizing agent that produces N_2 and combines with active Cr atoms to form Cr–N compounds [34]. This is consistent with the result of XRD.

Figure 3 shows the cross-section morphology of the chromized layer under different chromizing times. The microstructure exhibits distinct light–dark contrast, enabling clear differentiation into four distinct layers labeled A, B, C and D (as shown in Figure 3c). According to the EDS results shown in Table 4, the Cr content in the chromized layer gradually decreases and the Fe content gradually increases from the surface to the substrate, forming a gradual composition transition structure. The C content is higher than that of the substrate throughout the chromized layer due to the outward diffusion from the substrate. It is known that the diffusion coefficient of C atoms is three or four orders of magnitude greater than that of Cr atoms at high temperature. In the early stage of the chromizing process, Cr atoms first deposit on the surface and then diffuse inward; its penetration inside is stopped by a counter-diffusion of carbon as a result of the vigorous chemical affinity of these elements and formation of a carbide layer [39]. No N element is detected inside the chromized layer. A small amount of nitrogen atoms from the decomposition of NH_4Cl in the agent combine with Cr on the surface to form the chromium nitride phase sporadically dispersed on the surface.

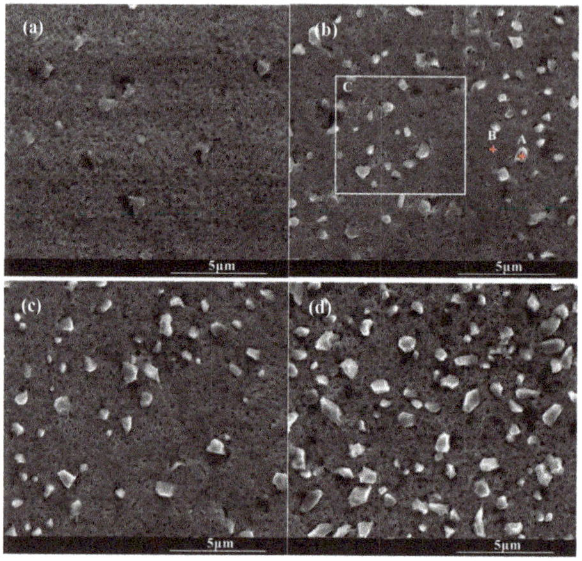

Figure 2. SEM surface morphology of the chromized layers under different chromizing time (**a**) CT6; (**b**) CT8; (**c**) CT10; (**d**) CT12.

Table 3. EDS elements composition in a different position on the surface of chromized layers in Figure 2.

Position	Elements (at.%)		
	Cr	C	N
A	60.60	17.96	21.43
B	72.40	27.60	/
C	71.87	28.13	/

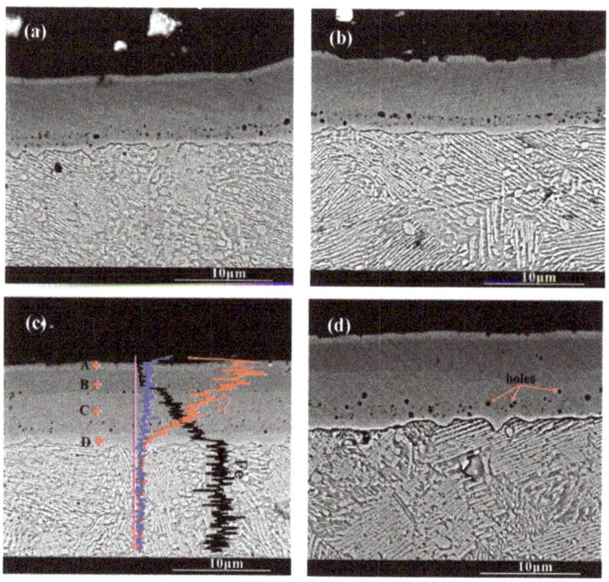

Figure 3. SEM cross-section morphology of the chromized layers with different chromizing time (**a**) CT6; (**b**) CT8; (**c**) CT10; (**d**) CT12.

Table 4. EDS elements composition in different positions of the cross-section chromized layer in Figure 3.

Position	Elements (at.%)		
	Cr	C	Fe
A	49.03	49.67	1.30
B	50.33	47.46	2.21
C	24.29	49.59	26.11
D	10.20	43.56	46.23

A certain number of pores distribute at the interface of the chromized layer/steel substrate shown in Figure 3. According to Figures 2 and 3, two types of pores are observed in the chromized layer: surface pores and interfacial pores. Previous studies have shown that the formation of pores on the surface of the chromized layer is caused by the mismatch diffusion of Cr atoms and Fe atoms in the chromized layer and the aggregation of vacancy migration at grain boundaries, as well as gas decomposition from the activator (NH_4Cl in the pack), etc. [30,40,41]. During the chromizing process, Cr atoms diffuse inwards and C atoms in the substrate diffuse outwards, meeting at the interface and reacting to form the chromium carbide phase. Once this carbide layer develops, surface-adsorbed Cr atoms must penetrate it to reach the substrate, significantly hindering inward diffusion and consequently limiting chromized layer growth. The initially formed chromium carbide phase (such as $Cr_{23}C_6$) can also serve as a Cr source for further inward diffusion. If the vacancies generated by the inward diffusion of Cr atoms cannot be timely, compensated by the inward diffusion of surface Cr atoms from external medium, a large number of vacancies will migrate and coalesce, leading to the formation of pores on the surface [22]. Wang H.F. et al. [30] observed that the formation of pores on the surface was caused by the differential Cr diffusion rates between grain boundaries and intragranular regions in the Cr-rich phase layer. The lower boundaries of most pores are connected to grain boundaries, while the upper boundaries of pores are located within the grains. The diffusion rate of atoms along the grain boundary is much higher than that through intragranular region, resulting in the formation of a large number of vacancies. Vacancies are constantly accumulating, and finally pores are formed. Similarly, the mechanism of pore formation and growth at the chromized layer/steel substrate interface is also the result of this inequality of atom diffusion. Additionally, the stress and relaxation, which are caused by the different specific volumes and coefficients of linear thermal expansion of the carbide layer and the steel substrate and which is accompanied by vacancy coagulation at the interface boundary, may also be causes of porosity at the interface of carbide-layer/steel [39]. Such interfacial porosity may reduce the bonding strength of the chromized layer to the substrate [42].

Figure 4 shows the thickness variation in the chromized layer under different chromizing times. As the chromizing time increases from 6 h to 12 h, the layer thickness exhibits a gradual growth from 6.97 μm to 8.67 μm. Previous research has shown that C content in the substrate significantly influences the thickness of the chromized layer. Liu T. et al. [22] proposed that the carbon content in the substrate of 0.2 wt.% is a cut-off point, which has an apparent influence on the structure and growth mechanism of the chromized layer. When the C content is $\leq$0.2 wt.%, the thickness of the chromized layer is significantly thicker, although the thickness of the permeable layer decreased slightly with the increase in C content in the substrate (ranging from 51.19 μm to 31.39 μm). In contrast, for substrates with higher carbon content (>0.2 wt.%), the chromized layer remains relatively thin (13.04 μm to 15.60 μm for C content of 0.3~0.45 wt.%) and shows only a slight increase in thickness with further elevation of C content. However, Lee J.W. et al. [28] indicated that the thickness of the chromized layer obeys the parabolic rate law: $X = K\sqrt{t}$ and increases with chromizing time and carbon contents in the substrate. The high carbon content (up to 1 wt.%) in GCr15 steel substrate promotes the rapid formation of the chromium carbide layer during the chromizing process. After the chromium carbide layer is formed due to the vigorous chemical affinity on the surface, Cr atoms adsorbed on the surface must pass through the carbides layer to reach the substrate, which increases the difficulty of inward diffusion, hindering the inward diffusion of Cr atoms, and thus limiting the growth rate of the chromized layer. As a result, the thickness of the chromized layer increases slightly with an increase in the chromizing time. Since the diffusion rate of C atoms at a high temperature is much higher than that of Cr, the outward diffusion of C atoms has become the main reason for the growth of the chromized layer. As indicated during chromized plating, about 70% of the thickness of the carbide layer is formed due to its outward growth, and only 30% due to inward growth into the depth of specimen [39].

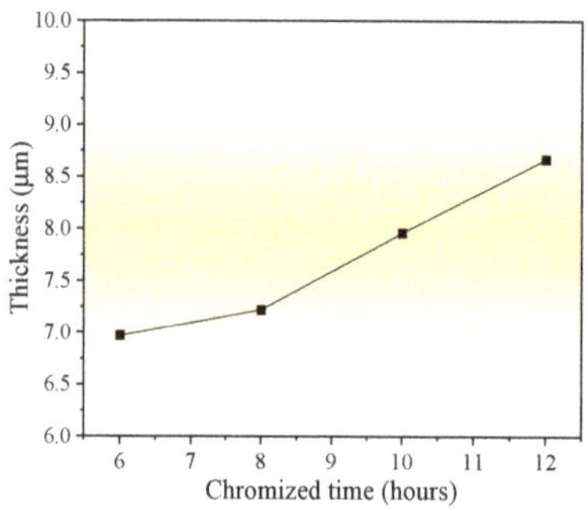

Figure 4. Influence of chromizing time on thickness of the chromized layers.

2.2. *Microhardness*

Figure 5 shows the surface microhardness of the chromized layer under different chromizing times. With the increase in chromizing time, the surface hardness of the chromized layer increases, varying from 1360.6 HV to 1588.2 HV. Previous studies indicate that the microhardness of the $(Cr,Fe)_{23}C_6$ phase is 1200 HV–1400 HV, 2000 HV–2500 HV for $(Cr,Fe)_7C_3$ phase, and 18GPa for $(Cr,Fe)_2N_{1-x}$ phase [28,40]. As indicated by the XRD results, the chromized layer is mainly composed of $(Cr,Fe)_{23}C_6$ and $(Cr,Fe)_7C_3$, meaning its overall hardness reflects a composite of these two phases. With the increase in chromizing time, the thickness of the chromized layer increases, while its microstructure and phase composition change slightly. Therefore, the increment of hardness within the chromizing time should be primarily attributed to the improvement of layer thickness. Compared with that of the GCr15 bearing steel substrate (412 HV), the chromizing treatment improves the surface hardness by more than three times.

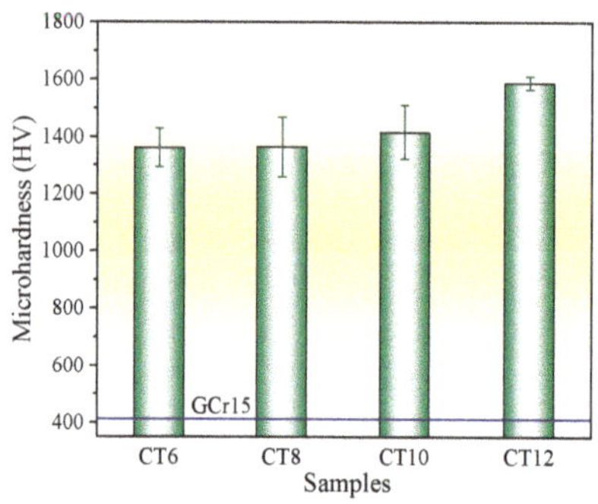

Figure 5. Microhardness of the chromized layers under different chromizing time.

2.3. Indentation

Figure 6 shows the SEM morphology of the indentation on the surface of the chromized layer. Numerous cracks exist within and around the indentation, where radial and circumferential cracks are observed inside the indentation zone, while radial cracks dominate the surrounding area. Radial cracks typically initiate inside the indentation, then propagate outward in a radial pattern through the indentation edge. Some radial cracks deflect and then continue to extend outward. Within the indentation zone, circumferential cracks appear to be discontinuous, terminating at radial cracks. That is, numerous discontinuous circumferential cracks exist between radial cracks, with coarser ones appearing at the indentation edge. The presence of circumferential cracks reflects fracture behavior in the material, indicating that the chromized layer inside the indentation has undergone fracturing, with the most severe fracture occurring and leading to the spalling around the indentation edge. Spalling tends to occur in enclosed regions formed by radial and circumferential cracks. The spalling regions surrounding the indention observed for CT6 and CT8 are relatively small, with minimal spalling at the indention edge. In contrast, CT10 and CT12 exhibited obviously larger spalling areas. The spalling zones surrounding the indentations are nearly completely interconnected, forming a continuous spalling circle. Furthermore, spalling at the indentation edge obviously increases and gradually propagates outward, merging with the outer spalling circle to form extensive spalling regions. This indicates that the brittleness of the chromized layer increases with prolonged chromizing time. Burnett P.J. et al. [43] revealed that the ratio of E_f/E_s (where E_f and E_s are the Young's modulus of coatings and substrate, respectively) significantly influences the indentation-cracking behavior of the coating, primarily depending on the deformation mode of the coating/substrate system. When the elastic modulus of the coating is close to that of the substrate, the coating deformation under stress is obstructed by the substrate, leading to cracking in the coating and typically in radial patterns. If a coating with a higher elastic modulus is deposited onto a relatively softer substrate, the high-hardness coating usually has higher yield stress, and the soft substrate tends to yield first under stress. In this case, the deformation of coating will be affected by the deformation of the substrate; that is, the plastic deformation zone of the substrate beneath the coating will increase the curvature of the coating, thus increasing the cracking displacement and generating frame-like cracks (circumferential cracks). Simulation results confirm that the material beneath the indenter undergoes plastic deformation under compressive stress during the indentation process, being extruded and accumulating/expanding around the indentation. The maximum radial tensile stress exists outside the indentation edge, which is the main reason for the circumferential cracking of the coating. As indentation depth increases, the maximum circumferential tensile stress increases and its position shifts toward the indentation edge [44,45].

The chromized layer forms a continuous metallurgical bond with the steel substrate, ensuring excellent bonding strength between them and preventing premature delamination during loading. During stress loading, the chromized layer bears most of the stress due to its high hardness. However, the grain boundaries and defects within the chromized layer are prone to becoming stress concentration points, where cracks first initiate and propagate. As shown in Figures 2 and 3, there are many pores on the surface and at the interface of the chromized layer/steel substrate. These pores may accelerate the propagation of cracks and lead to the spalling of layers. Furthermore, with the increase in chromizing time, the grains of the chromized layer are coarsened, causing the increment of the brittleness of the chromized layer and resulting in serious spalling around the indentation.

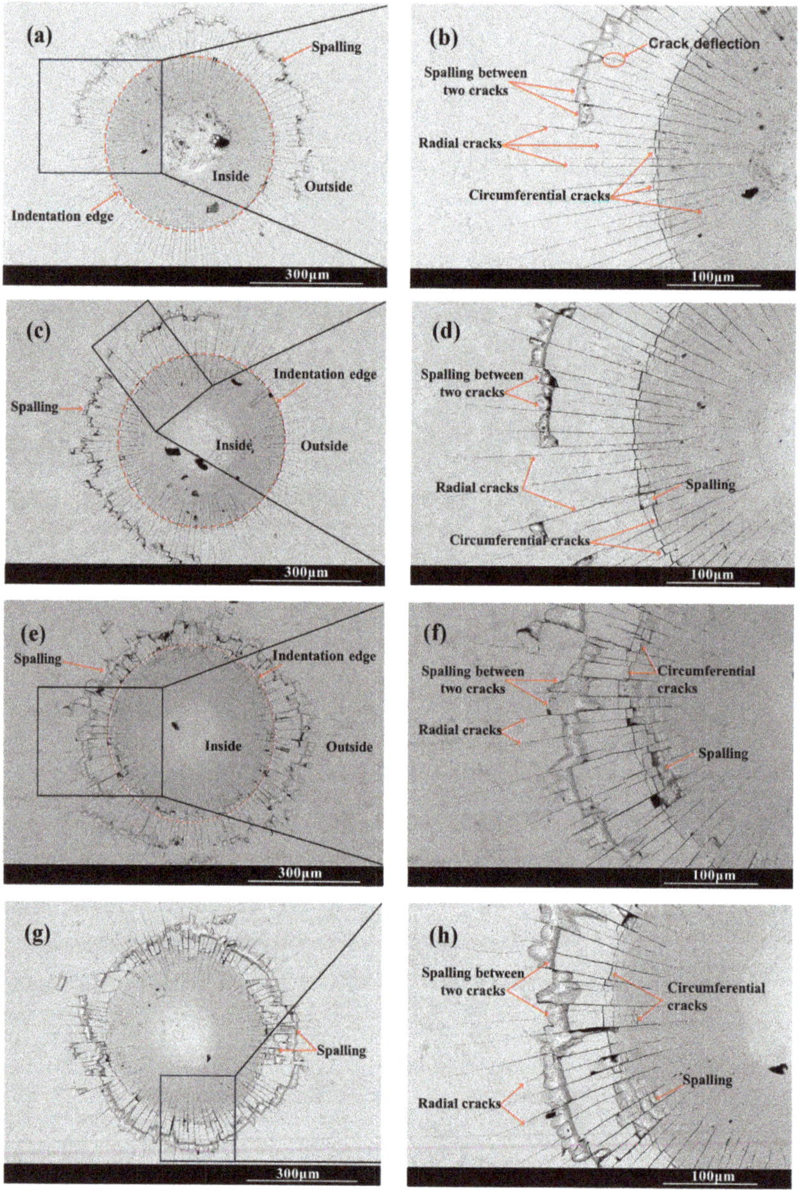

Figure 6. SEM indention morphology of the chromized layers (**a**,**b**) CT6; (**c**,**d**) CT8; (**e**,**f**) CT10; (**g**,**h**) CT12.

2.4. Tribological Properties

Figure 7 shows the friction and wear properties of the substrate and the chromized layer at room temperature. The average friction coefficients of the substrate, CT6, CT8, CT10 and CT12, are 0.69, 0.59, 0.62, 0.61 and 0.69, respectively. For CT6, CT8 and CT10, the average friction coefficient is lower than that of the substrate. However, with the increase in sliding time, their friction coefficient gradually increases and tends to be the same as the substrate at the end of the friction test. It is known that the main wear form of the GCr15 matrix is severe abrasive wear; thus, the friction coefficient is high, and the wear is serious [8]. After chromizing treatment, the high-hardness phase of chromium carbide is formed on the surface, and the resistance to abrasive wear is improved, resulting in a lower initial friction coefficient. However, with the increase in sliding time, the friction coefficient of the chromized layer gradually increases due to the continuous accumulation of hard abrasive particles at the friction interface. When the chromizing time is increased to 12 h (CT12 sample), coarse grains are formed on the surface of the chromized layer (see Figure 2), which can increase the surface roughness, resulting in a larger initial friction coefficient. As the coarse grains are gradually smoothed out with continued sliding, the friction coefficient decreased gradually. After approximately 50 min of wear, the variation trend of friction coefficient of the CT12 sample is similar to that of other chromized layers.

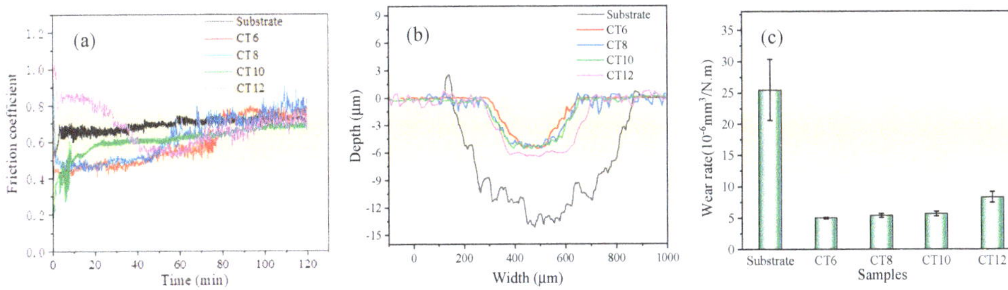

Figure 7. Wear properties of the substrate and the chromized layer (**a**) friction coefficient; (**b**) wear trace contour; (**c**) wear rate.

From the wear scar profile, the substrate displays a wear scar dimension with 726 μm in width and 14.29 μm in depth, whereas the chromized layer is markedly reduced in wear scar dimension with smaller width and depth. The wear depths of the chromized layer are lower than their thickness, indicating that the chromized layers have not been worn through. Figure 7c shows the wear rate, calculated based on the wear scar profile. The wear rate of the substrate is 25.47×10^{-6} $mm^3 \cdot N^{-1} \cdot m^{-1}$, which is about five times higher than that of the chromized layers. This suggests that the wear rate is reduced greatly by chromizing treatment. This improvement in wear resistance can be attributed to the high hardness of the chromized layer and excellent interface bonding strength between the substrate and coating. The wear rates of CT6, CT8 and CT10 samples exhibited minimal variations, while that of the sample CT12 has increased. This may be related to the larger friction coefficient of CT12 samples.

Figure 8 shows the wear scar morphologies of the substrate and CT10 chromized layer after the friction and wear test. There are a large number of scratches, along with torn-wear phenomena within the wear scar of the substrate. Some wear debris adhere in the scar and are crushed to form abrasive particles. The EDS result shown in Table 3 reveals that oxidation is not obvious in the smooth Area A, but is severe in the abrasive particle's Area B, which indicates that wear debris undergoes oxidation under repeated sliding of the load and forms hard oxide abrasive particles. The surface of steel substrate, mainly composed of Fe and a small amount of Cr, exhibits high oxygen affinity. During the repeated friction process, the interface temperature increases, causing these metals to combine with oxygen to form hard oxide particles. These hard particles embed in the wear interface, accelerating the abrasion of the soft substrate. Moreover, the affinity between the metal and the counter ball leads to adhesive wear, causing tearing and spalling of the material within the wear scar. Therefore, the wear mechanism of the GCr15 steel substrate is characterized by a combination of abrasive wear, adhesive wear and oxidative wear, collectively contributing to severe material wear.

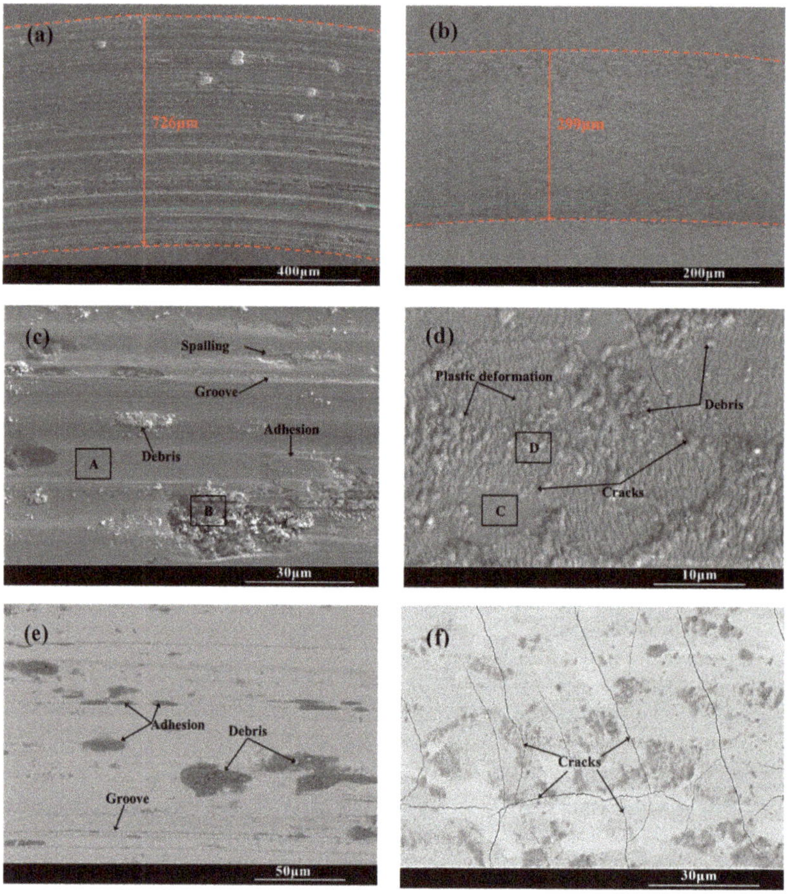

Figure 8. Wear trace morphology of the substrate and the chromized layer (**a**,**c**,**e**) substrate; (**b**,**d**,**f**) CT10.

The wear tracks of the chromized layer shown in Figure 8b,d,f exhibit relatively shallow profiles with minor superficial scratches, indicating that the superior wear is resistant to the chromized layer. Higher-magnification images (Figure 8d,f) reveal the presence of microcracks, plastic deformation zones and fine wear debris particles within the wear tracks. During the friction contact process, the material in the wear tracks is subjected to repeated shear stress and compressive stress, leading to plastic deformation as the material flows along the direction of the friction force. The thickness change in the deformation zone is inversely proportional to the hardness of the material. Generally, the material with large plastic deformation has lower hardness, while the material with higher hardness is more resistant to deformation. In the severe deformation zone, intense shear and contact stresses induce pronounced grain slip and flow, resulting in irregular deformation characteristics [46]. As observed in Figure 8d, it can be found that the plastic deformation zone within the wear scar of the chromized layer is shallow due to its high hardness. Under the cyclic contact stress and fatigue at the friction interface, micro-cracks nucleate at pores and grain boundary regions in the chromized layer. These micro-cracks gradually propagate under the repeated loading, eventually developing into macroscopic cracks within the wear tracks. As the crack density increases and propagates to form interlaced cracks, material spalling will be triggered.

It can be seen that although the substrate material undergoes extensive plastic deformation within the wear tracks due to its low hardness and high toughness, but no cracks are generated. However, the substrate is heavily ground due to its low hardness, resulting in poor wear resistance. In contrast, despite the presence of numerous cracks, the chromized layer demonstrates significantly enhanced wear resistance owing to its superior hardness and strong metallurgical bonding with the substrate.

Additionally, compared with that of the GCr15 steel substrate, no adhesive wear phenomenon is found within the wear tracks of the chromized layer. The chromized layer is mainly composed of hard ceramic phases: $(Cr,Fe)_{23}C_6$ and $(Cr,Fe)_7C_3$. The ceramic phases exhibit high chemical stability and poor affinity with the counter grinding ball. Therefore, the phenomena of adhesive wear and oxidative wear are significantly suppressed (as shown in Table 5). Thus, the main wear mechanism of the chromized layer is abrasive wear.

Table 5. EDS element composition of the different positions of wear trace in Figure 8.

Position	Elements (at.%)			
	Cr	C	Fe	O
A	1.45	1.34	94.16	2.19
B	0.85	2.67	48.81	47.05
C	49.05	8.95	34.96	7.03
D	47.86	7.25	31.32	13.57

3. Experimental Section

3.1. Materials and Process

GCr15 bearing steel (provided by Luoyang Bearing Research Institute CO., Ltd., Luoyang, China) with dimensions of Φ30 mm × 5 mm was used as the substrate material. The chemical composition of the GCr15 bearing steel is given in Table 6. The samples were polished to obtain a mirror finish and ultrasonically cleaned in acetone and alcohol, then dried before the samples were chromized.

Table 6. Chemical composition of the GCr15 bearing steel.

Element	C	Cr	Mn	Si	Al	Cu	S	P	Fe
wt.%	1.0	1.4	0.3	0.2	0.1	0.04	0.03	0.01	balance

The chromized layer was fabricated by a solid powder pack chromizing method. The chromizing powder consisted of 50 wt.% chromium powder, 40 wt.% Al_2O_3 powder, 7 wt.% LaO_2 powder and 3 wt.% NH_4Cl (chromium powder, Al_2O_3 powder and LaO_2 powder were provided by Zhongchun New Materials (Beijing) Technology Co., Ltd. China, Beijing, China, and NH_4Cl was provided by Guangzhou Chemical Reagent Factory, Guangzhou, China). Firstly, the cleaned samples were embedded in the mixture powder within a steel box. Then, the chromizing treatment was carried out using a KBF13Q atmosphere electrical resistance furnace (Nanjing Nanda Instrument Co., Ltd., Nanjing, China). Before heating, the furnace was pumped to a pressure below −0.1 MPa, followed by argon purging until the pressure reached 0 MPa (equivalent to atmospheric pressure). The vacuum-pumping and gas-filling process was repeated three times to ensure an inert environment. The chromizing parameters were as follows: the heating rate of 10 °C/min, the holding temperature of 840 °C, and the holding times of 6 h, 8 h, 10 h, and 12 h (designated as CT6, CT8, CT10, and CT12, respectively). After holding, the samples were furnace-cooled. Subsequently, quenching treatment was performed at 840 °C for 30 min, followed by oil quenching. Annealing was conducted at 160 °C for 2 h, and the samples were then furnace-cooled. Finally, the samples were ultrasonically degreased and cleaned in acetone and alcohol before drying.

3.2. Characterization

The surface and cross-sectional morphologies of the chromized layer were characterized using a NOVA NanoSEM430 field-emission scanning electron microscope (FE-SEM) (Thermo Fisher Scientific, Waltham, Massachusetts, USA), while the elemental composition was analyzed with energy-dispersive spectroscopy (EDS). For cross-sectional analysis, samples were first sectioned using wire electrical discharge machining, followed by resin embedding, polishing to a mirror finish, and etching with a 4% nitric acid–alcohol solution. Phase structure analysis was carried out using a Philips X'Pert Pro MPD X-ray diffractometer (XRD) (Philips Analytical) with Cu Kα radiation. Both conventional Bragg–Brentano geometry and grazing-incidence XRD (GIXRD) at an incident angle of 2° were employed. Surface microhardness was tested using an MH-5D microhardness tester (Hengyi Precision Instrument Co., Ltd, Shanghai, China) under a load of 50 g and holding for 15 s. The average value was obtained by repeating 5 times. The adhesion strength was evaluated by a Rockwell indentation tester (Laizhou Huayin Test Instrument Co., Ltd, Shandong, China), using a 150 kg load. Three indentations were performed for each sample and then SEM was used to analyze the indent morphology. Friction and wear properties of both the substrate and chromized layer were investigated using a UMT-tribolab tribometer (Bruker, Billerica, MA, USA) at room temperature. The tribological experimental test conditions were presented in Table 7. The experiment was repeated three times. Wear track profiles were measured using a Dektak XT 3D profilometer (Bruker, Billerica, MA, USA), and the wear rate was calculated according to Equation (13):

$$W = \frac{V}{LF_N} \tag{13}$$

where V is the wear volume (mm^3), L is the sliding distance (m) and F_N is the normal load (N).

Table 7. Tribological experimental test conditions.

Conditions	Value
Normal force (N)	10
Sliding speed (m/s)	0.105
Sliding distance (m)	756
Pin-end Diameter, spherical (mm)	6
Pin-end material	Al_2O_3
Environment (Dry)	Atmospheric environment
Track diameter (mm)	10
Temperature (°C)	26
Humidity (RH)	52%

4. Conclusions

The packing chromizing method was used to fabricate a chromized layer on the surface of GCr15 bearing steel. The microstructure, hardness, adhesion strength and tribological properties of the chromized layer were investigated.

The chromized layer exhibits a relatively smooth topography, with scattered protruding rhombic grains embedded on the surface. There are a certain number of fine pores existing on the surface and at the interface of the chromized layer/steel substrate. The content of the Cr element in the chromized layer gradually decreases from the surface to the interior, while the content of Fe element gradually increases, forming a good gradient structure. The content of C is high throughout the chromized layer. The chromized layer is mainly composed of $(Cr,Fe)_{23}C_6$ and $(Cr,Fe)_7C_3$ ceramic hard phase. With the increase in chromizing time, the microstructure of the chromized layer does not change obviously, except for the increase in thickness.

With the increase in chromizing time, the hardness of the chromized layer increases, ranging from 1360.6 HV to 1588.2 HV, which is more than threefold higher compared to that of the substrate material. A large number of radial cracks and circumferential cracks exist in and around the Rockwell indentation. As the chromizing time increases, the brittleness of the chromized layer increases and the spalling phenomenon of the chromized layer around the edge of the indentation is aggravated.

The wear mechanism of the GCr15 steel substrate is characterized by a combination of abrasive wear, adhesive wear and oxidative wear, collectively leading to severe material wear. In contrast, the wear behavior of the chromized layer is only slight abrasive wear, despite the occurrence of microcracks during the friction process. The average friction coefficient of the chromized layer has been reduced, and the wear rate is about one fifth of that of the substrate. This significant improvement in the wear resistance of the chromized layer is primarily attributed to its high hardness and strong metallurgical bonding with the substrate.

Author Contributions: Conceptualization, D.Y. and C.W.; Methodology, D.Y. and C.W.; Formal analysis, C.W.; Investigation, D.Y. and C.W.; Data curation, P.T. and S.H.; Writing—original draft, D.Y. and C.W.; Writing—review and editing, C.W., Q.S. and X.H.; Supervision, S.L.; Funding acquisition, C.W.; All authors have read and agreed to the published version of the manuscript.

Funding: This research was funded by the Guangdong Basic and Applied Basic Research Foundation Project, grant number 2022A1515240048 and 2024A1515010753; Funding Project for the Initiative to Establish a Comprehensive Industrial Technology Innovation Center by Guangdong Academy of Sciences, grant number 2022GDASZH-2022010201.

Institutional Review Board Statement: Not applicable.

Informed Consent Statement: Not applicable.

Data Availability Statement: Data is contained within the article.

Conflicts of Interest: The authors declare no conflicts of interest.

References

1. Zhao, Z.; Xie, X.; Tang, G.; Padhiar, M.A.; Xiao, J.; Liang, Z. Study on the effect of the strengthen grinding process surface coverage on the micro-morphology, micro-hardness, roughness, and residue stress of GCr15 bearing steels. *J. Manuf. Process.* **2023**, *99*, 362–372. [CrossRef]
2. Zhang, H.; Zheng, Z.; Qiao, Z.; Han, D.; Du, A.; Ma, R.; Fan, Y.; Zhao, X.; Cao, X. Novel process used to accelerate spheroidizing annealing of GCr15 bearing steel by modulating initial structure. *Mater. Today Commun.* **2025**, *44*, 112001. [CrossRef]
3. Zhang, H.; Jiao, F.; Lian, X.; Zhao, Y.; Niu, Y.; Tong, J. Surface quality in ultrasonic-electrolytic internal grinding of GCr15 steel. *Inte. J. Mech. Sci.* **2025**, *291–292*, 110188. [CrossRef]
4. Zhang, W.; Li, W.; Liu, H.; He, T.; Deng, S.; Tian, H.; Cao, W. Microstructure effects on improving rolling contact fatigue by a double-quenching heat treatment process for GCr15 steel. *Tribol. Int.* **2024**, *194*, 109519. [CrossRef]
5. Li, Y.; Ying, S.; Zhao, Y.; Li, Z.; Cao, J.; Wu, X.; Du, P.; Wu, T.; Yu, C.; Xie, G. Study on fretting friction properties based on GCr15 alloy and liquid epoxy resin protective coatings on GCr15 alloy. *Colloid. Surface. A* **2024**, *696*, 134364. [CrossRef]
6. Mei, S.; Xiao, Z.; Chen, Z.; Hu, Z.; Zou, X.; Xu, Q.; Zheng, Q. Prediction and optimization of Aluminum content in Ni-P-Al_2O_3 composite coatings on GCr15 Steel Based on PSO-BP Neural Network. *Surf. Interf.* **2025**, *64*, 106319. [CrossRef]
7. Xie, X.; Guo, Z.; Liang, Z.; Xiao, J.; Zhao, Z. Enhanced high-temperature wear resistance of GCr15 steel balls by generating a Ti + Nb diffusion layer via mechanical alloying and $NH_3.H_2O$ treatment. *Surf. Coat. Technol.* **2023**, *466*, 129632. [CrossRef]
8. Liu, S.; Li, Y.; Wang, Y.; Wei, Y.; Zhang, L.; Wang, J. High wear resistance WC-Co reinforced GCr15 bearing steel composite prepared via selective laser melting (SLM). *Int. J. Refract. Met. H.* **2022**, *109*, 105988. [CrossRef]

9. Wang, W.; Yuan, W.; Guo, Q.; Wang, N.; Chi, B. Effect of picosecond laser surface texturing under Babbitt coating mask on friction and wear properties of GCr15 bearing steel surface. *Eng. Fail. Anal.* **2024**, *157*, 107878. [CrossRef]
10. Xiao, J.; Xie, X.; Liu, X.; Li, D.; Zhou, Y. Fabrication of gradient structure for enhancing wear resistance of GCr15 bearing steel friction shim. *Wear* **2024**, *552–553*, 205449. [CrossRef]
11. Li, Y.; Chen, C.; Fan, Z.; Jiang, D.; Shuai, S.; Tu, T. Effects of cooling rate on isothermal spheroidizing annealing of hot-rolled GCr15 bearing steel. *J. Mater. Res. Technol.* **2024**, *31*, 329–337. [CrossRef]
12. Maskavizan, A.J.; Quintana, J.P.; Dalibón, E.L.; Márquez, A.B.; Brühl, S.P.; Farina, S.B. Evaluation of wear and corrosion resistance in acidic and chloride solutions of Cathodic Arc PVD chromium nitride coatings on untreated and plasma nitrided AISI 4140 steel. *Surf. Coat. Technol.* **2024**, *494*, 131476. [CrossRef]
13. Novák, M.; Novotný, R.; Valtr, J.; Dašek, D.; Cvrček, L.; Krejčí, J.; Vrtílková, V.; Macák, J. The effect of chromium-based coatings on corrosion behavior of alloy Zr1Nb in 70ppm Li^+ water environment. *J. Nucl. Mater.* **2024**, *599*, 155245. [CrossRef]
14. Bai, C.Y.; Wen, T.M.; Hou, K.H.; Ger, M.D. The bipolar plate of AISI 1045 steel with chromized coatings prepared by low-temperature pack cementation for proton exchange membrane fuel cell. *J. Power Sources* **2010**, *195*, 779–786. [CrossRef]
15. Jathar, S.; Honnali, S.K.; Farhadizadeh, A.; Febvier, A.l.; Odén, M.; Eklund, P. Influence of nitrogen and niobium incorporation in bcc-chromium coatings on microstructure and mechanical properties. *Surf. Coat. Technol.* **2024**, *494*, 131489. [CrossRef]
16. Iliescu, I.; Gazal, Y.; Michau, A.; Addou, F.; Duguet, T.; Monsifrot, E.; Schuster, F.; Maury, F. Low temperature Direct Liquid Injection MOCVD of amorphous CrCx coatings in large-scale reactors: An original route to nanostructured multilayer coatings. *Surf. Coat. Technol.* **2021**, *416*, 127174. [CrossRef]
17. Roshan, K.; Jayakrishnan, R.; Thirumalai, K.S. Corrosion measurement of thermally sprayed carbide coatings on stainless steel pipes. *Results Surf. Interfaces* **2024**, *17*, 100320.
18. Beck, K.; Ulrich, A.S.; Thor, N.; Oskay, C.; Galetz, M.C. Chromium diffusion coatings for improving the oxidation behavior of refractory metals at intermediate temperatures. *Int. J. Refract. Met. H. Mat.* **2024**, *121*, 106626. [CrossRef]
19. Ye, J.L.; Ye, F. Effects of strong carbide-forming elements on low temperature salt-bath chromizing. *Adv. Mater. Res.* **2011**, *214*, 646–650. [CrossRef]
20. Ganji, O.; Sajjadi, S.A.; Yang, Z.G.; Mirjalili, M. Tribological properties of duplex coatings of chromium-vanadium carbide produced by thermo-reactive diffusion (TRD). *Ceram. Int.* **2022**, *48*, 7475–7490. [CrossRef]
21. Najari, M.R.; Sajjadi, S.A.; Ganji, O. Microstructural evolution and wear properties of chromium carbide coating formed by thermo-reactive diffusion (TRD) process on a cold-work tool steel. *Results Surf. Interfaces* **2022**, *8*, 100059. [CrossRef]
22. Liu, T.; Wang, C.; Meng, Q.; Song, Q.; Xue, B.; Zhang, Y.; Cheng, H.; Wang, Y. Effect of steel matrix with different C content on the growth mechanism, microstructure and properties of the chromized layer. *Surf. Coat. Technol.* **2024**, *478*, 130446. [CrossRef]
23. Wei, C.B.; Lin, S.S.; Dai, M.J.; Shi, Q.; Su, Y.; Tang, P. Structure and Performance Stability of Chromized Coatings by Pack Cementation. *Rare Met. Mat. Eng.* **2021**, *50*, 1706–1712.
24. Dong, Z.; Zhou, T.; Liu, J.; Zhang, X.; Shen, B.; Hu, W.; Liu, L. Effects of pack chromizing on the microstructure and anticorrosion properties of 316L stainless steel. *Surf. Coat. Technol.* **2019**, *366*, 86–96. [CrossRef]
25. Dong, Z.; Zhou, T.; Liu, J.; Zhang, X.; Shen, B.; Hu, W.; Liu, L. Performance of surface chromizing layer on 316L stainless steel for proton exchange membrane fuel cell bipolar plates. *Int. J. Hydrog. Energ.* **2019**, *44*, 22110–22121. [CrossRef]
26. Lee, S.B.; Cho, K.H.; Lee, W.G.; Jang, H. Improved corrosion resistance and interfacial contact resistance of 316L stainless-steel for proton exchange membrane fuel cell bipolar plates by chromizing surface treatment. *J. Power Sources* **2009**, *187*, 318–323. [CrossRef]
27. Liu, S.; Yang, J.; Liang, X.; Sun, Y.; Zhao, X.; Cai, Z. Investigation of the preparation, corrosion inhibition, and wear resistance of the chromized layer on the surfaces of T9 and SPCC steels. *Materials* **2022**, *15*, 7902. [CrossRef]
28. Lee, J.W.; Duh, J.G. Evaluation of microstructures and mechanical properties of chromized steels with different carbon contents. *Surf. Coat. Technol.* **2004**, *177–178*, 525–531. [CrossRef]
29. Tan, C.; Zong, X.; Zhou, W.; Cao, H.; Wang, J.J.; Wang, C.; Peng, J.; Li, Y.; Li, H.; Wang, J.S.; et al. Insights into the microstructure characteristics, mechanical properties and tribological behaviour of gas-phase chromized coating on GCr15 bearing steel. *Surf. Coat. Technol.* **2022**, *443*, 128605. [CrossRef]
30. Wang, H.F.; Wang, C.M.; Liu, T.T.; Song, Q.; Xue, Y.L.; Ma, Q.C.; Cui, H.Z. Growth Mechanism and Pore Formation of Solid Chromizing Layer of GCr15 Pin. *Rare Met. Mat. Eng.* **2023**, *52*, 1809–1818.
31. Zong, X.; Jiang, W.; Fan, Z. Evaluation of chromium carbide coatings on AISI 52100 steel obtained by thermo-reactive diffusion technique. *Mater. Sci.* **2019**, *25*, 36–41.
32. Vidakis, N.; Antoniadis, A.; Bilalis, N. The VDI 3198 indentation rest evaluation of a reliable qualitative control for layered compounds. *J. Mater. Process. Technol.* **2003**, *143–144*, 481–485. [CrossRef]
33. Wu, C.; Hong, Y.; Chen, W.; Chen, J.; Yuan, M.; Liao, X. A double strengthened surface layer fabricated by nitro-chromizing on carbon steel. *Surf. Coat. Technol.* **2016**, *298*, 83–92. [CrossRef]

34. Guo, S.; Liu, L.; He, F.; Wang, S. Preparation of Cr-N coatings on 316H stainless steel via pack chromizing and gas nitriding, and their resistance to liquid metal corrosion in early stages. *Surf. Coat. Technol.* **2024**, *481*, 130665. [CrossRef]
35. Li, L.B.; Li, Z.X.; Liu, L.T.; He, F. Effect of reaction temperature and time on microstructure of chromizing layer on austenitic stainless steel matrix. *Rare Met. Mat. Eng.* **2021**, *50*, 1743–1752.
36. Hu, J.; Zhang, Y.; Yang, X.; Li, H.; Xu, H.; Ma, C.; Dong, Q.; Guo, N.; Yao, Z. Effect of pack-chromizing temperature on microstructure and performance of AISI 5140 steel with Cr-coatings. *Surf. Coat. Technol.* **2018**, *344*, 656–663. [CrossRef]
37. He, Y.; Huang, G.; Liu, Y.; Tang, Y.; Fu, X.; Bao, H.; Ge, M.; Huang, H.; Zhang, R.; Liu, H.; et al. A simple and feasible method for preparing chromium carbide coating on graphite through binary NaCl-KCl molten salt. Mater. *Today Commun.* **2024**, *40*, 109610.
38. Iorga, S.; Cojocaru, M.; Chivu, A.; Ciuca, S.; Burdusel, M.; Badica, P.; Leuvrey, C.; Schmerber, G.; Ulhaq-bouillet, C.; Colis, S. Influence of the Carbo-Chromization Process on the Microstructural, Hardness, and Corrosion Properties of 316L Sintered Stainless Steel. *Metall. Mater. Trans. A* **2014**, *45A*, 3088–3096. [CrossRef]
39. Arzamasov, B.N.; Mel'nikov, R.A. Investigation of the pore-formation process during the chromizing of steel 40 Kh by the circulation method. *Met. Sci. Heat Treat.* **1994**, *36*, 237–243. [CrossRef]
40. Lin, N.; Guo, J.; Xie, F.; Zou, J.; Tian, W.; Yao, X.; Zhang, H.; Tang, B. Comparison of surface fractal dimensions of chromizing coating and P110 steel for corrosion resistance estimation. *Appl. Surf. Sci.* **2014**, *311*, 330–338. [CrossRef]
41. Cao, H.; Luo, C.P.; Liu, J.; Zou, G. Phase transformations in low-temperature chromized 0.45 wt.% C plain carbon steel. *Surf. Coat. Technol.* **2007**, *201*, 7970–7977. [CrossRef]
42. Meng, T.X.; Guo, Q.; Xi, W.; Ding, W.Q.; Liu, X.Z.; Lin, N.M.; Yu, S.W.; Liu, X.P. Effect of surface etching on the oxidation behavior of plasma chromizing-treated AISI440B stainless steel. *Appl. Surf. Sci.* **2018**, *433*, 855–861. [CrossRef]
43. Burnett, P.J.; Rickerby, D.S. The mechanical properties of wear-resistant coatings I: Modelling of hardness behaviour. *Thin Solid Film.* **1987**, *148*, 41–50. [CrossRef]
44. Fu, K.; Yin, Y.; Chang, L.; Shou, D.; Zheng, B.; Ye, L. Analysis on multiple ring-like cracks in thin amorphous carbon film on soft substrate under nanoindentation. *J. Phys. D Appl. Phys.* **2013**, *46*, 505314. [CrossRef]
45. Liu, J.N.; Xing, Z.G.; Wang, H.D.; Cui, X.F.; Jin, G.; Xu, B.S. Microstructure and fatigue damage mechanism of Fe-Co-Ni-Al-Ti-Zr high-entropy alloy film by nanoscale dynamic mechanical analysis. *Vacuum* **2019**, *159*, 516–523. [CrossRef]
46. Borodin, E.; Bushuev, O.; Bratov, V.; Jivkov, A.P. Discrete model for discontinuous dynamic recrystallisation applied to grain structure evolution inside adiabatic shear bands. *J. Mater. Res. Technol.* **2024**, *30*, 2125–2139. [CrossRef]

molecules

MDPI

Article

Improving Anti-Corrosion and Conductivity of NiTi Alloy Bipolar Plate Used for PEMFCs via Nb Alloying

Ziyang Niu [1,†], Yingping Li [1,†], Yuanyuan Li [1,2,*], Xiaofen Wang [3,4], Yumin Pan [3], Zhuo He [3], Guohong Zhang [2], Zhen Wang [2] and Qiongyu Zhou [3,*]

1 The State Key Laboratory of Refractories and Metallurgy, Wuhan University of Science and Technology, Wuhan 430081, China; niuziyang@wust.edu.cn (Z.N.); liyingping@wust.edu.cn (Y.L.)
2 Analytical and Testing Center, Wuhan University of Science and Technology, Wuhan 430081, China; zhangguohong@wust.edu.cn (G.Z.); wangzhen@wust.edu.cn (Z.W.)
3 Key Laboratory of Green Surface Technology and Functional Coatings for Materials, China National Light Industry, Foshan University, Foshan 528000, China; wangxiaofen@shu.edu.cn (X.W.); 20230580130@stu.fosu.edu.cn (Y.P.); 20220580212@stu.fosu.edu.cn (Z.H.)
4 School of Materials Science and Engineering, Shanghai University, Shanghai 200444, China
* Correspondence: liyuanyuan@wust.edu.cn (Y.L.); zhouzhouqiongyu@fosu.edu.cn (Q.Z.); Tel.: +86-135-12741496 (ext. 430081) (Y.L.); +86-159-70962829 (ext. 528000) (Q.Z.)
† These authors contributed equally to this work.

Abstract: NiTi alloy has emerged as a promising bipolar plate (BP) material for proton exchange membrane fuel cells (PEMFCs), combining Ti-like corrosion resistance with Ni-like electrical conductivity through its intermetallic characteristics. However, its performance faces greater challenges under aggressive operating conditions (70 °C, F^--containing acidic solution with air bubbling). This study demonstrates that Nb alloying effectively enhances NiTi while preserving its balanced properties. The developed NiTiNb alloy exhibits improved performance with 26% lower corrosion current density (i_c) and 29% reduced interfacial contact resistance (ICR) compared to conventional NiTi, effectively overcoming the conventional corrosion–conductivity trade-off in metallic BPs. The alloy also shows superior electrochemical stability and microhardness relative to pure Ti and Ni. These enhancements stem from a unique dual-phase microstructure comprising a NiTi (B2) matrix with continuous β-Nb grain boundary networks. During operation, this structure enables in situ formation of protective TiO_2-Nb_2O_5 films while maintaining conductive Nb/Nb_2O_5 pathways and metallic Ni domains. The findings establish Nb alloying as a viable optimization strategy for NiTi-based BP substrate in demanding PEMFC applications.

Keywords: NiTiNb alloy; corrosion resistance; interfacial contact resistance; bipolar plate; PEMFC

1. Introduction

Hydrogen energy has emerged as a pivotal clean energy solution to address the dual crises of fossil fuel depletion and escalating greenhouse gas emissions [1]. Among various hydrogen utilization technologies, fuel cells (FCs) have garnered significant attention for their ability to convert chemical energy from hydrogen directly into electricity with exceptional efficiency and zero greenhouse gas emissions [2]. Proton exchange membrane fuel cells (PEMFCs), in particular, are renowned for their high energy efficiency, near-zero emissions, and rapid cold-start capability, making them one of the most viable power sources for next-generation electric vehicles and portable devices [3,4]. Within PEMFC stacks, bipolar

https://doi.org/10.3390/molecules30173658

plates (BPs) serve as one of the most critical components, accounting for 60–80% of the stack weight and 20–30% of the total cost [5]. To perform their essential functions (such as gas distribution, current collection, membrane electrode support, and heat discharge), BPs must satisfy stringent requirements, including excellent corrosion resistance in acidic environments, low interfacial contact resistance, and high mechanical robustness [6,7].

BP materials, including graphite, composites, and metals, have been extensively investigated over the past three decades [8]. Graphite, the earliest BP material developed, displays advantages such as low density, superior electrical conductivity, and outstanding chemical inertness [9]. However, its inherent brittleness and poor manufacturability significantly limit their application in compact mobile systems [9,10]. Composite BPs, which combine good corrosion resistance, high mechanical strength, and ease of molding, are regarded as ideal candidates and a promising future direction for BP development. Nevertheless, their practical application is currently hindered by challenges related to high production cost, non-uniform conductivity, and complex fabrication processes [11,12]. In comparison, metal BPs are more suitable for large-scale manufacturing owing to their low cost, low gas permeability, and satisfactory balance between electrical and thermal conductivity. However, a major drawback is their susceptibility to metal dissolution and surface passivation, both of which can deteriorate cell performance. Metal dissolution under cathodic operation potentials of PEMFCs (e.g., +0.6 V vs. SCE (saturated calomel electrode)) can lead to the degradation of the proton exchange membrane through the migration of dissolved metal ions, while surface passivation results in drastically increased interfacial contact resistance (ICR) due to the formation of semiconducting or insulating oxide films [13,14].

To address these issues, numerous studies have shown that utilizing suitable coatings can effectively improve the corrosion resistance and electrical conductivity of metal BPs [2]. Regrettably, limited research has been conducted on the durability of these coatings under actual production and operational conditions. Coating degradation is inevitable during long-term corrosion, which can result in localized corrosion of the underlying metal BPs near damaged coating areas. The corrosion resistance and service life of the coating are also closely related to the properties of the metallic substrate [15,16]. Meanwhile, recent advances have highlighted alloy design as a promising approach to mitigate the corrosion–ICR dilemma [17]. Our previous work revealed that NiTi alloy exhibits significantly decreased ICR values compared to commercial Ti BPs. In addition, the NiTi alloy showed fairly low current densities (1.28 $\mu A\ cm^{-2}$) during 8-h potentiostatic tests, only slightly above the U.S. Department of Energy (DOE) standard ($\leq 1\ \mu A\ cm^{-2}$) [18]. H. Zhu et al. [19] investigated the effects of cold rolling, annealing time, solution temperature, and fluoride concentration in simulated PEMFC environments on the corrosion behavior of Ti-Nb-Ni foil, which consisted of β-Ti and $NiTi_2$ phases. Their findings emphasized the beneficial role of traceable Nb, deposited via high-power impulse magnetron sputtering technology, in improving both electrical conductivity and corrosion resistance in PEMFC-like conditions. Furthermore, Ishikawa et al. [20] reported that the $Nb_{19}Ti_{40}Ni_{41}$ alloy, featuring a cube-on-cube structure composed of β-Nb and B2-TiNi phases, exhibited high hydrogen permeability.

Inspired by these studies, this paper proposes a strategy to further improve the performance of NiTi alloy via Nb alloying in a simulated PEMFC environment. A novel NiTiNb alloy, comprising dispersed NiTi (B2) phase and continuous reticular β-Nb phase, was prepared by vacuum arc melting. The corrosion resistance and conductivity of the designed alloy were systematically investigated through electrochemical experiments, ICR measurements, and detailed characterization.

2. Results and Discussion

To explore the structural advantages of Nb-alloyed NiTi alloy, the phase composition and microstructure of the NiTiNb alloy, which are critical factors influencing its performance as BPs in PEMFCs, were systematically analyzed, as shown in Figure 1. The optical micrograph (OM) reveals a homogeneous as-cast microstructure characterized by fine, cellular-like dendrites (Figure 1a). This morphology closely resembles that of unalloyed NiTi, owing to the identical vacuum arc melting preparation method used for both alloys. During solidification, the relatively small ingot size (~20 mm) establishes a thermal gradient that promotes directional grain growth along the heat flow [21]. In comparison, pure Ti and pure Ni exhibit much coarser primary equiaxed grains, with a small number of twins observed in pure Ti (Figure S1 in Supplementary Materials). The X-ray diffraction (XRD) pattern (Figure 1b) confirms that the NiTiNb alloy mainly consists of the NiTi (B2) phase (JCPDS No. 96-901-4020) and the β-Nb phase (JCPDS No. 96-151-2525) [22,23] (both featuring a body-centered cubic (BCC) crystal structure), consistent with the phase composition typically reported in the literature for the $Ni_{47}Ti_{44}Nb_9$ alloy [24]. The introduction of Nb leads to a eutectic-like microstructure, as evidenced by the scanning electron microscopy (SEM) image in Figure 1c and further supported by the corresponding energy dispersive spectroscopy (EDS) in Figure 1d–f. The β-Nb phase is continuously distributed along the grain boundaries, where a small amount of NiTi (B2) phase is also present, appearing as brighter areas in the SEM image. The grain interiors are predominantly composed of the NiTi (B2) matrix. Additionally, Nb atoms are partially substituted into the NiTi (B2) lattice. Given the larger atomic radius of Nb, this substitution induces lattice expansion, which manifests as a leftward shift of the diffraction peaks in the XRD pattern. Such a refined and homogeneous microstructure, along with the presence of a supersaturated solid solution and a continuous β-Nb phase network at grain boundaries, contributes to improved mechanical strength and enhanced workability of the NiTiNb alloy. These characteristics are particularly advantageous for BP applications in PEMFCs.

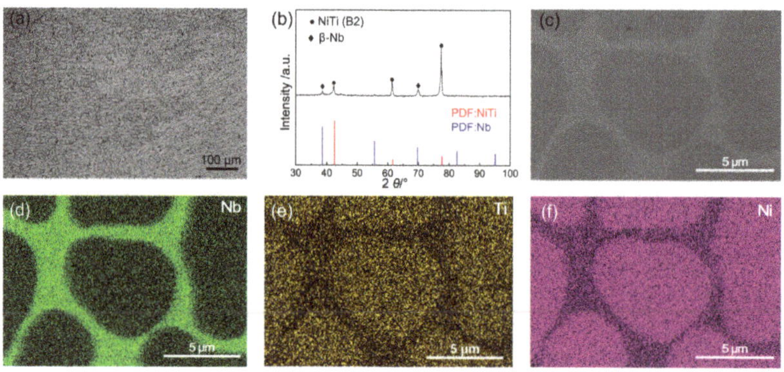

Figure 1. (**a**) Optical micrograph (OM), (**b**) X-ray diffraction (XRD) pattern, and (**c**) scanning electron microscopy (SEM) image of the NiTiNb alloy, along with the corresponding elemental distribution maps for (**d**) Nb, (**e**) Ti, and (**f**) Ni.

The open circuit potential versus time (E_{OCP}-t) curves and potentiodynamic polarization curves of the NiTiNb alloy, NiTi alloy, pure Ti, and pure Ni are shown in Figure 2, with key electrochemical parameters summarized in Table 1. Clearly, both the E_{OCP}-t curves of NiTi alloy and pure Ti display fluctuations and downward drifts before stabilizing. This behavior can be attributed to localized dissolution of their native passive films during initial immersion in the halide-containing (e.g., F^-) solution and the resulting instability during the dissolution–passivation process [25]. Pure Ni, on the other hand, exhibits a continuous decrease in open circuit potential (E_{OCP}), implying persistent surface degradation and poor corrosion resistance. In comparison, the NiTiNb alloy demonstrates a brief initial rise in E_{OCP} followed by rapid stabilization into a smooth plateau, indicating the admirable stability of the formed passive films, which offers an effective barrier to improve corrosion resistance. All E_{OCP} values tend to stabilize after long-term immersion (3600 s). At the steady state, the E_{OCP} of NiTiNb alloy (0.13 V) is quite close to that of NiTi alloy (0.14 V). Both values are markedly nobler than those of pure Ni (−0.03 V) and pure Ti (−0.42 V), confirming the strong corrosion resistance tendency of NiTiNb alloy.

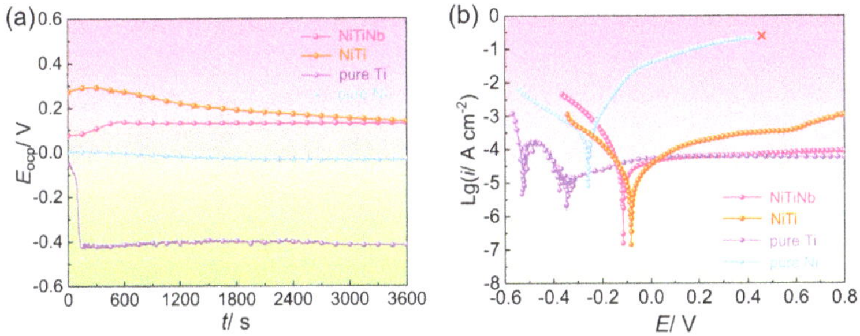

Figure 2. (**a**) Open circuit potential versus time (E_{OCP}-t) curves and (**b**) potentiodynamic polarization curves of the NiTiNb alloy, NiTi alloy, pure Ti, and pure Ni.

Table 1. The electrochemical parameters extracted from the E_{OCP}-t curves and the potentiodynamic polarization curves.

Samples	E_{OCP} at Steady State (V)	E_{OCP} Standard Deviation During 2400–3600 s (mV)	E_{corr} (V)	i_{corr} (μA cm^{-2})
NiTiNb	0.13	0.21	−0.12	5.01
NiTi	0.14	8.98	−0.08	6.51
Pure Ti	−0.42	4.76	−0.53	75.95
Pure Ni	−0.03	1.43	−0.26	221.23

Potentiodynamic polarization curves shown in Figure 2b further support these observations. The relative order of corrosion potentials (E_{corr}) among the four materials aligns well with their E_{OCP} values. Both NiTiNb and NiTi alloys exhibit similarly low corrosion current densities (i_{corr}), approximately one order of magnitude lower than that of pure Ti and two orders lower than pure Ni. Notably, no passivation feature is observed in the anodic branch of pure Ni, with the anodic current density continuously rising until reaching the measurement limit (0.1 A) at 0.42 V. Pure Ti undergoes an unstable sequence involving initial passivation, transpassive dissolution, and re-passivation before reaching a quasi-stable state, accompanied by pronounced current fluctuations. The dissolution process can also be observed at the initial stage of immersion in the E_{OCP}-t curve of pure Ti (Figure 2a). In contrast, anodic branches of NiTi and NiTiNb alloys change directly into the passive region without the active-to-passive transition, suggesting their superior passivation characteristics, as the passive films form rapidly and almost spontaneously at E_{corr} [26]. However, compared to ambient conditions [18], the presence of F^- and elevated temperature under air bubbling significantly reduce the stability of passive films on NiTi surfaces, leading to gradually increasing passive current density (i_{pass}). Remarkably, the NiTiNb alloy maintains a consistently low i_{pass} under harsh conditions, a value quite close to that of pure Ti. This result highlights its excellent electrochemical stability and corrosion resistance, underscoring the strong potential of NiTiNb alloy as a durable and protective BP material for PEMFC applications.

Figure 3 presents the core properties of metal materials as BPs for PEMFC applications, including the corrosion durability, ICR, and mechanical hardness. The potentiostatic polarization curves, measured in a simulated PEMFC cathodic environment at 70 °C with air bubbling, reveal significant performance variations among the four tested materials. Pure Ni exhibits obviously higher cathodic current density (i_c) than the other materials and turns into complete failure within 4.5 h due to severe dissolution. Pure Ti shows the lowest i_c value of 11.2 μA cm^{-2} but suffers from considerable electrochemical instability. The NiTiNb alloy emerges as the most promising candidate, demonstrating not only superior corrosion resistance (29.8 μA cm^{-2}, representing a 26% improvement over the NiTi alloy) but also exceptional electrochemical stability. This advantage becomes particularly evident under ambient temperature and non-aerated conditions, where NiTiNb achieves a low current density of 0.90 μA cm^{-2} (Figure S2 in Supplementary Materials), significantly surpassing both the DOE target (<1 μA cm^{-2}) and the NiTi alloy performance (1.28 μA cm^{-2}) by 30% [18,27]. Importantly, even under accelerated corrosion conditions at elevated temperatures with aeration, the NiTiNb alloy consistently outperforms most uncoated metallic materials while maintaining the most stable electrochemical behavior across all test conditions [28–31].

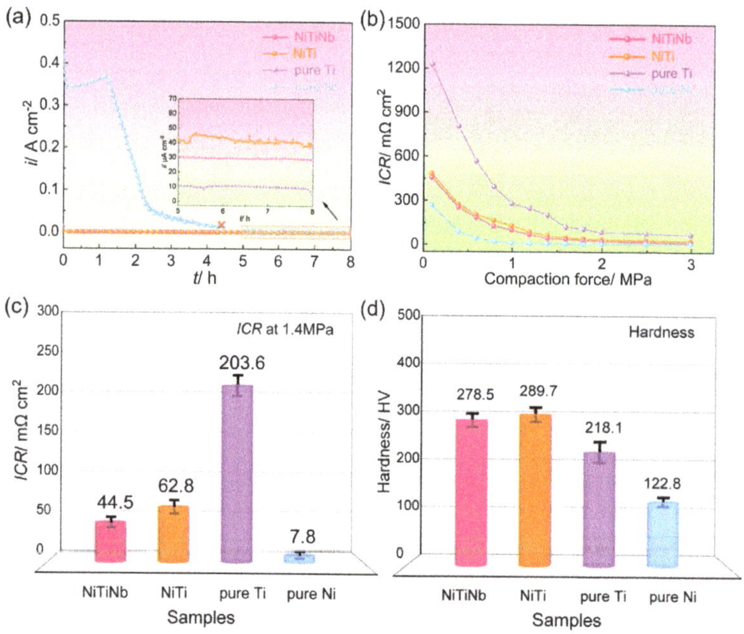

Figure 3. (**a**) Potentiostatic polarization curves, (**b**) interfacial contact resistance (ICR) values under varying compaction force and (**c**) 1.4 MPa, and (**d**) hardness of the NiTiNb alloy, NiTi alloy, pure Ti, and pure Ni.

Figure 3b,c illustrate the variations in ICR of the four tested materials as a function of compressive stress, with particular emphasis on 1.4 MPa, a typical assembly pressure in PEMFCs [32,33]. All materials show a characteristic decline in ICR as compressive stress increases, featuring an initial rapid decrease followed by gradual stabilization due to the enhanced effective contact area between the BP and the gas diffusion layer (GDL) [34]. The ICR ranking remains consistent across all pressure conditions: pure Ti > NiTi alloy > NiTiNb alloy > pure Ni. At 1.4 MPa, pure Ti demonstrates exceptionally high ICR (203.6 mΩ cm^2), exceeding NiTi (62.8 mΩ cm^2) and NiTiNb (44.5 mΩ cm^2) by one order of magnitude and surpassing pure Ni (7.8 mΩ cm^2) by two orders. This trend highlights the inherent trade-off between corrosion resistance and electrical conductivity in metallic BPs [7]. Pure Ti achieves the lowest i_c due to its pronounced passivation, but this comes at the cost of severely compromised conductivity. Conversely, pure Ni's inability to form stable passive films results in poor corrosion resistance but excellent electrical contact. Our previous work has shown that NiTi intermetallic can effectively balance these two properties by combining the advantages of Ti and Ni [18]. The present results reveal that Nb alloying further optimizes this balance, with NiTiNb achieving both enhanced corrosion resistance and 29% lower ICR compared to NiTi.

Furthermore, the NiTiNb alloy exhibits comparable microhardness to the NiTi alloy while surpassing both pure Ti and pure Ni (Figure 3d), demonstrating excellent mechanical properties suitable for PEMFC BP applications. In summary, Nb alloying of NiTi alloy achieves simultaneous enhancement in both corrosion resistance and electrical conductivity while maintaining superior mechanical strength. These combined advantages position NiTiNb as a highly promising candidate material for metallic BP substrates in practical PEMFC applications.

Electrochemical impedance spectroscopy (EIS), SEM, and X-ray photoelectron spectroscopy (XPS) were employed to investigate the surface states of different BP materials in a simulated PEMFC environment, which are critical factors determining both corrosion resistance and electrical conductivity. Figure 4 exhibits the comparative EIS results obtained before and after potentiostatic polarization. As shown in Figure 4a,c, the impedance magnitude (|Z|) before potentiostatic polarization follows a clear descending order: NiTiNb alloy > NiTi alloy > pure Ti > pure Ni. This trend confirms the superior intrinsic corrosion resistance of NiTiNb alloy at the initial stage compared to other BP materials, consistent with the results of i_{corr} obtained from potentiodynamic polarization curves (Figure 2b). After potentiostatic polarization, pure Ti shows the highest |Z| (Figure 4b,d) due to the semiconductive passive film formation, which can offer corrosion resistance but simultaneously result in undesirably high ICR. The phase angle plots reveal distinct electrochemical responses (Figure 4c,d). Pure Ni shows only one single time constant at medium frequencies (~10^3 Hz) both before and after potentiostatic polarization, indicating its metallic surface state is devoid of effective passivation, which explains its poor corrosion resistance and low ICR. In contrast, other BP materials exhibit an additional time constant at higher frequencies that merges with the metallic characteristic time constant, leading to relatively stable phase angles across the medium–low frequencies. Nevertheless, all tested materials display high and slanted impedance modulus values, confirming the presence of the protective passive films formed on the surfaces [26,35].

Based on the characteristic description of EIS plots, the equivalent circuit proposed by A. K. Iversen [36] was used to fit the pure Ni data (Figure 5a), reflecting its charge transfer behavior at the metal/electrolyte interface without stable passivation. For other materials, the equivalent circuit from C. Boissy et al. [37] was applied (Figure 5b), accounting for both charge transfer and the capacitive response from the passive film/electrolyte interface. In these models, R_s represents the solution resistance. C_b and R_b correspond to the double-layer capacitance and charge transfer resistance, respectively. C_p and R_p denote the capacitance and resistance of the passive film. The "n" is the CPE exponent (0–1), reflecting deviation from ideal capacitance. The fitted parameters are summarized in Table 2, and the fitting lines are also shown in Figure 4. As shown in Table 2, the obtained chi-square (χ^2) values are on the order of 10^{-3}, indicating that the used equivalent circuits appropriately represent the actual electrochemical processes that occurred on the sample surfaces [38].

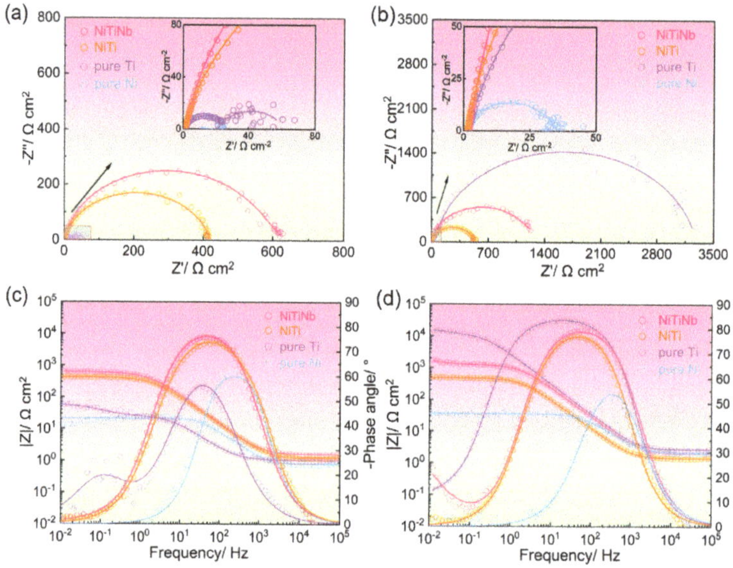

Figure 4. The electrochemical impedance spectroscopy (EIS) results for NiTiNb alloy, NiTi alloy, pure Ti, and pure Ni, presented as (**a**,**b**) Nyquist plots and (**c**,**d**) Bode-phase plots, measured (**a**,**c**) before and (**b**,**d**) after potentiostatic polarization.

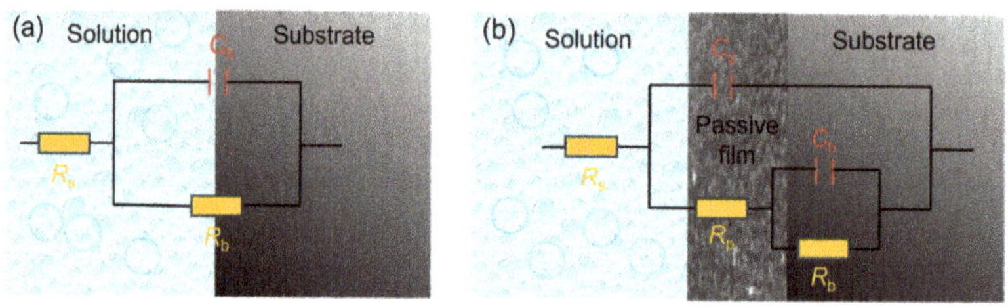

Figure 5. Equivalent electrical circuits used for EIS fitting: (**a**) one-time constant model; (**b**) two-time constant model.

Table 2. Electrochemical parameters extracted from equivalent circuit fitting of EIS data.

	Elements	R_s (Ω cm^2)	C_b (Ω^{-1} cm^{-2} s^n)	n_1	R_b (Ω cm^2)	C_p (Ω^{-1} cm^{-2} s^n)	n_2	R_p (Ω cm^2)	$\Sigma\chi^2$
Before	NiTiNb	1.59	7.02×10^{-3}	0.53	137.6	1.88×10^{-4}	0.91	528.2	5.34×10^{-3}
	NiTi	1.20	7.99×10^{-5}	0.88	7.66	2.05×10^{-4}	0.77	413.7	8.67×10^{-3}
	Pure Ti	1.03	6.23×10^{-2}	0.82	32.11	1.82×10^{-3}	0.85	24.97	9.62×10^{-3}
	Pure Ni	0.79	2.68×10^{-4}	0.94	20.7	-	-	-	7.24×10^{-3}
	Elements	R_s (Ω cm^2)	C_b (Ω^{-1} cm^{-2} s^n)	n_1	R_b (Ω cm^2)	C_p (Ω^{-1} cm^{-2} s^n)	n_2	R_p (Ω cm^2)	$\Sigma\chi^2$
After	NiTiNb	1.97	1.18×10^{-3}	0.58	376.7	8.26×10^{-5}	0.94	1030	3.18×10^{-3}
	NiTi	1.38	1.02×10^{-4}	0.93	363.6	1.49×10^{-4}	0.80	520.2	6.55×10^{-3}
	Pure Ti	1.71	2.05×10^{-4}	0.89	346.8	9.81×10^{-5}	0.99	2935	9.28×10^{-3}
	Pure Ni	1.67	1.60×10^{-4}	0.89	32.9	-	-	-	9.08×10^{-3}

As quantified in Table 2, pure Ni consistently exhibits low R_b both before and after potentiostatic polarization, implying its poor corrosion resistance. In contrast, pure Ti shows a significant increase in R_b and R_p coupled with a remarkable decrease in C_b and C_p due to the formation of a semiconductive passive film barrier. It is worth noting that after potentiostatic polarization, the NiTiNb alloy shows a much higher R_p and lower C_p compared to the NiTi alloy, confirming enhancement of anti-corrosion performance. Importantly, while maintaining comparable C_p values to pure Ti, the NiTiNb alloy shows significantly smaller R_p, demonstrating an optimal balance between protective barrier effects (enhancing corrosion resistance) and improved electrical conductivity (beneficial for ICR). These results demonstrate that the Nb alloying strategy successfully overcomes the typical trade-off between corrosion protection and electrical conductivity in metal BP materials.

Figure 6 presents the SEM image and corresponding elemental mapping of NiTiNb alloy after potentiostatic polarization. The surface reveals remarkable corrosion resistance as evidenced by the absence of visible corrosion damage and the preservation of its original as-cast cellular-like dendrite microstructure. Notably, EDS analysis demonstrates distinct Nb enrichment along the continuous intergranular network. Complementary observations from Figure S3 in Supplementary Materials show that NiTi alloy develops initial pitting corrosion under more aggressive heated and aerated conditions, while pure Ti maintains its corrosion-free surface. In stark contrast, pure Ni suffers from severe intergranular corrosion damage. These morphological characteristics correlate well with the electrochemical analysis results.

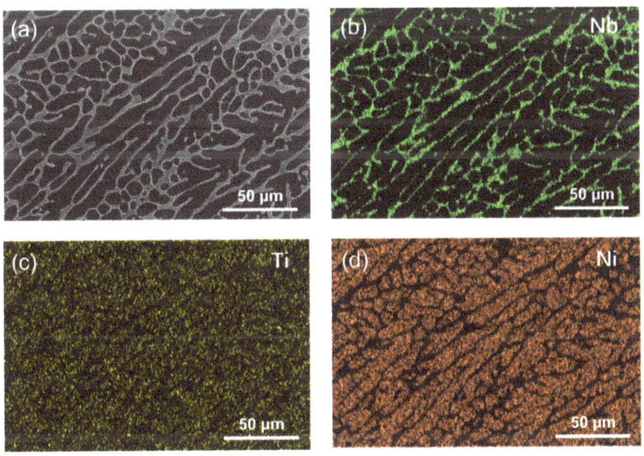

Figure 6. SEM image of (**a**) NiTiNb alloy after potentiostatic polarization and (**b**–**d**) corresponding elemental distribution maps.

Figure 7 presents the XPS spectra from the surfaces of NiTiNb alloy, NiTi alloy, pure Ti, and pure Ni after potentiostatic polarization, with detailed peak deconvolution revealing distinct surface chemical states. The Ni 2p spectrum (Figure 7a) exhibits three characteristic components, namely $Ni(OH)_2$ at 855.6 eV, NiO at 853.7 eV, and metallic Ni at 852.6 eV, while the Ti 2p spectrum (Figure 7b) shows TiO_2 at 458.5 eV and metallic Ti at 454.1 eV. Additionally, the Nb 3d spectrum of the NiTiNb alloy (Figure 7c) demonstrates the coexistence of Nb_2O_5 (207.1 eV) and metallic Nb (202.4 eV). Notably, the O 1s spectrum (Figure 7d) reveals three major contributions, namely lattice oxygen (O^{2-} at 530.0 eV), hydroxyl groups (OH^- at 531.6 eV), and absorbed water (H_2O at 532.3 eV), with O^{2-}/OH^- serving as a critical indicator of passive film quality. The presence of bound water facilitates the capture of dissolved metal cations on the electrode surface, thereby promoting the development of denser passivation film to mitigate corrosion damage [25]. Quantitative analysis of surface chemical species for all materials is provided in Table 3.

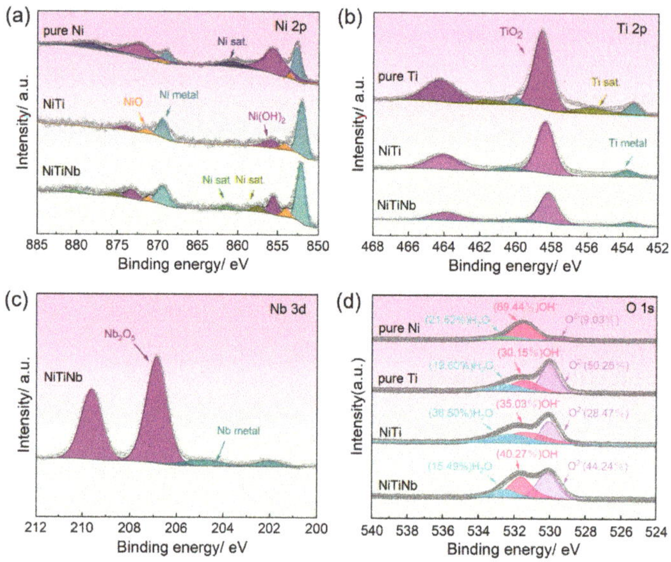

Figure 7. X-ray photoelectron spectroscopy (XPS) analysis results of NiTiNb alloy, NiTi alloy, pure Ti, and pure Ni: (**a**) Ni 2p, (**b**) Ti 2p, (**c**) Nb 3d, (**d**) O 1s.

Table 3. The estimated ratios of different chemical states at the surface of the NiTiNb alloy, NiTi alloy, pure Ti, and pure Ni.

Samples	Ni 2p (%)		Ti 2p (%)		Nb 3d (%)		O^{2-}/OH^-
	$NiO + Ni(OH)_2$	Metallic Ni	TiO_2	Metallic Ti	Nb_2O_5	Metallic Nb	
NiTiNb	23.87	76.11	91.37	8.63	92.53	7.47	1.10
NiTi	28.06	71.94	90.89	9.11	/	/	0.81
Pure Ti	/	/	82.31	17.69	/	/	1.67
Pure Ni	87.65	12.35	/	/	/	/	0.13

XPS spectral analysis reveals fundamental differences in surface chemical states that govern the performance of these BP materials. Pure Ni surfaces are dominated by $Ni(OH)_2$ formation, exhibiting a remarkably low O^{2-}/OH^- ratio of 0.13 that reflects severe corrosion and ineffective passivation. In contrast, pure Ti develops a TiO_2-rich passive film with a high O^{2-}/OH^- ratio of 1.67, suggesting excellent corrosion resistance at the expense of electrical conductivity. The NiTiNb alloy demonstrates an optimal compromise with a moderate O^{2-}/OH^- ratio of 1.1 (versus 0.81 for NiTi), maintaining 76.11% metallic Ni content (compared to 71.94% in NiTi). This high metallic content implies the presence of a thin yet protective oxide layer, further supported by a TiO_2-Nb_2O_5 composite passive film. This unique surface composition explains the material's superior performance. On the one hand, the TiO_2-Nb_2O_5 composite passive film provides corrosion resistance comparable to pure Ti. On the other hand, the continuous Nb/Nb_2O_5 network along grain boundaries establishes conductive pathways, effectively reducing ICR. These results are consistent with previous findings demonstrating that Nb/Nb_2O_5 coatings significantly improve both corrosion resistance and electrical conductivity in 316L stainless steel as well as Ti-based BPs [39], with similar enhancements reported for Nb_2O_5-modified Ti-6Al-4V alloys [40]. As a result, the present study demonstrates that bulk Nb alloying in NiTi achieves comparable benefits through in situ formation of a multifunctional surface structure during polarization, effectively overcoming the conventional trade-off between corrosion resistance and conductivity in metallic BP materials.

3. Materials and Methods

3.1. Materials

In this study, a NiTiNb alloy with a nominal atomic ratio of Ni:Ti:Nb = 47:44:9, representative of typical commercial compositions, was selected as the primary research material. For comparison, a binary NiTi alloy with a nominal atomic ratio of Ni:Ti = 51:49, along with pure Ti and pure Ni, which were previously studied as PEMFC bipolar plate (BP) materials under ambient conditions, were also included [18]. Both NiTiNb and NiTi alloys were fabricated using the vacuum arc melting process (WK-I, Physcience Opto-electronics Co., Ltd., Beijing, China). To ensure compositional homogeneity, small ingots (~10 mm in diameter) were first remelted six times with flipping between melts. These ingots were then consolidated and subjected to three additional remelting cycles, also with flipping, to form the final ingots (~25 mm in diameter). High-purity raw materials were used, including electrolytic nickel (purity > 99.9%), titanium sponge (purity > 99.7%), and niobium particle (purity > 99.9%). Pure Ti and pure Ni samples were purchased from Shengshida Metal Materials Co. Ltd., Xingtai, China. All materials were cut into thin plates with dimensions of 20 mm × 10 mm × 1 mm using wire electrical discharge machining. The sample surfaces were mechanically ground using progressively finer silicon carbide papers up to 2000 grit, polished with a 1 μm diamond suspension, and sequentially cleaned in anhydrous ethanol and deionized water under ultrasonic agitation. These prepared samples were used for subsequent electrochemical tests, microstructural characterization, and phase identification.

3.2. Electrochemical Measurements

Electrochemical measurements were carried out using a CHI660E electrochemical workstation (Chenhua, Shanghai, China) in a conventional three-electrode cell. The cell configuration was composed of a saturated calomel electrode (SCE) as the reference, a platinum electrode as the counter, and the prepared sample as the working electrode. All measured potentials (V_{SCE}) were converted to values relative to the standard hydrogen electrode (SHE) using the following relationship: $V_{SHE} = V_{SCE} + 0.244$ V.

Prior to each test, the working electrode was stabilized under the open circuit condition for 3600 s to establish a steady open circuit potential (E_{OCP}). Potentiodynamic polarization measurements were conducted by scanning the potential from $E_{OCP} - 0.25$ V to $E_{OCP} + 1.00$ V at a rate of 1 mV/s. Potentiostatic polarization was carried out at a simulated cathodic potential of 0.85 V_{SHE} for 8 h [41]. Electrochemical impedance spectroscopy (EIS) was performed at the E_{OCP} over a frequency range from 10 mHz to 100 kHz with a sinusoidal perturbation of 10 mV [42]. All electrochemical tests were conducted in a simulated PEMFC environment, consisting of 0.5 M H_2SO_4 containing 2 ppm F^- at 70 °C, with continuous air bubbling [43]. Each test was repeated at least three times under consistent conditions to ensure parameter suitability and data accuracy.

3.3. ICR Measurements

Interfacial contact resistance (ICR) measurements were conducted after potentiostatic polarization tests, following the method proposed by Wang et al. [44], using a DC low-resistance meter (TH2516B, Tonghui, Changzhou, China). The ICR was calculated using Equation (1):

$$ICR = (R_2 - R_1)/2 \quad (1)$$

where R_2 is the measured total resistance, R_1 is the resistance attributed to the gas diffusion layer (GDL) with carbon paper, and A_c is the contact area between the sample and the carbon paper. The procedure for determining R_1 and R_2 is illustrated in the schematic diagram shown in Figure 8. Compaction pressures ranging from 0.1 to 3.0 MPa were applied, and the corresponding ICR value under each pressure was obtained by averaging three separate measurements taken at different locations on the sample surface to minimize local variability.

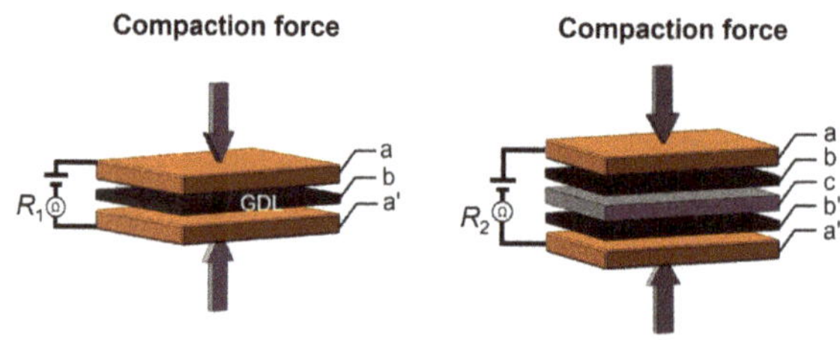

Figure 8. The schematic diagram for testing the R_1 and R_2 values for calculating the ICR value.

3.4. Characterization

The microstructure of the alloys was characterized using optical microscopy (OM, DM2700 M, Leica, Wetzlar, Germany) and field-emission scanning electron microscopy (FE-SEM, Apreo S Hivac, Thermo Fisher, Waltham, MA, USA). Prior to OM observation, the samples were etched with a mixed solution of $HF:HNO_3:H_2O$ in a volume ratio of 1:4:5 to reveal the microstructure. Elemental distributions were analyzed by energy dispersive spectroscopy (EDS, Ultim Live 100X, Oxford, Abingdon, UK) integrated with FE-SEM. Phase identification was performed using X-ray diffraction (XRD, SmartLab SE, Rigaku, Tokyo, Japan) with CuKα radiation. The chemical compositions and elemental valence states of the passive film formed on the sample surfaces were analyzed by X-ray photoelectron spectroscopy (XPS, AXIS SUPRA+, Shimadzu, Kyoto, Japan) equipped with an AlKα X-ray source (1486.6 eV, probing depth < 10 nm).

4. Conclusions

This study systematically evaluates the performance of NiTiNb alloy as a potential BP material in simulated PEMFC environments (70 °C, 0.5 M H_2SO_4 + 2 ppm F^- with air bubbling). Results reveal that Nb alloying maintains the intrinsic balance between corrosion protection and electrical conductivity of NiTi alloy while significantly enhancing both properties. After 8 h of polarization, the NiTiNb alloy demonstrates a 26% current density reduction and a 29% ICR decrease compared to the NiTi alloy, effectively overcoming the conventional performance trade-off in metallic BPs. The alloy also displays superior electrochemical stability and significantly enhanced microhardness relative to pure Ti and Ni. These exceptional properties originate from the alloy's unique dual-phase microstructure, featuring an NiTi (B2) matrix with a continuous β-Nb network along grain boundaries. The in situ formation of TiO_2-Nb_2O_5 composite passive film enhances corrosion resistance, while the interconnected Nb/Nb_2O_5 network and preserved metallic Ni domains synergistically improve electrical conductivity. These findings establish NiTiNb as a highly promising BP substrate for PEMFC and provide an effective optimization strategy through compositional modification.

Supplementary Materials: The following supporting information can be downloaded at https://www.mdpi.com/article/10.3390/molecules30173658/s1, Figure S1. Optical micrographs of (a) NiTi alloy, (b) pure Ti, and (c) pure Ni; Figure S2. Potentiostatic polarization curve of the NiTiNb alloy tested under ambient conditions with no air bubbling; Figure S3. Scanning electron microscopy (SEM) images of (a) NiTi alloy, (b) pure Ti, and (c) pure Ni after potentiostatic polarization.

Author Contributions: Conceptualization, Y.L. (Yuanyuan Li) and Q.Z.; methodology, Z.N., Y.L. (Yingping Li) and X.W.; formal analysis, Z.N., Y.L. (Yingping Li) and Y.P.; funding acquisition, Y.L. (Yuanyuan Li); investigation, Z.N., Y.L. (Yingping Li), Z.H., G.Z. and Z.W.; project administration, Y.L. (Yuanyuan Li) and Q.Z.; writing—original draft, Z.N. and Y.L. (Yingping Li); writing—review and editing, Y.L. (Yuanyuan Li) and Q.Z. All authors have read and agreed to the published version of the manuscript.

Funding: This research was supported by the National Natural Science Foundation of China under Grant No. 12102310 and the Guangdong Basic and Applied Basic Research Foundation under Grant No. 2020A1515110818.

Institutional Review Board Statement: Not applicable.

Informed Consent Statement: Not applicable.

Data Availability Statement: The data presented in this study are available on request from the corresponding author.

Conflicts of Interest: The authors declare no conflicts of interest.

References

1. Ang, T.-Z.; Salem, M.; Kamarol, M.; Das, H.S.; Nazari, M.A.; Prabaharan, N. A Comprehensive Study of Renewable Energy Sources: Classifications, Challenges and Suggestions. *Energy Strategy Rev.* **2022**, *43*, 100939. [CrossRef]
2. Niu, Z.; Wang, X.; Li, Y.; He, Z.; Huang, F.; Deng, Q.; Hu, F.; Zhou, Q. Review of Conductive Polymer Coatings for Metallic Bipolar Plates in Proton Exchange Membrane Fuel Cells. *J. Polym. Mater.* **2025**, *42*, 57–80. [CrossRef]
3. Atapour, M.; Rajaei, V.; Trasatti, S.; Casaletto, M.P.; Chiarello, G.L. Thin Niobium and Niobium Nitride PVD Coatings on AISI 304 Stainless Steel as Bipolar Plates for PEMFCs. *Coatings* **2020**, *10*, 889. [CrossRef]
4. Zhao, J.; Tu, Z.; Chan, S.H. Carbon Corrosion Mechanism and Mitigation Strategies in a Proton Exchange Membrane Fuel Cell (PEMFC): A Review. *J. Power Sources* **2021**, *488*, 229434. [CrossRef]
5. Ha, H.; Gong, M.; Choi, B.; Na, Y. Design of Stiffened Ribs for Improving Forming Defects of Metal Bipolar Plate for PEMFC. *Int. J. Precis. Eng. Manuf.-Green Tech.* **2025**, *12*, 935–950. [CrossRef]
6. Xiong, K.; Wu, W.; Wang, S.; Zhang, L. Modeling, Design, Materials and Fabrication of Bipolar Plates for Proton Exchange Membrane Fuel Cell: A Review. *Appl. Energy* **2021**, *301*, 117443. [CrossRef]
7. Liu, Q.; Lan, F.; Zeng, C.; Chen, J.; Wang, J. A Review of Proton Exchange Membrane Fuel Cell's Bipolar Plate Design and Fabrication Process. *J. Power Sources* **2022**, *538*, 231543. [CrossRef]
8. Gao, X.; Chen, J.; Xu, R.; Zhen, Z.; Zeng, X.; Chen, X.; Cui, L. Research Progress and Prospect of the Materials of Bipolar Plates for Proton Exchange Membrane Fuel Cells (PEMFCs). *Int. J. Hydrogen Energy* **2024**, *50*, 711–743. [CrossRef]
9. Wu, X.; Xing, S.; Luo, J.; Wang, H.; Huang, F.; Zhao, C. Progress and Challenges on Air-Cooled Open-Cathode Proton Exchange Membrane Fuel Cells: Materials, Structures, and Systems. *Energy Rev.* **2025**, *4*, 100130. [CrossRef]
10. Jeong, K.I.; Oh, J.; Song, S.A.; Lee, D.; Lee, D.G.; Kim, S.S. A Review of Composite Bipolar Plates in Proton Exchange Membrane Fuel Cells: Electrical Properties and Gas Permeability. *Compos. Struct.* **2021**, *262*, 113617. [CrossRef]
11. Maheshwari, P.H.; Mathur, R.B.; Dhami, T.L. Fabrication of High Strength and a Low Weight Composite Bipolar Plate for Fuel Cell Applications. *J. Power Sources* **2007**, *173*, 394–403. [CrossRef]
12. Li, W.; Xie, Z.; Qiu, S.; Zeng, H.; Liu, M.; Wu, G. Improved Performance of Composite Bipolar Plates for PEMFC Modified by Homogeneously Dispersed Multi-Walled Carbon Nanotube Networks Prepared by In Situ Chemical Deposition. *Nanomaterials* **2023**, *13*, 365. [CrossRef]
13. Leng, Y.; Ming, P.; Yang, D.; Zhang, C. Stainless Steel Bipolar Plates for Proton Exchange Membrane Fuel Cells: Materials, Flow Channel Design and Forming Processes. *J. Power Sources* **2020**, *451*, 227783. [CrossRef]
14. Chen, Z.; Zhang, G.; Yang, W.; Xu, B.; Chen, Y.; Yin, X.; Liu, Y. Superior Conducting Polypyrrole Anti-Corrosion Coating Containing Functionalized Carbon Powders for 304 Stainless Steel Bipolar Plates in Proton Exchange Membrane Fuel Cells. *Chem. Eng. J.* **2020**, *393*, 124675. [CrossRef]

15. Gong, J.; Wei, H.; Hao, P.; Li, S.; Zhao, X.; Tang, Y.; Zuo, Y. Study on the Influence of Metal Substrates on Protective Performance of the Coating by EIS. *Materials* **2024**, *17*, 378. [CrossRef]
16. Saikia, M.; Dutta, T.; Jadhav, N.; Kalita, D.J. Insights into the Development of Corrosion Protection Coatings. *Polymers* **2025**, *17*, 1548. [CrossRef]
17. Yuan, Z.; Chu, H.; Pan, Y.; Li, F.; Wang, X.; Li, L.; Hu, P.; Du, B.; Hayat, M.D. Enhanced Conductivity Behavior of Titanium-Based Bipolar Plate Oxide Films through Microalloying: First-Principles Calculations and Experimental Investigations. *Ceram. Int.* **2024**, *50*, 4101–4108. [CrossRef]
18. Li, Y.; Wang, X.; Li, Y.; He, Z.; Zhang, G.; Wang, Z.; Wang, S.; Hu, F.; Zhou, Q. Corrosion and Interfacial Contact Resistance of NiTi Alloy as a Promising Bipolar Plate for PEMFC. *Molecules* **2024**, *29*, 3696. [CrossRef] [PubMed]
19. Zhu, H.; Wang, X.; Meng, W.; Ming, P.; Kong, F. Effect of Annealing Time on Microstructure and Properties of Ti-Nb-Ni Sheets as Bipolar Plates Substrates. *Int. J. Hydrogen Energy* **2024**, *57*, 1263–1272. [CrossRef]
20. Ishikawa, K.; Saeki, Y.; Miyajima, Y.; Watanuki, T.; Machida, A. Effects of Rolling and Annealing on Hydrogen Permeability and Crystal Orientation in Nb-TiNi Two-Phase Alloys. *Int. J. Hydrogen Energy* **2019**, *44*, 23101–23106. [CrossRef]
21. Liu, L.; Huang, T.; Qu, M.; Liu, G.; Zhang, J.; Fu, H. High Thermal Gradient Directional Solidification and Its Application in the Processing of Nickel-Based Superalloys. *J. Mater. Process. Technol.* **2010**, *210*, 159–165. [CrossRef]
22. Sitepu, H. Texture and Structural Refinement Using Neutron Diffraction Data from Molybdite (MoO_3) and Calcite ($CaCO_3$) Powders and a Ni-Rich $Ni_{50.7}Ti_{49.30}$Alloy. *Powder Diffr.* **2009**, *24*, 315–326. [CrossRef]
23. Lejaeghere, K.; Van Speybroeck, V.; Van Oost, G.; Cottenier, S. Error Estimates for Solid-State Density-Functional Theory Predictions: An Overview by Means of the Ground-State Elemental Crystals. *Crit. Rev. Solid State Mater. Sci.* **2014**, *39*, 1–24. [CrossRef]
24. Zhang, Y.H.; Li, H.; Yang, Z.W.; Liu, X.; Gu, Q.F. Microstructure Evolution and Shape Memory Behaviors of $Ni_{47}Ti_{44}Nb_9$ Alloy Subjected to Multistep Thermomechanical Loading with Different Prestrain Levels. *J. Mater. Sci. Technol.* **2024**, *171*, 80–93. [CrossRef]
25. Cheng, H.; Luo, H.; Wang, X.; Dong, C.; Li, X. Effect of Fluoride Ion Concentration and Fluctuating Conditions on Titanium Bipolar Plate in PEM Water Electrolyser Environment. *Corros. Sci.* **2023**, *222*, 111414. [CrossRef]
26. Zhou, Q.; Sheikh, S.; Ou, P.; Chen, D.; Hu, Q.; Guo, S. Corrosion Behavior of Hf0.5Nb0.5Ta0.5Ti1.5Zr Refractory High-Entropy in Aqueous Chloride Solutions. *Electrochem. Commun.* **2019**, *98*, 63–68. [CrossRef]
27. Asri, N.F.; Husaini, T.; Sulong, A.B.; Majlan, E.H.; Daud, W.R.W. Coating of Stainless Steel and Titanium Bipolar Plates for Anticorrosion in PEMFC: A Review. *Int. J. Hydrogen Energy* **2017**, *42*, 9135–9148. [CrossRef]
28. Jayaraj, J.; Kim, Y.C.; Kim, K.B.; Seok, H.K.; Fleury, E. Corrosion Studies on Fe-Based Amorphous Alloys in Simulated PEM Fuel Cell Environment. *Sci. Technol. Adv. Mater.* **2005**, *6*, 282–289. [CrossRef]
29. Wang, Y.; Northwood, D. An Investigation of the Electrochemical Properties of PVD TiN-Coated SS410 in Simulated PEM Fuel Cell Environments. *Int. J. Hydrogen Energy* **2007**, *32*, 895–902. [CrossRef]
30. Lee, H.-Y.; Lee, S.-H.; Kim, J.-H.; Kim, M.-C.; Wee, D.-M. Thermally Nitrided Cu–5.3Cr Alloy for Application as Metallic Separators in PEMFCs. *Int. J. Hydrogen Energy* **2008**, *33*, 4171–4177. [CrossRef]
31. Lee, C.-H.; Lee, Y.-B.; Kim, K.-M.; Jeong, M.-G.; Lim, D.-S. Electrically Conductive Polymer Composite Coating on Aluminum for PEM Fuel Cells Bipolar Plate. *Renew. Energy* **2013**, *54*, 46–50. [CrossRef]
32. Bi, F.; Yi, P.; Zhou, T.; Peng, L.; Lai, X. Effects of Al Incorporation on the Interfacial Conductivity and Corrosion Resistance of CrN Film on SS316L as Bipolar Plates for Proton Exchange Membrane Fuel Cells. *Int. J. Hydrogen Energy* **2015**, *40*, 9790–9802. [CrossRef]
33. Mi, B.; Wang, Q.; Qi, T.; Qin, Z.; Chen, Z.; Wang, H. Performance and Structure of Ti-Doped Amorphous Carbon/CrN/Ti Multilayer Coating Deposited on 316L Stainless Steel for Use as Bipolar Plate in Proton Exchange Membrane Fuel Cell. *J. Alloys Compd.* **2023**, *943*, 169080. [CrossRef]
34. Ma, T.; Guo, H.; Tian, Y.; Qi, J.; Yao, N. TiZrC-Coated 316 L Stainless Steel Bipolar Plates for Proton Exchange Membrane Fuel Cells. *Mater. Today Commun.* **2025**, *42*, 111170. [CrossRef]
35. Rosero-Navarro, N.C.; Pellice, S.A.; Durán, A.; Aparicio, M. Effects of Ce-Containing Sol–Gel Coatings Reinforced with SiO2 Nanoparticles on the Protection of AA2024. *Corros. Sci.* **2008**, *50*, 1283–1291. [CrossRef]
36. Iversen, A.K. Stainless Steels in Bipolar Plates—Surface Resistive Properties of Corrosion Resistant Steel Grades during Current Loads. *Corros. Sci.* **2006**, *48*, 1036–1058. [CrossRef]
37. Boissy, C.; Ter-Ovanessian, B.; Mary, N.; Normand, B. Correlation between Predictive and Descriptive Models to Characterize the Passive Film—Study of Pure Chromium by Electrochemical Impedance Spectroscopy. *Electrochim. Acta* **2015**, *174*, 430–437. [CrossRef]
38. Zhao, L.; Dai, H.; Pei, F.; Ming, P.; Wei, X.; Zhou, J. A Comparative Study of Equivalent Circuit Models for Electro-Chemical Impedance Spectroscopy Analysis of Proton Exchange Membrane Fuel Cells. *Energies* **2022**, *15*, 386. [CrossRef]

39. Cao, C.; Liang, C.; Huang, N. Electrochemical Behavior of Niobium Electrodeposited 316 Stainless Steel Bipolar Plate for PEMFC in Choline Chloride Based Ionic Liquids. *J. Wuhan Univ. Technol.-Mater. Sci. Ed.* **2015**, *30*, 1061–1067. [CrossRef]
40. Nascimento, J.P.L.D.; Ferreira, M.O.A.; Gelamo, R.V.; Sampaio, R.M.D.S.; Slade, N.B.L.; Moreto, J.A.; Fernandes, J.C.S. Corrosion and Solid-State Properties of Nb_2O_5 Coating Deposited on Ti-6Al-4V Alloy via DC Reactive Sputtering. *Electrochim. Acta* **2025**, *530*, 146323. [CrossRef]
41. Zhang, D.; Duan, L.; Guo, L.; Wang, Z.; Zhao, J.; Tuan, W.-H.; Niihara, K. TiN-Coated Titanium as the Bipolar Plate for PEMFC by Multi-Arc Ion Plating. *Int. J. Hydrogen Energy* **2011**, *36*, 9155–9161. [CrossRef]
42. Pugal Mani, S.; Srinivasan, A.; Rajendran, N. Effect of Nitrides on the Corrosion Behaviour of 316L SS Bipolar Plates for Proton Exchange Membrane Fuel Cell (PEMFC). *Int. J. Hydrogen Energy* **2015**, *40*, 3359–3369. [CrossRef]
43. Wang, L.; Tao, Y.; Zhang, Z.; Wang, Y.; Feng, Q.; Wang, H.; Li, H. Molybdenum Carbide Coated 316L Stainless Steel for Bipolar Plates of Proton Exchange Membrane Fuel Cells. *Int. J. Hydrogen Energy* **2019**, *44*, 4940–4950. [CrossRef]
44. Wang, H. Stainless Steel as Bipolar Plate Material for Polymer Electrolyte Membrane Fuel Cells. *J. Power Sources* **2003**, *115*, 243–251. [CrossRef]

molecules

MDPI

Article

Influence of Bi^{3+} Doping on Electrochemical Properties of $Ti/Sb\text{-}SnO_2/PbO_2$ Electrode for Zinc Electrowinning

Jia Wu [1,*], Xuanqi Kang [1,2], Shuangwen Xu [1], Zhen Wei [1], Shangyuan Xu [1], Kang Liu [1], Qing Feng [1], Bo Jia [1] and Yunhai Wang [2]

[1] Xi'an Taijin New Energy & Materials Sci-Tech Co., Ltd., Xi'an 710016, China; kangsiqiqi@163.com (X.K.); xu_shuangwen@163.com (S.X.); wzgryx1991@163.com (Z.W.); xushangyuanx@163.com (S.X.); liukang5410@163.com (K.L.); fengqing0715@126.com (Q.F.); ppqtmyy@163.com (B.J.)

[2] State Key Laboratory of Multiphase Flow in Power Engineering, Department of Environmental Science and Engineering, Xi'an Jiao tong University, Xi'an 710049, China; wang.yuanhai@mail.xjtu.edu.cn

* Correspondence: lhwu123456@126.com; Tel.: +86-15802900957; Fax: +86-29-86968411

Abstract: Bi^{3+} doped $Ti/Sb\text{-}SnO_2/PbO_2$ electrode materials were fabricated by electrodeposition to improve their electrochemical performance in zinc electrowinning. The surface morphology, chemical composition, and hydrophilicity of the as-prepared electrodes were characterized using scanning electron microscopy (SEM), energy dispersive X-ray spectroscopy (EDS), X-ray diffraction (XRD), X-ray photoelectron spectroscopy (XPS), and contact angle. An electrochemical measurement and an accelerated lifetime experiment were also conducted to investigate the electrocatalytic performance and stability of the electrodes. The results show that the Bi^{3+} modification electrode has an important effect on the coating morphology, the crystal structure, the surface hydrophilicity, the electrocatalytic activity, and the stability. The electrode prepared from the solution containing 2 $mmol \cdot L^{-1}$ $Bi(NO_3)_3$ (marked as the $Ti/Sb\text{-}SnO_2/2Bi\text{-}PbO_2$ electrode) exhibits the best hydrophilicity performance (θ = 21.6°) and the longest service life (1196 h). During the electrochemical characterization analysis, the $Ti/Sb\text{-}SnO_2/2Bi\text{-}PbO_2$ electrode showed the highest oxygen evolution activity, which can be attributed to it having the highest electroactive surface ($q_T{}^*$ = 21.20 $C \cdot cm^{-2}$) and the best charge-transfer efficiency. The DFT calculation demonstrated that the doping of Bi^{3+} leads to a decrease in the OER reaction barrier and an increase in the DOS of the electrode, which further enhances the catalytic activity and the conductivity of the electrode. Moreover, the simulated zinc electrowinning experiment demonstrated that the $Ti/Sb\text{-}SnO_2/2Bi\text{-}PbO_2$ electrode consumes less energy than other electrodes. Therefore, it is expected that the Bi^{3+} modified electrode will become a very promising electrode material for zinc electrowinning in the future.

Keywords: $Ti/Sb\text{-}SnO_2/PbO_2$ electrode; Bi^{3+} doping; electrocatalytic activity; zinc electrowinning

1. Introduction

Zinc is widely used in mechanical processing, chemical industry, medicine, aerospace, electronics, and other fields [1–3]. Currently, approximately 85% of the zinc used in different industries is obtained by hydrometallurgy [4–6]. Compared to pyrometallurgy, hydrometallurgy has the advantages of a large production capacity, high efficiency, and low environmental pollution. Zinc electrowinning is considered an important part of the hydrometallurgy process; it occupies more than 80% of the total energy of zinc hydrometallurgy [7–9]. The electrode material employed in zinc electrowinning has a significant impact on various parameters, including current efficiency, energy consumption, production cost, quality of the zinc, etc. [10–12]. At present, lead-based electrodes have been widely used for zinc electrowinning [13–15]. However, there are some problems which arise when using this anode, including poor corrosion resistance, easy distortion, high energy consumption, and the cathode zinc being susceptible to Pb contamination [16–18]. Therefore, novel anode materials (Ti-based metal oxide, Al/PbO_2, SS/PbO_2) with suitable stability, high electrical conductivity, and good electrocatalytic activity have attracted much attention [19–21]. Among the anode materials mentioned above, Ti-PbO_2 anodes exhibit superior properties, including strong size stability, long service life, excellent corrosion resistance, low production cost, and high electrocatalytic activity [22–24]. However, the Ti-PbO_2 electrode is not ideal for actual application. The main drawbacks of the Ti-PbO_2 electrode for zinc electrowinning are the high oxygen evolution reaction (OER), the large internal stress, and the short service life. To solve these problems, the doping of various elements and composites in electrolytes including Ce^{4+}, Ag^+, CeO_2, Co_3O_4, and RuO_2 has been employed in the PbO_2 coating, which has also been demonstrated to be an effective approach [25–30]. With the rapid progress of technology, green and sustainable development has become the target of zinc hydrometallurgy. Thus, researchers have given much interest to the anode materials with long lifetimes, high activity, and low energy consumption.

The non-toxic Bi^{3+} (103 pm) and Pb^{2+} (119 pm) are known to have very similar ionic radii. Therefore, their oxides easily form solid solutions that improve electrode surface adhesion and increase the electrode conductivity [31–33]. It is worth noting that the presence of Bi^{3+} ions influences the kinetics of PbO_2 electrodeposition, resulting in the morphological modification of the PbO_2 coating. Moreover, the phase composition is also influenced by the incorporation of Bi^{3+} ions [34,35]. Therefore, in this work, Bi^{3+} ions were introduced into the Ti/Sb-SnO_2/PbO_2 electrode by electrodeposition to improve the anode stability and decrease the OER of the electrode. The principal diagram of the Bi^{3+} modification electrode is shown in Figure 1. Meanwhile, the effect of Bi^{3+} ions with various concentrations on the crystal structure and surface morphology was also studied in detail. The electrochemical properties of the as-fabricated anodes were investigated using cyclic voltammetry (CV), linear sweep voltammetry (LSV), and electrochemical impedance spectroscopy (EIS).

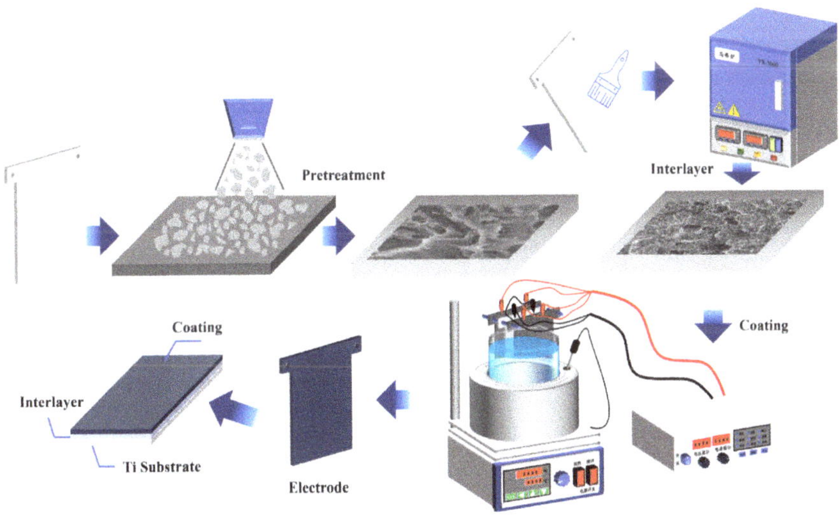

Figure 1. A schematic diagram of electrode preparation.

2. Results and Discussion

2.1. Surface Morphology and Crystal Structure

The surface morphologies of Ti/Sb-SnO_2/Bi-PbO_2 electrodes prepared at different Bi^{3+} concentrations are shown in Figure 2. It can be seen that the addition of Bi^{3+} into a PbO_2 coating significantly affects the coating morphology. The film becomes denser and the particles become smaller with the introduction of Bi^{3+}. We suggest that the doping of Bi^{3+} into the PbO_2 coating could improve the coating structure effectively. This can be attributed to the formation of heterogeneous nuclei of bismuth oxide, which increases the number of crystal nuclei and hinders the growth of particles during the crystallization of the coating [36]. The particles on the electrode surface exhibit the smallest size when the $Bi(NO_3)_3$ concentration is 2 mmol·L^{-1}. However, a further increase in $Bi(NO_3)_3$ concentration is not beneficial for the decrease in the size of the particles. The corresponding EDS mapping images of the Bi, O, and Pb elements are shown in Figure 3. We found that the Bi element is not detected in Ti/Sb-SnO_2/PbO_2, Ti/Sb-SnO_2/1Bi-PbO_2, Ti/Sb-SnO_2/2Bi-PbO_2, or Ti/Sb-SnO_2/3Bi-PbO_2, due to the detection limits of the instrument. However, the Bi element is evenly distributed on the Ti/Sb-SnO_2/5Bi-PbO_2 electrode surface.

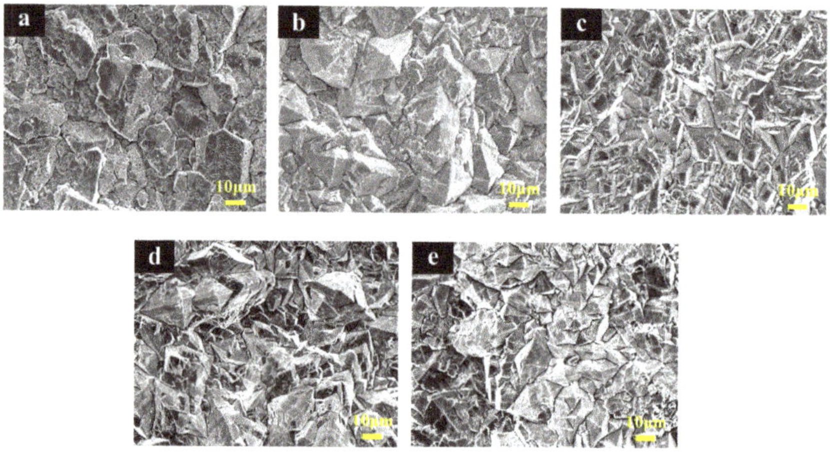

Figure 2. SEM of the electrodes: (**a**) Ti/Sb-SnO_2/PbO_2; (**b**) Ti/Sb-SnO_2/1Bi-PbO_2; (**c**) Ti/Sb-SnO_2/2Bi-PbO_2; (**d**) Ti/Sb-SnO_2/3Bi-PbO_2; and (**e**) Ti/Sb-SnO_2/5Bi-PbO_2.

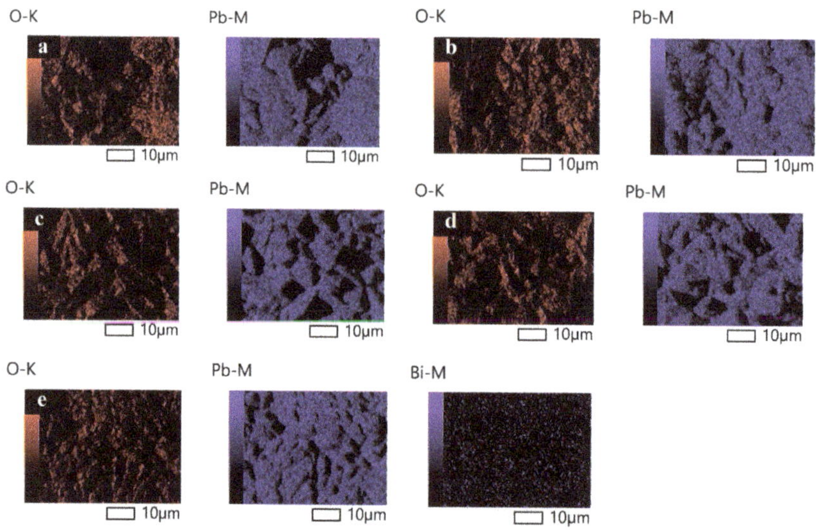

Figure 3. EDS Mapping of the electrodes: (**a**) Ti/Sb-SnO_2/PbO_2; (**b**) Ti/Sb-SnO_2/1Bi-PbO_2; (**c**) Ti/Sb-SnO_2/2Bi-PbO_2; (**d**) Ti/Sb-SnO_2/3Bi-PbO_2; and (**e**) Ti/Sb-SnO_2/5Bi-PbO_2.

Figure 4 shows the XRD patterns of the as-prepared electrodes with a variation Bi^{3+}concentration. As depicted in Figure 4, the diffraction peaks observed at 2θ = 25.4°, 32.0°, 36.2°, 52.1°, 58.9°, 62.5°, 74.4° were assigned to the (110), (101), (200), (220), (310),

(301), and (321) planes of β-PbO_2 (JCPDS#41-1492). The diffraction peaks observed at 2θ = 28.6° were assigned to the (111) plane of α-PbO_2 (JCPDS#45-1416). It was found that all the electrodes exhibited reflections of β-PbO_2 and α-PbO_2. No peaks corresponding to bismuth oxide were detected in the XRD pattern of Ti/Sb-SnO_2/xBi-PbO_2, indicating that the doping amount of bismuth is small and lower than the detective limit of XRD. It should be noted that the crystalline of α-PbO_2 and the main plane of β-PbO_2 varied with the concentration of Bi^{3+} doping. The main planes of Ti/Sb-SnO_2/PbO_2 were (110) and (220). After doping with a low-concentration $Bi(NO_3)_3$ ($\leq$2 mmol·L^{-1}), the diffraction peaks of crystal planes (101) and (301) were enhanced, indicating that the directions of crystal planes (101) and (301) were preferentially selected. However, $Bi(NO_3)_3$ content which was too high was not favorable for the growth of crystal planes (101) and (301). When the concentration of $Bi(NO_3)_3$ was 5 mmol·L^{-1}, the main plane of Ti/Sb-SnO_2/5Bi-PbO_2 was (220). These results imply that the doping concentration of $Bi(NO_3)_3$ affects the selective growth of β-PbO_2 crystal planes.

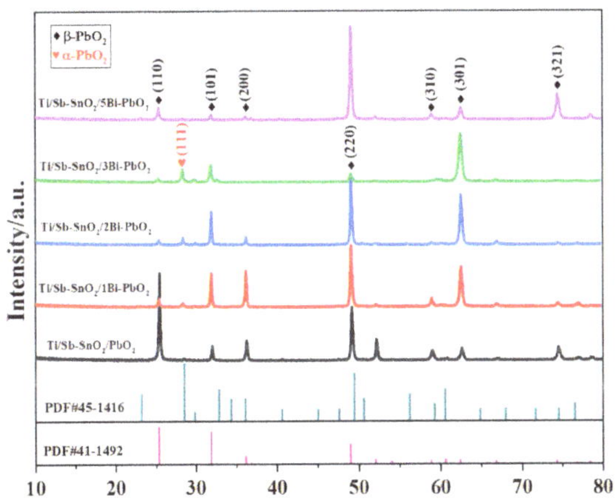

Figure 4. XRD of the as−prepared electrodes with different $Bi(NO_3)_3$ concentrations.

2.2. XPS Analysis

An XPS measurement is performed to further analyze the chemical states of the elements supported on the as-synthesized electrodes. Figure 5a reveals the survey spectrum of the as-prepared electrodes, which exhibit Pb, O, and C elemental signals. Moreover, Bi^{3+} elemental peaks can be observed for Ti/Sb-SnO_2/xBi-PbO_2 (x = 1, 2, 3, 5), suggesting that Bi ions are successfully doped in the PbO_2 coating. Figure 5b displays the 4f core-level spectra recorded for Bi. The peaks at 158.7 and 163.8 eV correspond to Bi $4f_{7/2}$ and Bi $4f_{5/2}$, respectively, which confirms the presence of the Bi^{3+} species. Figure 5c shows the O 1s core level of the as-prepared electrodes, which can be deconvoluted into three characteristic peaks: lattice oxygen species (~529.2 eV for O_L), surface oxygen vacancies and adsorbed oxygen (530.5–531.7 eV for $O_{d\text{-}ad}$), and surface adsorbed molecular water (~532.9 eV for $O_{s\text{-}ad}$), as listed in Table 1. It should be noted that the formation of surface oxygen vacancies is closely related to the highly oxidative oxygen species and is active for catalyzing OER [37]. From Table 1, it can be found that the proportions of $O_{d\text{-}ad}$ present on the electrode surface first increase with the Bi^{3+} doping content and then decrease with a further increase of the Bi^{3+} doping content. The maximum proportion of $O_{d\text{-}ad}$ is obtained when the concentration of $Bi(NO_3)_3$ is 2 mmol·L^{-1}, indicating the best OER activity.

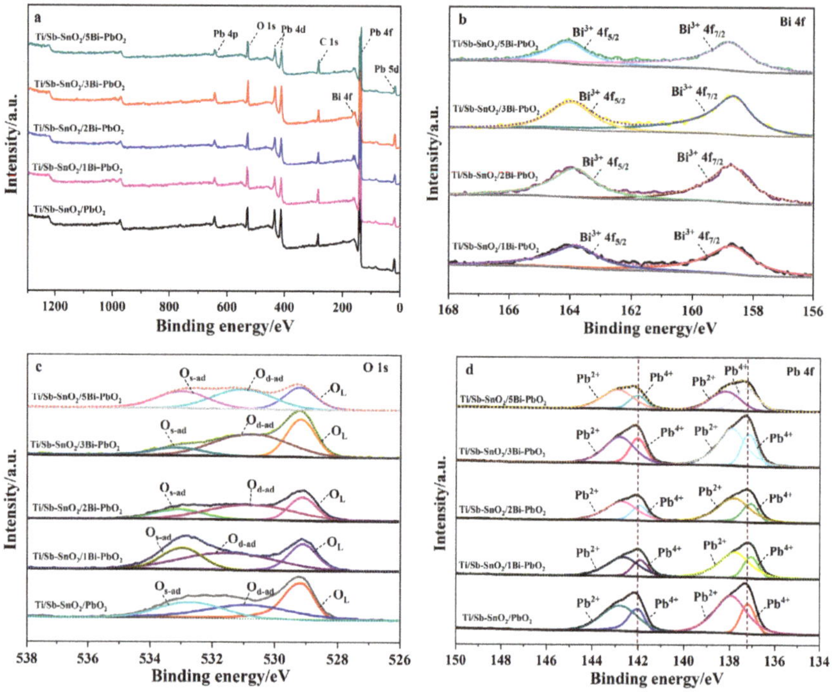

Figure 5. XPS spectra of the as−prepared electrode: (**a**) survey; (**b**) Pb 4f; (**c**) O 1s; and (**d**) Bi 4f.

Table 1. The binding energy of O 1s and its proportions on each electrode.

Electrode	O_L/eV	O_L/%	$O_{d\text{-}ad}$/eV	$O_{d\text{-}ad}$/%	$O_{s\text{-}ad}$/eV	$O_{s\text{-}ad}$/%
Ti/Sb-SnO_2/PbO_2	529.16	36.4	530.85	35.8	532.65	27.8
Ti/Sb-SnO_2/1Bi-PbO_2	529.11	26.6	531.48	47.3	532.95	26.1
Ti/Sb-SnO_2/2Bi-PbO_2	529.12	28.5	530.74	51.5	533.07	20.0
Ti/Sb-SnO_2/3Bi-PbO_2	529.18	32.8	530.68	50.3	533.12	16.8
Ti/Sb-SnO_2/5Bi-PbO_2	529.22	25.4	531.04	41.6	532.97	33.0

To elucidate the impact of doping on PbO_2 electrodeposition, the Pb 4f spectrum was performed, as shown in Figure 5d. In the Ti/Sb-SnO_2/PbO_2 electrode, peaks of 142.01 eV and 137.20 eV correspond to the Pb^{4+}, while peaks of 142.82 eV and 137.96 eV correspond to Pb^{2+}, as shown in Table 2. The simultaneous presence of Pb^{4+} and Pb^{2+} in the electrode indicates the formation of non-stoichiometric PbO_2 during the electrodeposition. After doping with Bi^{3+}, the binding-energy peak of Pb^{4+} shifted, suggesting a strong interaction between Bi^{3+} and Pb [38]. This converts some PbO_2 into a lower-valence compound, promoting the proportion of non-stoichiometric PbO_2, thus improving the electrode conductivity.

Table 2. The binding energy of Pb 4f on each electrode.

Electrode	$Pb^{4+}4f_{7/2}$/eV	$Pb^{4+}4f_{5/2}$/eV	$Pb^{2+}4f_{7/2}$/eV	$Pb^{2+}4f_{5/2}$/eV
Ti/Sb-SnO_2/PbO_2	137.20	142.01	137.96	142.82
Ti/Sb-SnO_2/1Bi-PbO_2	137.07	141.92	137.80	142.67
Ti/Sb-SnO_2/2Bi-PbO_2	137.06	141.96	137.79	142.74
Ti/Sb-SnO_2/3Bi-PbO_2	137.15	142.04	137.88	142.83
Ti/Sb-SnO_2/5Bi-PbO_2	137.29	142.06	138.17	142.92

2.3. Hydrophilicity Analysis

The hydrophilicity of the surfaces has been shown to promote the charge-transfer rate between the electrolytes and the electrodes, enhancing the oxygen evolution reaction activity. Therefore, the contact angle of the as-synthesized electrodes was measured to evaluate the surface wettability. As shown in Figure 6, the contact angles (θ) in Ti/Sb-SnO_2/PbO_2, Ti/Sb-SnO_2/1Bi-PbO_2, Ti/Sb-SnO_2/2Bi-PbO_2, Ti/Sb-SnO_2/3Bi-PbO_2, and Ti/Sb-SnO_2/5Bi-PbO_2 were 60.3°, 29.1°, 21.6°, 27.1°, and 27.7°, respectively. It can be found that much smaller θ are observed for the Ti/Sb-SnO_2/xBi-PbO_2 electrode compared to pure Ti/Sb-SnO_2/PbO_2 (60.3°), indicating the enhanced compatibility and affinity of the electrode for Bi^{3+} doping, which in turn leads to enhancement of their catalytic activity in the OER. When the concentration of $Bi(NO_3)_3$ further increases, the contact angles of Ti/Sb-SnO_2/3Bi-PbO_2 and Ti/Sb-SnO_2/5Bi-PbO_2 are slightly higher than that of Ti/Sb-SnO_2/2Bi-PbO_2, which may be ascribed to the change of crystal planes for the electrode.

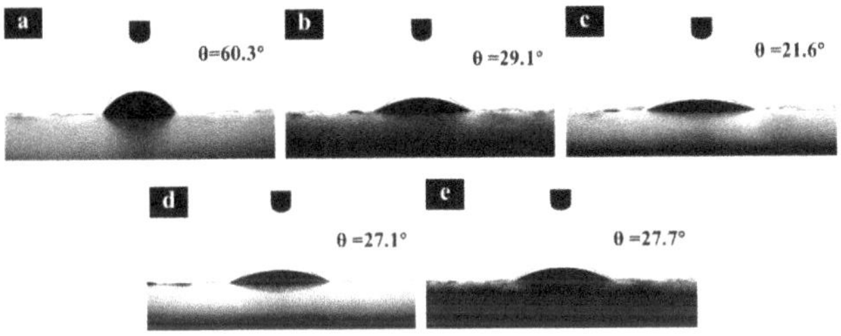

Figure 6. Contact angles (θ) of the as−prepared electrode: (**a**) Ti/Sb-SnO_2/PbO_2; (**b**) Ti/Sb-SnO_2/1Bi-PbO_2; (**c**) Ti/Sb-SnO_2/2Bi-PbO_2; (**d**) Ti/Sb-SnO_2/3Bi-PbO_2; and (**e**) Ti/Sb-SnO_2/5Bi-PbO_2.

2.4. Electrochemical Property

An LSV test was performed to obtain the oxygen evolution potentials (OEP) of the as-prepared electrodes, as displayed in Figure 7. The results show that the OEP of the electrode decreases with the addition of $Bi(NO_3)_3$. The onset potentials for OEP on Ti/Sb-SnO_2/PbO_2, Ti/Sb-SnO_2/1Bi-PbO_2, Ti/Sb-SnO_2/2Bi-PbO_2, Ti/Sb-SnO_2/3Bi-PbO_2, and Ti/Sb-SnO_2/5Bi-PbO_2 are 1.99 V, 1.84 V, 1.80 V, 1.87 V, and 1.93 V (vs. Ag/AgCl), respectively. This implies that the introduction of Bi^{3+} provides more active sites for the oxygen evolution reaction, and that the electrocatalytic activity gradually increased. Furthermore, as shown in Figure 7, the potential of the Ti/Sb-SnO_2/2Bi-PbO_2 electrode is found to be 89 mV lower than that of the Ti/Sb-SnO_2/PbO_2 electrode (500 A/m^2), indicating that the addition of Bi^{3+} can effectively improve the electrocatalytic activity of the PbO_2 electrode. However, when the concentration was greater than 2 mmol·L^{-1}, there was little difference in potential, and the increase of Bi^{3+} resulted in the excessive increase of PbO_2 crystal grain. This may cause a decrease in surface roughness and a decrease in the active regions involved in the reaction.

EIS measurements were conducted to further explain the influence of $Bi(NO_3)_3$ on the oxygen evolution capacity of the PbO_2 coatings. Figure 8 exhibits the electrochemical impedance of PbO_2 electrodes with an applied potential of 1.90 V vs. Ag/AgCl, which corresponds to the oxygen evolution zone. In the equivalent circuits, R_s, R_{ct}, and constant phase angle element (CPE) represent the ohmic resistance, the charge-transfer resistance, and the double-layer capacitance. The R_s include the electrolyte resistance and the active material resistance. R_{ct} reflects the oxygen release reaction activity. The CPE was employed to substitute the capacitance when n is between 0.9 and 1. The simulation data for each parameter are shown in Table 3. The R_{ct} value of the Ti/Sb-SnO_2/2Bi-PbO_2 electrode is 7.30 ohm, which is lower than those of the other electrodes. The Ti/Sb-SnO_2/PbO_2 electrode exhibits the largest R_{ct} value of 20.80 ohm. Furthermore, the R_{ct} value gradually reduced with the addition of $Bi(NO_3)_3$. The minimum value was obtained when the $Bi(NO_3)_3$ concentration was 2 mmol·L^{-1}. On the contrary, the R_{ct} value increased slightly when the $Bi(NO_3)_3$ concentration was further increased, indicating that the modification of Bi^{3+} could significantly increase the catalytic activity of the coating, and that the optimum $Bi(NO_3)_3$ concentration is 2 mmol·L^{-1}. It can be inferred that high oxygen evolution activity and low R_{ct} are in favor of the catalytic activity. Earlier research has observed that oxygen evolution activity depends on the active site of the PbO_2 coating, and that large active sites imply good reactivity [39].

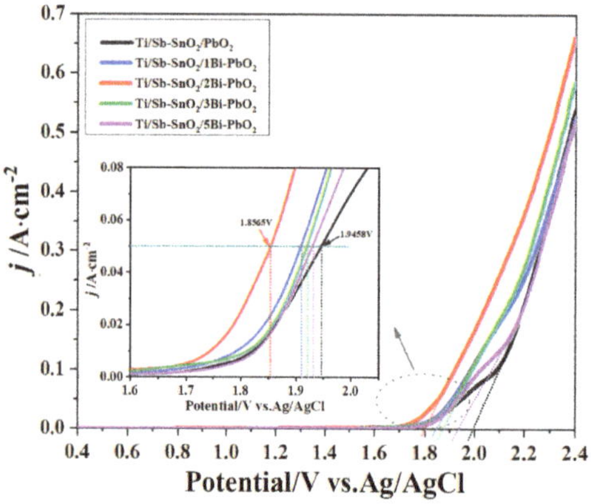

Figure 7. LSV curves of the as−prepared electrode.

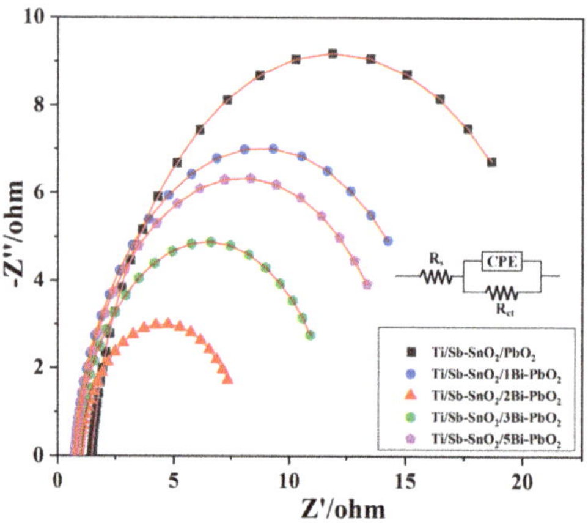

Figure 8. EIS curves of the as-prepared electrode.

Table 3. The equivalent circuit parameters of the as-prepared electrodes.

Materials	R_s/ohm	R_{ct}/ohm	CPE/S·sec^n·cm^{-2}	n
$Ti/Sb-SnO_2/PbO_2$	1.521	20.80	0.0306	0.921
$Ti/Sb-SnO_2/1Bi-PbO_2$	0.797	16.07	0.0369	0.914
$Ti/Sb-SnO_2/2Bi-PbO_2$	0.952	7.30	0.0599	0.905
$Ti/Sb-SnO_2/3Bi-PbO_2$	0.854	11.29	0.0388	0.908
$Ti/Sb-SnO_2/5Bi-PbO_2$	0.823	14.37	0.0350	0.920

The oxygen evolution activity can be determined by comparing the oxygen evolution overpotentials. Figure 9 displays the Tafel curve of the as-prepared electrodes. The oxygen evolution overpotential (η) was employed to investigate the oxygen evolution overpotential (η) of the electrode material and it was obtained according to Formula (2) [40]:

$$\eta = E + 0.224 - 1.241 - iR_s \quad (1)$$

where E represents the potential for the reference electrode (Ag/AgCl), measured by the anodic polarization curve; 1.241 V is the standard potential in the measurement system (50 $g{\cdot}L^{-1}$ Zn^{2+} + 150 $g{\cdot}L^{-1}$ H_2SO_4); 0.224V is the potential of the Ag/AgCl relative to the standard hydrogen electrode; and R_s is the electrolyte resistance between the reference electrode and the working electrode.

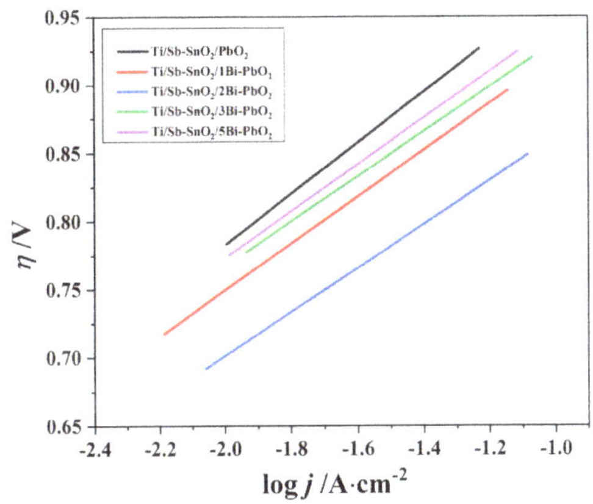

Figure 9. Tafel curves of the as−prepared electrode.

The η and the current density j exhibit a semi-logarithmic relationship, as shown in Formulas (3)–(5):

$$\eta = a + b\lg j; \quad (2)$$

$$a = -2.3(RT/\beta nF)\lg J_0; \quad (3)$$

$$b = 2.3RT/\beta nF, \quad (4)$$

where a and b are Tafel parameters during anodic polarization and j is the measured current density at the relative potential.

The corresponding overpotentials and kinetic parameters for the as-synthesized electrodes are given in Table 4. We found that under the conditions of 500 and 1000 $A{\cdot}m^{-2}$, the Ti/Sb-SnO_2/2Bi-PbO_2 electrode exhibited the lowest η values (0.814 V and 0.862 V), indicating the high probability of oxygen evolution reaction. When the concentration of $Bi(NO_3)_3$ increased from 0 to 2 $mmol{\cdot}L^{-1}$, the η gradually decreased, and the η began to increase at 3 $mmol{\cdot}L^{-1}$. The electrocatalytic performance of the electrode is closely related to the Tafel parameters. Usually, parameter a reflects the cell voltage and the cell voltage decreases with the decreasing value, which means a small energy consumption. Table 4 shows that the values of the electrodes Ti/Sb-SnO_2/PbO_2 and Ti/Sb-SnO_2/2Bi-PbO_2 are 1.141 V and 1.023 V, respectively. The variation of a value and η follows the same tendency. Moreover, the b value represents the overpotential of the material. The smaller b value denotes high catalytic activity. The electrode fabricated at 2 $mmol{\cdot}L^{-1}$ $Bi(NO_3)_3$ exhibited the lowest b value, indicating the highest catalytic activity for an oxygen evolution reaction. This consequence is in accord with the EIS results.

Table 4. Overpotential and kinetic parameters for oxygen evolution on different samples.

Materials	a/V	b/V	η	
			500 A·m^{-2}	1000 A·m^{-2}
Ti/Sb-SnO_2/PbO_2	1.141	0.187	0.898	0.954
Ti/Sb-SnO_2/1Bi-PbO_2	1.091	0.171	0.869	0.920
Ti/Sb-SnO_2/2Bi-PbO_2	1.023	0.161	0.814	0.862
Ti/Sb-SnO_2/3Bi-PbO_2	1.096	0.164	0.882	0.931
Ti/Sb-SnO_2/5Bi-PbO_2	1.114	0.171	0.892	0.944

The amount of voltammetric charge (q^*) is related to the actual surface area and number of active sites, which can reflect the electrochemical activity of the electrode [41]. It is found that the q^* depends on the potential scan rate (ν). Therefore, q^* can be obtained by integrating the CV curve. Figure 10 displays the CV curve of the as-prepared electrodes under various potential scan rates. According to the literature, total voltammetric charge (q_T^*), outer voltammetric charge (q_o^*), and inner voltammetric charge (q_i^*) are calculated by the Equations (6)–(8), respectively.

$$q^* = q_o^* + k_1 \nu^{-1/2}; \quad (5)$$

$$q^{*-1} = q_T^{*-1} + k_2 \nu^{1/2}; \quad (6)$$

$$q_T^* = q_o^* + q_i^*, \quad (7)$$

where k_1 and k_2 are constants; ν is the scan rate. It is reported that q_T^* and q_i^* are related to the entire electroactive surface and to less accessible electroactive sites, respectively. The electrochemical porosity can be obtained by computing q_i^*/q_T^*. Figure 11a shows that q^* is linearly related to $\nu^{-1/2}$, and q_o^* can be obtained from the extrapolation to $\nu^{-1/2} = 0$. Moreover, q^{*-1} is also linearly related to $\nu^{1/2}$, and extrapolation of these straight lines to $\nu^{1/2} = 0$ gives q_T^*, as shown in Figure 11b.

Table 5 summarizes the total, outer, and inner charges obtained for the as-prepared electrodes. The q_T^* of Ti/Sb-SnO_2/PbO_2, Ti/Sb-SnO_2/1Bi-PbO_2, Ti/Sb-SnO_2/2Bi-PbO_2, Ti/Sb-SnO_2/3Bi-PbO_2, and Ti/Sb-SnO_2/5Bi-PbO_2 is 4.44 C·cm^{-2}, 7.25 C·cm^{-2}, 21.20 C·cm^{-2}, 11.49 C·cm^{-2}, and 11.01 C·cm^{-2}, respectively. It can be seen that the doping of Bi^{3+} could greatly improve the electroactive surface of the electrodes. The electrode prepared at 2 mmol·L^{-1} $Bi(NO_3)_3$ exhibited the highest q_T^*, suggesting a higher electroactive surface. However, excessive doping resulted in a decrease in q_T^*, which is not beneficial for the increase in the electroactive surface. Moreover, the ratio of q_i^*/q_T^* for Ti/Sb-SnO_2/2Bi-PbO_2 electrodes is higher than the other electrodes, which is in favor of an increase in the active site. This result is consistent with the Tafel tendency.

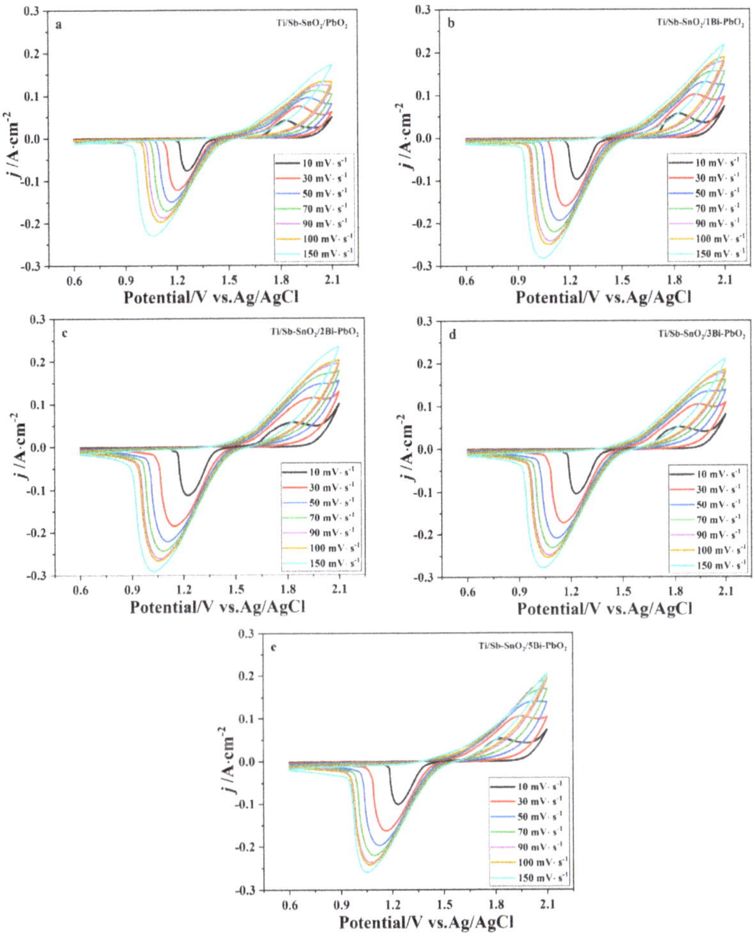

Figure 10. CV curves of the as-prepared electrode under different scan rates: (**a**) Ti/Sb-SnO_2/PbO_2; (**b**) Ti/Sb-SnO_2/1Bi-PbO_2; (**c**) Ti/Sb-SnO_2/2Bi-PbO_2; (**d**) Ti/Sb-SnO_2/3Bi-PbO_2; and (**e**) Ti/Sb-SnO_2/5Bi-PbO_2.

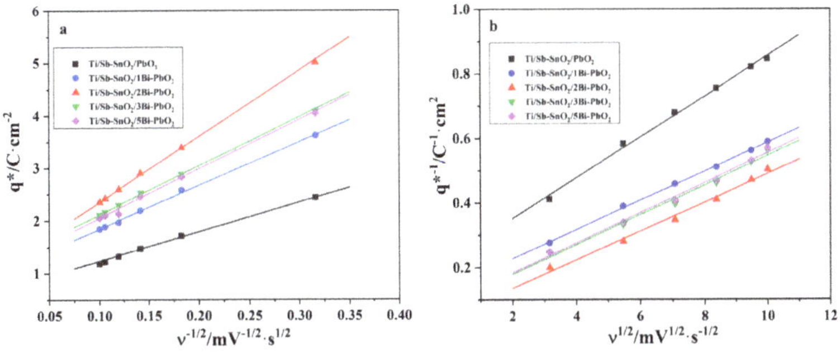

Figure 11. (**a**) Extrapolation of q_o* for the as−prepared electrode from the representation of q* versus $\nu^{-1/2}$. (**b**) Extrapolation of q_T* for the as−prepared electrode from the representation of $(q^*)^{-1}$ versus $\nu^{1/2}$.

Table 5. The detailed voltammetric charge (q*) of different samples.

Materials	q_T*/C·cm^{-2}	q*$_o$/C·cm^{-2}	q*$_i$/C·cm^{-2}	q*$_i$/q_T*
Ti/Sb-SnO_2/PbO_2	4.44	0.68	3.76	84.7%
Ti/Sb-SnO_2/1Bi-PbO_2	7.25	1.01	6.24	86.1%
Ti/Sb-SnO_2/2Bi-PbO_2	21.20	1.09	20.11	94.9%
Ti/Sb-SnO_2/3Bi-PbO_2	11.49	1.18	10.31	89.8%
Ti/Sb-SnO_2/5Bi-PbO_2	11.01	1.11	9.90	89.9%

2.5. DFT Calculations

To verify the effect of Bi^{3+} doping on PbO_2, the Density of State (DOS) and the OER activity of the Bi atom were studied using DFT calculations (Figure 12). Figure 12a,b displays the Schematic diagram of the structures. Figure 12c,d shows the DOS of the PbO_2 and Bi-PbO_2 electrodes. We observe that Bi doping could slightly improve the value of the DOS at the Fermi level. Thus, the introduction of Bi can improve electrode conductivity. The Gibbs free energies of the PbO_2 and Bi-PbO_2 at each step of their OER reactions are also calculated (Figure 12e). According to the literature [42–44], OER is a four-electron step process. Oxygen intermediates include *OH, *O, and *OOH. The largest Gibbs free energy is usually considered as the rate-determining step (RDS), which determines the electrode OER electrocatalytic activity. As shown in Figure 12e, the conversion of *O to *OOH is the RDS of the PbO_2 and Bi-PbO_2. In the RDS procedure, the energy barrier of PbO_2 and Bi-PbO_2 is 2.32 eV and 2.23 eV, respectively. Therefore, the electrocatalytic activity of the Bi-PbO_2 electrodes is higher than that of the PbO_2 electrode. The difference in energy barrier can be ascribed to the introduction of the Bi atom and the enhanced catalytic sites. Combined with theoretical calculations, the Bi doping leads to a decrease in the OER reaction barrier and an increase in the specific surface area of the electrode, which further improves the activity of the catalytic layer. These findings are consistent with the experimental results.

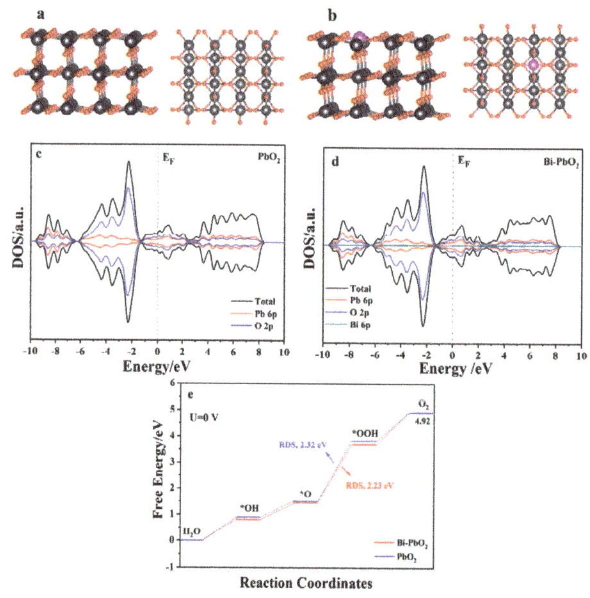

Figure 12. (**a**,**b**) Structural diagrams of PbO_2 and Bi−PbO_2, respectively (black, red, and purple spheres represent Pb, O, and Bi atoms, respectively); (**c**,**d**) Density of States for PbO_2 and Bi−PbO_2, respectively; (**e**) free−energy diagrams for OER in PbO_2 and Bi−PbO_2, respectively (* represent the ctive site on the surface model).

2.6. Electrochemical Stability Test

The accelerated life tests were carried out to evaluate the electrodes' stability. Figure 13 displays the time course of cell potential in the accelerated life measurement for the as-prepared electrodes. It is observed that the Ti/Sb-SnO_2/2Bi-PbO_2 electrode exhibits the longest lifetime of 1196 h, which is 2.79 times that of the Ti/Sb-SnO_2/PbO_2 electrode (428 h). The lifetime of Ti/Sb-SnO_2/2Bi-PbO_2 electrodes is longer than those of the other electrodes. This can be attributed to the growth of the PbO_2 crystal grain, which reduces the gap between the grain boundaries, making the electrode surface denser. The SEM results show that the PbO_2 particle size becomes small after Bi^{3+} doping. The decrease in the PbO_2 particle size could reduce the defect density of the electrode surface and make a compact and fine surface layer. The compact electrode surface could not only baffle the penetration of the electrolyte through cracks and pores but also prevent an increase in pressure inside the electrode, caused by the internal oxygen evolution. The lifetime of the Ti/Sb-SnO_2/3Bi-PbO_2 electrode is smaller than that of the Ti/Sb-SnO_2/2Bi-PbO_2 electrode. This is probably because the nucleation rate is larger than the growth rate of the grain, thereby increasing grain addition. Moreover, the internal stress of the crystal grain is increased, and the surface is cracked. Therefore, the corrosion resistance of the electrode is reduced. The SEM of the as-prepared deactivated electrodes was measured to study the deactivated behavior, as shown in Figure 14. It is observed that the coating of all deactivated electrodes disappeared, and some granular agglomeration was on the electrode. The EDS and EDS-mapping results of the undoped Ti/Sb-SnO_2/PbO_2 electrode and the Ti/Sb-SnO_2/2Bi-PbO_2 electrode are displayed in Figure 15. It can be seen that Ti, Sn, and Sb elements were detected in both of the deactivated electrodes, indicating that the coating fell off from the electrode surface and that the Sb-SnO_2 intermediate layer and the Ti substrate were exposed. Moreover, Pb, S, and O elements were also measured. This can be ascribed to the transformation of PbO_2 to $PbSO_4$ during the long-time electrolysis process. From these results, we can speculate that the deactivation of the electrode is due to the detachment of the active layer.

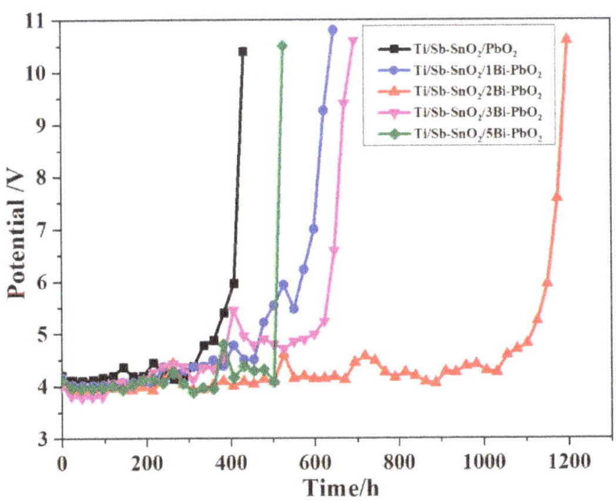

Figure 13. Variation of cell potential with time in the accelerated life test for different electrodes.

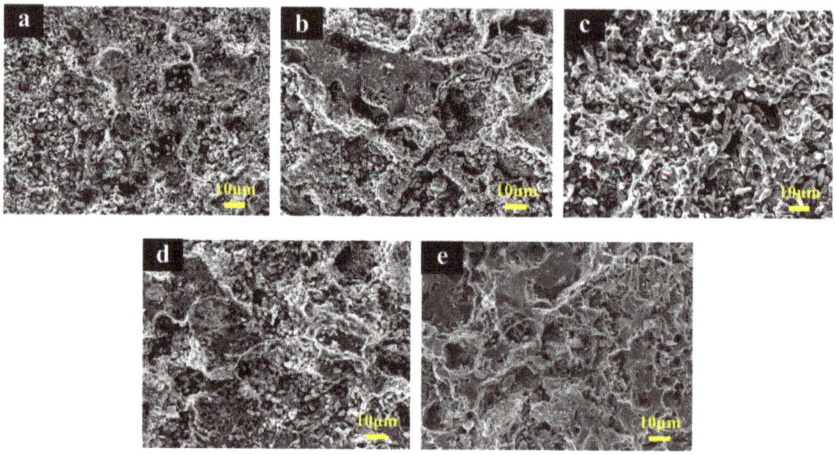

Figure 14. SEM of the deactivated electrodes: (**a**) Ti/Sb-SnO_2/PbO_2; (**b**) Ti/Sb-SnO_2/1Bi-PbO_2; (**c**) Ti/Sb-SnO_2/2Bi-PbO_2; (**d**) Ti/Sb-SnO_2/3Bi-PbO_2; (**e**) Ti/Sb-SnO_2/5Bi-PbO_2.

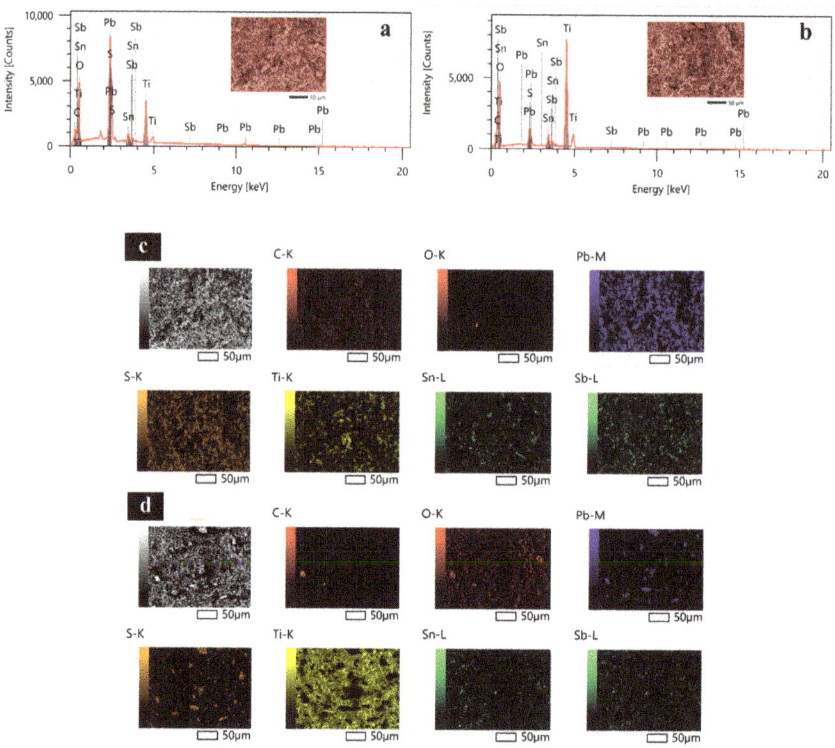

Figure 15. (**a**) the EDS of the deactivated Ti/Sb-SnO_2/PbO_2 electrode; (**b**) the EDS of the deactivated Ti/Sb-SnO_2/2Bi-PbO_2 electrode; (**c**) EDS-mapping of the deactivated Ti/Sb-SnO_2/PbO_2 electrode; and (**d**) EDS-mapping of the deactivated Ti/Sb-SnO_2/2Bi-PbO_2 electrode.

2.7. Simulated Zinc Electrowinning Experiment

The simulated zinc electrowinning experiment was carried out to study the voltage change of the electrode. Figure 16 shows the cell voltage of the as-prepared electrodes during the 3 h zinc electrowinning process. The cell potential was calculated using the following equation:

$$U = (E_1 - E_2) + IR \quad (8)$$

where $E_1 - E_2$ is the electrode polarization potential, including the theoretical decomposition voltage and overpotential of $ZnSO_4$. R is the total resistance, including the electrode resistance, electrolyte resistance, and contact resistance. The average cell potentials of Ti/Sb-SnO_2/PbO_2, Ti/Sb-SnO_2/1Bi-PbO_2, Ti/Sb-SnO_2/2Bi-PbO_2, Ti/Sb-SnO_2/3Bi-PbO_2, and Ti/Sb-SnO_2/5Bi-PbO_2 are 3.32 V, 3.24 V, 2.96 V, 3.16 V, and 3.18 V, respectively. It can be seen that the doping of Bi^{3+} reduces the cell voltage of the electrodes. From the Tafel curve, it can be found that the b value decreases with the introduction of Bi^{3+}, indicating a small overpotential of the electrode. According to Equation (9), the cell voltage decreases with the decrease in the polarization potential. Moreover, it should be noted that the polarization potential has a positive relationship with the overpotential. Therefore, the decrease in overpotential results in a decrease in voltage drop. Commonly, the small cell potential is favorable for the reduction in energy consumption. The smallest cell voltage was obtained by the Ti/Sb-SnO_2/2Bi-PbO_2 electrode, suggesting less energy consumption during zinc electrowinning.

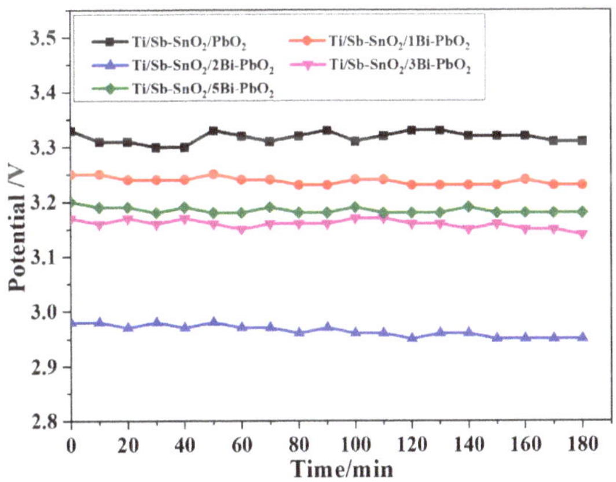

Figure 16. Cell voltage changes at 450 $A{\cdot}m^{-2}$ during the simulated zinc electrowinning in 3h.

3. Experimental Details

3.1. Pretreatment of Ti Sheet

The Ti foil (TA1), with a 1.5 mm thickness, was cut into pieces with a dimension of 80 mm × 100 mm before the experiments. The surface of the Ti sheet was subjected to pretreatment to obtain a gray surface with uniform roughness. The pre-treatment contained degreasing, polishing, and etching for 2 h in boiling oxalic acid. Finally, the specimens were washed with distilled water.

3.2. Preparation of Sb-SnO_2 Interlayer

The sol-gel technique was employed to deposit the Sb-SnO_2 interlayer to prevent the passivation of the Ti substrate. The citric acid (CA) was dissolved in ethylene glycol (EG) to form citric acid ester, which was then added to $SnCl_4$ and $SbCl_3$ to obtain sol-gel precursor solutions. The molar ratio of Sn: Sb was 10:1 and the molar ratio of CA: EG was 1:4. The precursor solutions were brushed on the acid-etched Ti sheet and then placed in an oven at 130 °C and annealed at 500 °C. This procedure was repeated several times.

3.3. Synthesis of Bi^{3+} Doped Ti/Sb-SnO_2/PbO_2 Electrode

Bi^{3+} doped Ti/Sb-SnO_2/PbO_2 electrodes were prepared using an anodic electrodeposition approach with a Ti sheet as the cathode. The deposition solutions contained $Pb(NO_3)_2$, $Cu(NO_3)_2$, HNO_3, some additives, and 0–5 mmol·L^{-1} $Bi(NO_3)_3$. The current density was controlled at 200 A·m^{-2} for 2 h and the bath temperature was 60 °C. Magnetic stirring was employed to enhance the electrolytic diffusion property. The as-prepared electrodes were rinsed thoroughly with deionized water and were denoted as Ti/Sb-SnO_2/xBi-PbO_2, where x (x = 1, 2, 3, 5) presents the Bi^{3+} concentration. In addition, the blank sample was also prepared and was marked as Ti/Sb-SnO_2/PbO_2.

3.4. Characterization and Electrochemical Measurements

The surface morphology and composition of the as-prepared electrodes were characterized using scanning electron microscopy (SEM, JSM-IT200, JEOL, Tokyo, Japan), and the microscope was equipped with an energy-dispersive X-ray spectroscopy (EDS, JEOL, Tokyo, Japan) detector. The crystal pattern of the electrodes was detected by an X'pert PRO MRD diffractometer (XRD, D8 Advance, Bruker, Germany). The chemical states of Pb, O, and Bi in the as-prepared electrodes were determined using X-ray photoelectron spectroscopy (XPS, ESCALAB 250Xi, ThermoFisher, Waltham, MA, USA) on an Ultra DLD Electron Spectrometer (Al Ka radiation; hν = 1486.71 eV). The binding energy of each spectrum was calibrated with C1s (284.8 eV) and was fitted using commercial software (Thermo Avantage 5.9931). Powereach JC2000D was used to measure the contact angle (θ) of the electrode surface. To avoid the effect of the droplet's gravity on the contact angle, all contact angles were tested 10 s after the droplets were dropped onto the facet. The experiment was conducted at room temperature (25 °C).

An electrochemical workstation (Corrtest CS2350) was adopted to conduct electrochemical research using a traditional three-electrode system. The working electrode was the as-prepared electrode connected to a reference electrode (SCE). The Pt sheet served as the counter electrode. The simulated zinc electrowinning electrolyte was composed of 50 g·L^{-1} Zn^{2+} and 150 g·L^{-1} H_2SO_4. The linear sweep voltammetry (LSV) characterization was measured at the scan rate of 10 mV·s^{-1}. Cyclic voltammetry (CV) was performed at the scan rate between 10 and 100 mV·s^{-1}. The voltammetric charge (q*) was evaluated by integrating cyclic voltammograms. EIS experiments were carried out at 1.9 V in the frequency range of 100 kHz–0.1 Hz with an amplitude of 5 mV. The potential values in this work are all quoted with respect to Ag/AgCl. Accelerated life tests were carried out to research the stability and lifetime of the as-prepared electrodes in 15% H_2SO_4 aqueous solutions with a current density of 10,000 A·m^{-2}. In this procedure, the experiment is supposed to be finished when the cell voltage is higher than 10 V. Moreover, to study the anodic performance and durability of the composited electrodes, galvanostatic polarization was tested under the simulated zinc electrowinning condition at a current density of 450 A·m^{-2}.

3.5. Theoretical Calculation

First-principle calculations were conducted on the Vienna Ab initio Simulation Package (VASP (6.X.X.)) [45]. The exchange-correlation influences are described by the Perdew-Burke-Ernzerhof (PBE) functional within the generalized gradient approximation (GGA) method [46]. The core-valence interactions were accounted for using the projected augmented wave (PAW) method [47]. The simulation model was built based on Visualization for electronic and structural analysis (VESTA (3.5.8)) software. The simulation model for Bi-PbO_2 was built by uniformly doping Bi atoms into PbO_2 (110) fragments. The energy cutoff for plane wave expansions was set to 480 eV, and the 3 × 3 × 1 Monkhorst-Pack grid k-points were selected to sample the Brillouin zone integration. The vacuum space was adopted 15 Å above the surfaces to avoid periodic interactions. The structural optimization was completed for energy and force convergence set at 1.0×10^{-4} eV and 0.02 eV Å^{-1}, respectively.

The Gibbs free energy change (ΔG) of each step was calculated using Formula (1)

$$\Delta G = \Delta E + \Delta ZPE - T\Delta S \quad (9)$$

where ΔE is the electronic energy difference directly obtained from DFT calculations; ΔZPE is the zero point energy difference; T is the room temperature (298.15 K); and ΔS is the entropy change.

4. Conclusions

In this study, we studied the role of $Bi(NO_3)_3$ doping in enhancing the electrochemical performance of $Ti/Sb\text{-}SnO_2/PbO_2$ electrodes in zinc electrowinning. The XPS measurement confirms that Bi^{3+} is successfully doped into the electrode. The addition of $Bi(NO_3)_3$ could refine the surface grain, change crystal structure, improve the hydrophilicity, and enhance the electrocatalytic activity of the $Ti/Sb\text{-}SnO_2/PbO_2$ electrode. We found that the $Bi(NO_3)_3$ content has a significant influence on the coating morphology, the crystal orientation, the surface hydrophilicity, the oxygen evolution reaction kinetics, the charge-transfer efficiency, the stability, and the reactivity. The hydrophilicity, the OEP, and the R_{ct} exhibit a declining trend when there is an increase in the $Bi(NO_3)_3$ concentration. The optimum electrocatalytic activity of the $Ti/Sb\text{-}SnO_2/Bi\text{-}PbO_2$ electrode is achieved when the concentration of $Bi(NO_3)_3$ is 2 $mmol \cdot L^{-1}$. The $Ti/Sb\text{-}SnO_2/2Bi\text{-}PbO_2$ electrode illustrates the largest $q_T{}^*$ of 21.20 $C \cdot cm^{-2}$ and the lowest overpotential of 0.814V (500 $A \cdot m^{-2}$), suggesting that it has the more active site and the higher oxygen evolution activity. Our DFT calculations demonstrate that the introduction of Bi^{3+} leads to a decrease in the OER reaction barrier and an increase in the DOS of the electrode, which further enhances the catalytic activity and conductivity of the electrode. This theoretical calculation is consistent with the experimental results. Furthermore, the accelerated life measurement indicates that the introduction of Bi^{3+} can enhance the stability of the electrode, and that the $Ti/Sb\text{-}SnO_2/2Bi\text{-}PbO_2$ electrode possesses the highest accelerated lifetime of 1196 h, which is 2.79 times that of the undoped electrode. The reason for the failure of the electrode is that the active coating gradually detaches from the electrode surface as the electrolysis time increases, until the active layer disappears completely. Furthermore, compared with the undoped electrode, the cell voltage of the $Ti/Sb\text{-}SnO_2/2Bi\text{-}PbO_2$ electrode is reduced, indicating less energy consumption. In summary, the proper doping of Bi^{3+} is beneficial for improving the electrocatalytic activity of the $Ti/Sb\text{-}SnO_2/PbO_2$ electrode. These attractive results show that the potential future applications of this electrode material should be strongly anticipated in zinc electrowinning.

Author Contributions: Conceptualization, J.W. and Z.W.; Methodology, J.W. and K.L.; Software, J.W., S.X. (Shangyuan Xu) and K.L.; Validation, Y.W.; Formal analysis, S.X. (Shuangwen Xu) and B.J.; Investigation, X.K., Z.W., S.X. (Shangyuan Xu) and Q.F.; Resources, X.K., Q.F. and B.J.; Data curation, S.X. (Shuangwen Xu); Writing—original draft, J.W.; Writing—review & editing, J.W. and Y.W. All authors have read and agreed to the published version of the manuscript.

Funding: This research was funded by the National Natural Science Foundation of China (Grant No. 21878242); Xi'an Talent Program, China (No. 20240020); and the Young Talent Fund of Association for Science and Technology in Shaanxi, China (No. 20230521) and the APC was funded by Xi'an Talent Program, China (No. 20240020).

Institutional Review Board Statement: Not applicable.

Informed Consent Statement: Not applicable.

Data Availability Statement: The data is unavailable due to privacy.

Conflicts of Interest: Authors J.W., X.K., S.X., Z.W., S.X., K.L., Q.F. and B.J. were employed by the company Xi'an Taijin New Energy & Materials Sci-Tech Co., Ltd. The remaining authors declare that the research was conducted in the absence of any commercial or financial relationships that could be construed as a potential conflict of interest.

References

1. Ortiz, I.Y.; Santos, A.R.; Costa, A.M.; Mavropoulos, E.; Tanaka, M.N.; Silva, M.H.P.; Camargo, S.S., Jr. In vitro assessment of zinc apatite coatings on titanium surfaces. *Ceram. Int.* **2016**, *42*, 15502–15510. [CrossRef]
2. Li, S.; Chen, W.; Huang, X.; Ding, L.; Ren, Y.; Xu, M.; Zhu, J.; Miao, Z.; Liu, H. Enabling wasted A4 papers as a promising carbon source to construct partially graphitic hierarchical porous carbon for high-performance aqueous Zn-ion storage. *ACS Appl. Mater. Interfaces* **2024**, *16*, 10126–10137. [CrossRef] [PubMed]
3. Wan, Z.; Wang, H.; Li, J.; Solomon, J.; Zhu, Y.; Carlson, B. Novel measures for spatter prediction in laser welding of thin-gage zinc-coated steel. *Int. J. Heat Mass Tran.* **2021**, *167*, 120830. [CrossRef]
4. Rudnik, E. Recovery of zinc from zinc ash by leaching in sulphuric acid and electrowinning. *Hydrometallurgy* **2019**, *188*, 256–263. [CrossRef]
5. Xu, X.; Li, D.; Chen, L.; Liu, M.; Liu, C.; Jia, J. Improve the energy efficiency: Effects of additives on longtime zinc electrowinning. *Hydrometallurgy* **2020**, *193*, 105326. [CrossRef]
6. Karbasi, M.; Alamdari, E.K.; Dehkordi, E.A. Electrochemical performance of Pb-Co composite anode during Zinc electrowinning. *Hydrometallurgy* **2019**, *183*, 51–59. [CrossRef]
7. Skaftun, S.M.; Sunde, S.; Haarberg, G.M.; Seland, F. Deposition and removal of manganese oxide in zinc electrowinning. *Electrochim. Acta* **2024**, *484*, 144040. [CrossRef]
8. Xue, Y.; Hao, X.; Liu, X.; Zhang, N. Recovery of zinc and iron from steel mill dust-an overview of available technologies. *Materials* **2022**, *15*, 4127. [CrossRef] [PubMed]
9. Valdrez, I.V.; Almeida, M.F.; Dias, J.M. Direct recovery of Zn from wasted alkaline batteries through selective anode's separation. *J. Environ. Manage.* **2022**, *321*, 115979. [CrossRef]
10. Kim, H.K.; Jang, H.; Jin, X.; Kim, M.G.; Hwang, S.J. A crucial role of enhanced Volmer-Tafel mechanism in improving the electrocatalytic activity via synergetic optimization of host, interlayer, and surface features of 2D nanosheets. *Appl. Catal. B Environ.* **2022**, *312*, 121391. [CrossRef]
11. Abedini, A.; Valmoozi, A.A.E.; Afghahi, S.S.S. Anodized graphite as an advanced substrate for electrodeposition of PbO_2. *Mater. Today Commun.* **2022**, *31*, 103464. [CrossRef]
12. Sangeetha, V.; Sivakumar, V.; Sudha, A.; Kannan, K. Electrochemical degradation of sago wastewater using Ti/PbO_2 electrode: Optimisation using response surface methodology. *Int. J. Electrochem. Sci.* **2014**, *10*, 1506–1516. [CrossRef]
13. Tao, Y.; Luo, K.; Chang, L.; Chen, B.; Huang, H.; He, Y.; Guo, Z. The influence of rare earth La on properties of lead-based alloy anode for zinc electrowinning. *J. Solid State Electr.* **2022**, *26*, 2555–2564. [CrossRef]
14. Li, H.; Yuan, T.; Li, R.; Wang, W.; Zheng, D.; Yuan, J. Electrochemical properties of powder-pressed Pb-Ag-PbO_2 anodes. *Trans. Nonferr. Met. Soc.* **2019**, *29*, 2422–2429. [CrossRef]
15. Ye, W.; Xu, F.; Jiang, L.; Duan, N.; Li, J.; Zhang, F.; Zhang, G.; Chen, L. A novel functional lead-based anode for efficient lead dissolution inhibition and slime generation reduction in zinc electrowinning. *J. Clean. Prod.* **2021**, *284*, 124767. [CrossRef]
16. Zhou, Y.; Chen, J.; Jiang, L. Industrial tests of composite porous Pb-Ag anode in zinc electrowinning plant. *Ionics* **2023**, *29*, 377–385. [CrossRef]
17. Wang, W.; Yuan, T.; Li, R.; Zhu, X.; Li, H.; Lin, W.; Li, L.; Zheng, D. Electrochemical corrosion behaviors of Pb-Ag anodes by electric current pulse assisted casting. *J. Electroanal. Chem.* **2019**, *847*, 113250. [CrossRef]
18. Wang, S.; Zhou, X.; Yuan, M.; Long, B.; Wang, H.; Tang, J.; Yang, J. Electrochemical properties of Pb-0.6wt% Ag powder-pressed alloy in sulfuric acid electrolyte containing Cl^-/Mn^{2+} ions. *Hydrometallurgy* **2018**, *177*, 218–226. [CrossRef]
19. Wang, X.; Wang, J.; Jiang, W.; Chen, C.; Xu, R. Facile synthesis $MnCo_2O_4$ modifying PbO_2 composite electrode with enhanced OER electrocatalytic activity for zinc electrowinning. *Sep. Purif. Technol.* **2021**, *272*, 118916. [CrossRef]
20. Tang, C.; Lu, Y.; Wang, F.; Niu, H.; Xue, J. Influence of a MnO_2-WC interlayer on the stability and electrocatalytic activity of titanium-based PbO_2 anodes. *Electrochim. Acta* **2020**, *331*, 135381. [CrossRef]
21. Zhang, F.; Zuo, J.; Jin, W.; Xu, F.; Jiang, L.; Xi, D.; Wen, Y.; Li, J.; Yu, Z.; Li, Z.; et al. Size effect of γ-MnO_2 precoated anode on lead-containing pollutant reduction and its controllable fabrication in industrial-scale for zinc electrowinning. *Chemosphere* **2022**, *287*, 132457. [CrossRef]
22. Chen, S.; Li, J.; Liu, L.; He, Q.; Zhou, L.; Yang, T.; Wang, X.; He, P.; Zhang, H.; Jia, B. Fabrication of Co/Pr co-doped Ti/PbO_2 anode for efficiently electrocatalytic degradation of β-naphthoxyacetic acid: Efficiency and mechanism. *Chemosphere* **2020**, *256*, 127139.
23. Duan, P.; Qian, C.; Wang, X.; Jia, X.; Jiao, L.; Chen, Y. Fabrication and characterization of Ti/polyaniline-Co/PbO_2-Co for efficient electrochemical degradation of cephalexin in secondary effluents. *Environ. Res.* **2022**, *214*, 113842. [CrossRef] [PubMed]
24. Shao, D.; Li, W.; Wang, Z.; Yang, C.; Xu, H.; Yan, W.; Yang, L.; Wang, G.; Yang, J.; Feng, L.; et al. Variable activity and selectivity for electrochemical oxidation wastewater treatment using a magnetically assembled electrode based on Ti/PbO_2 and carbon nanotubes. *Sep. Purif. Technol.* **2022**, *301*, 122008. [CrossRef]
25. Zhang, C.; Liu, J.; Chen, B. Effect of $Ce(NO_3)_4$ on the electrochemical properties of Ti/PbO_2-TiO_2-$Ce(NO_3)_4$ electrode for zinc electrowinning. *Appl. Phys. A-Mater.* **2019**, *125*, 150. [CrossRef]
26. Chen, S.; Chen, B.; Wang, S.; Yan, W.; He, Y.; Guo, Z.; Xu, R. Ag doping to boost the electrochemical performance and corrosion resistance of Ti/Sn-Sb-RuO_x/α-PbO_2/β-PbO_2 electrode in zinc electrowinning. *J. Alloys Compd.* **2020**, *815*, 152551. [CrossRef]
27. Zhang, C.; Liu, J.; Chen, B. Effect of CeO_2 and graphite powder on the electrochemical performance of Ti/PbO_2 anode for zinc electrowinning. *Ceram. Int.* **2018**, *44*, 19735–19742. [CrossRef]

28. Dan, Y.; Lu, H.; Liu, X.; Lin, H.; Zhao, J. Ti/PbO_2+nano-Co_3O_4 composite electrode material for electrocatalysis of O_2 evolution in alkaline solution. *Int. J. Hydrog. Energ.* **2011**, *36*, 1949–1954. [CrossRef]
29. He, S.; Xu, R.; Hu, G.; Chen, B. Electrosynthesis and performance of WC and Co_3O_4 co-doped α-PbO_2 electrodes. *RSC Adv.* **2016**, *6*, 3362–3371. [CrossRef]
30. Hakimi, F.; Rashchi, F.; Ghalekhani, M.; Dolati, A.; Astaraei, F.R. Effect of a synthesized pulsed electrodeposited Ti/PbO_2-RuO_2 nanocomposite on zinc electrowinning. *Ind. Eng. Chem. Res.* **2021**, *60*, 11737–11748. [CrossRef]
31. Dolatabadi, M.; Ehrampoush, M.H.; Pournamdari, M.; Ebrahimi, A.A.; Fallahzadeh, H.; Ahmadzadeh, S. Simultaneous electrochemical degradation of pesticides from the aqueous environment using Ti/SnO_2-Sb_2O_3/PbO_2/Bi electrode; process modeling and mechanism insight. *Chemosphere* **2023**, *311*, 137001. [CrossRef]
32. Figueredo-Sobrinho, F.A.A.; Souza Lucas, F.W.; Fill, T.P.; Rodrigues-Filho, E.; Mascaro, L.H.; Silva Casciano, P.N.; Lima-Neto, P.; Correia, A.N. Insights into electrodegradation mechanism of tebuconazole pesticide on Bi-doped PbO_2 electrodes. *Electrochim. Acta* **2015**, *154*, 278–286. [CrossRef]
33. Rahmani, A.; Ansari, A.; Seid-mohammadi, A.; Leili, M.; Nematollahi, D.; Shabanloo, A. Bismuth-doped 3D carbon felt/PbO_2 electrocatalyst for degradation of diuron herbicide and improvement of pesticide wastewater biodegradability. *J. Environ. Chem. Eng.* **2023**, *11*, 109118. [CrossRef]
34. Chang, L.; Zhou, Y.; Duan, X.; Liu, W.; Xu, D. Preparation and characterization of carbon nanotube and Bi co-doped PbO_2 electrode. *J. Taiwan Inst. Chem. Eng.* **2014**, *45*, 1338–1346. [CrossRef]
35. Gao, G.; Zhang, X.; Wang, P.; Ren, Y.; Meng, X.; Ding, Y.; Zhang, T.; Jiang, W. Electrochemical degradation of doxycycline hydrochloride on Bi/Ce co-doped Ti/PbO_2 anodes: Efficiency and mechanism. *J. Environ. Chem. Eng.* **2022**, *10*, 108430. [CrossRef]
36. Shmychkova, O.; Luk'yanenko, T.; Yakubenko, A.; Amadelli, R.; Velichenko, A. Electrooxidation of some phenolic compounds at Bi-doped PbO_2. *Appl. Catal. B-Environ.* **2015**, *162*, 346–351. [CrossRef]
37. Zhu, Y.; Zhou, W.; Yu, J.; Chen, Y.; Liu, M.; Shao, Z. Enhancing electrocatalytic activity of perovskite oxides by tuning cation deficiency for oxygen reduction and evolution reactions. *Chem. Mater.* **2016**, *28*, 1691–1697. [CrossRef]
38. Wan, C.; Zhao, L.; Wu, C.; Lin, L.; Liu, X. Bi^{5+} doping improves the electrochemical properties of Ti/SnO_2-Sb/PbO_2 electrode and its electrocatalytic performance for phenol. *J. Clean. Prod.* **2022**, *380*, 135005. [CrossRef]
39. Yan, Z.; Zhao, Y.; Zhang, Z.; Li, G.; Li, H.; Wang, J.; Feng, Z.; Tang, M.; Yuan, X.; Zhang, R.; et al. A study on the performance of IrO_2-Ta_2O_5 coated anodes with surface treated Ti substrates. *Electrochim. Acta* **2015**, *157*, 345–350. [CrossRef]
40. Xu, R.; Huang, L.; Zhou, J.; Zhan, P.; Guan, Y.; Kong, Y. Effects of tungsten carbide on electrochemical properties and microstructural features of Al/Pb-PANI-WC composite inert anodes used in zinc electrowinning. *Hydrometallurgy* **2012**, *12*, 8–15. [CrossRef]
41. Chen, Y.; Hong, L.; Xue, H.; Han, W.; Wang, L.; Sun, X.; Li, J. Preparation and characterization of TiO_2-NTs/SnO_2-Sb electrodes by electrodeposition. *J. Electroanal. Chem.* **2010**, *648*, 119–127. [CrossRef]
42. Yang, X.; Cheng, J.; Li, H.; Xu, Y.; Tu, W.; Zhou, J. Self-supported N-doped hierarchical Co_3O_4 electrocatalyst with abundant oxygen vacancies for acidic water oxidation. *Chem. Eng. J.* **2023**, *465*, 142745. [CrossRef]
43. Yang, H.; Gao, S.; Rao, D.; Zhang, C.; Zhou, X.; Yan, X. The regulation mechanism of cationic substitution in morphology-controlled oxy-spinel for oxygen evolution reaction. *J. Catal.* **2022**, *407*, 221–231. [CrossRef]
44. Peng, Y.; Hajiyani, H.; Pentcheva, R. Influence of Fe and Ni doping on the OER performance at the Co_3O_4(001) surface: Insights from DFT+U calculations. *ACS Catal.* **2021**, *11*, 5601–5613. [CrossRef]
45. Dziri, F.; Djaballah, Y.; Belbacha, E.; Bouzida, A.B. Thermodynamic reassessment of the Bi-Rb system supported by Ab-Initio Calculations. *J. Phase Equilibria Diffus.* **2022**, *43*, 382–392. [CrossRef]
46. Shi, B.; Nicholls, R.J.; Yates, J.R. Accurate and efficient structure factors in ultrasoft pseudopotential and projector augmented wave DFT. *Phys. Rev. B* **2023**, *108*, 115112. [CrossRef]
47. Grimme, S.; Antony, J.; Ehrlich, S.; Krieg, H. A consistent and accurate ab initio parametrization of density functional dispersion correction (DFT-D) for the 94 elements H-Pu. *J. Chem. Phys.* **2010**, *132*, 154104. [CrossRef] [PubMed]

molecules

MDPI

Article

NiMoS-Modified Carbon Felt Electrode for Improved Efficiency and Stability in a Neutral S/Fe Redox Flow Battery

Dan Mei [1], Bowen Liu [1,2], Haiqing Ma [2], Zhaoguo Zhang [1,2], Fan Wu [2], Yanan Chen [1,2,*], Jawad Ali [1,2], Futang Xing [1] and Liangbin Xiong [3,*]

1 College of Resources and Environmental Engineering, Wuhan University of Science and Technology, Wuhan 430081, China; meidan@wust.edu.cn (D.M.); liubowen@wust.edu.cn (B.L.)
2 School of Environment and Biological Engineering, Wuhan Technology and Business University, Wuhan 430065, China
3 School of Optoelectronic Engineering, Guangdong Polytechnic Normal University, Guangzhou 510665, China
* Correspondence: chenyanan@wtbu.edu.cn (Y.C.); xiongliangbin@gpnu.edu.cn (L.X.)

Abstract: Polysulfide-ferricyanide redox flow batteries (PFRFBs) are gaining significant attention in long-duration energy storage for their abundant availability and environmental benignity. However, the sluggish kinetics of the polysulfide redox reactions have tremendously constrained their performances. To address this issue, we developed a NiMoS catalyst-modified carbon felt (NiMoS-CF) electrode, which significantly accelerates the electrochemical reaction rates and enhances the cycling stability of PFRFB. Our PFRFB system, integrated with the NiMoS-CF electrode, exhibited an energy efficiency of 70% and a voltage efficiency of 87%, with a remarkable doubling of its cycle life as opposed to the pristine carbon felt (CF) electrode at a current density of 40 mA cm^{-2}. Notably, during 2500 cycles of charge–discharge testing, we achieved an average coulombic efficiency exceeding 99%. These improvements in PFRFB performance can be attributed to the NiMoS-CF electrode's large surface area, low resistance, and robust redox activity. This study offerings a novel approach for enhancing the electrochemical reaction kinetics and cycling stability in PFRFBs, laying a scientific foundation in the applications of practical PFRFBs for next-generation energy storage.

Keywords: polysulfide-ferricyanide redox flow battery; NiMoS catalyst; carbon felt electrode; electrochemical kinetics; cycling stability

1. Introduction

As environmental pollution escalates and fossil fuel reserves diminish, there is a growing focus on harnessing eco-friendly and sustainable energy alternatives, such as solar and wind energy. To achieve the successful adoption of these intermittent resources, developing the long-duration energy storage systems is critically important [1]. Redox flow batteries (RFBs) have become an ideal solution for long-term energy storage due to their versatile design, high efficiency, durable cycle life, and excellent deep discharge abilities [2–4]. The current widely used flow battery is all-vanadium redox flow batteries (VRFBs), it still faces low energy density and cost fluctuations due to low vanadium solubility and limited resources [5–8]. Optimizing electrochemical performance, increasing energy and power density, lowering internal resistance, and improving electrolyte cycle stability are all critical for the development of RFBs for long-term energy storage.

Current research and development focuses on low-cost, high-energy density, and long-term RFBs [9]. Emerging neutral polysulfide-ferricyanide redox flow batteries (PFRFBs) are gaining popularity because of their high energy density (260 Wh L^{-1}) and inexpensive cost (<19.26 $ kW h^{-1}). Iron is abundant and affordable for large-scale energy storage [10–12]; Sulfur's multielectron redox reaction and high theoretical energy density, when paired with iron, boost active ingredient usage and electrochemical performance [13,14], especially in S/Fe RFBs [15,16]. However, the battery performances of S/Fe RFBs including the voltage efficiency, energy efficiency, power density, and cycle life, are constrained by the sluggish kinetics of the polysulfide redox reactions. In the charging process, the large overpotential (>500 mV) and low energy efficiency (<50%) are resulted from the electrochemical reduction of polysulfide even with the moderate current densities (10–20 mA cm^{-2}). This is a common challenge in polysulfide-based battery systems, which can be found in other sodium and lithium-sulfur batteries.

To facilitate the reduction of polysulfide, adding chemicals [17–20], electrolytes [21–23], etc. can increase its electrochemical characteristics and stability. Notably, redox processes that involve electrochemically active couples predominantly occur at the electrodes of RFBs. The dynamics of these redox reactions, which directly affect the performance of the battery, are affected by various factors associated with the physical and chemical characteristics of the electrodes, as well as their electrochemical behavior. These factors include surface roughness, wettability, durability, surface area, electrical conductivity, and the electrochemical activity. Several approaches have been investigated to enhance the kinetics of polysulfide ions. For example, Lan [13] improved the mass transfer and redox behavior of aqueous polysulfides by employing a nanostructured, self-assembled, hierarchically porous Co and N dual-doped carbon (OHP-Co/NC) as an electrocatalyst in an electrocatalytic reactor, resulting in a high power density of 110 mW cm^{-2} and 99.7% capacity retention, but the interfacial force between the carbon carrier and the active center in the catalyst is weak, which may result in the loss or inactivation of its active components, compromising the catalyst's long-term stability. Sulfides have higher electrical conductivity than oxides, which allows for more efficient charge transfer [24], hance various metal sulfides like mesocrystalline NiS_2 [25], CoS_2/CoS [26,27], CuS [28] have been employed to boost the kinetics of polysulfide ions. Motivated by these pioneered studies, this work aims to develop novel S/Fe RFB system with accelerated kinetics of polysulfide redox reactions and long cycling life. It has not been reported, though, that NiMoS—a catalyst with strong stability and activity in electrocatalytic water splitting and catalytic hydrogenation and oxidation reactions [26,29,30]—is used in polysulfide-ferricyanide redox flow batteries. Here, we introduce NiMoS catalysts modified carbon felt (NiMoS-CF) as an anode electrode, capitalizing on the catalytic properties of metal sulfides to enhance polysulfide redox processes. NiMoS-CF demonstrated good charge transfer and catalytic performance for polysulfide redox processes. By directly growing electrocatalytic materials on carbon felt (CF), we created modified electrodes without binders, avoiding the negative impacts of adhesives. The PFRFB, equipped with the NiMoS-CF electrode, exhibited improved electrochemical reaction rates, energy density and cycling stability, with a 70% energy efficiency, a coulombic efficiency of over 99% and the cycle life of 2500 cycles at 40 mA cm^{-2}. In contrast, the PFRFB with pristine CF as the negative electrode ended after 1300 cycles, leading to a faster capacity decline.

2. Results and Discussion

Figure 1a illustrates the configuration of S/Fe redox flow battery, in which K_2S and $K_3[Fe(CN)]_6$ are utilized as the anolyte and catholyte, respectively. The catholyte and anolyte are composed of the redox couples $[Fe(CN)_6]^{4-/3-}$ and S_2^{2-}/S^{2-}, respectively, with KCl as the supporting electrolyte. As demonstrated in the preparation process of the NiMoS-CF electrode, a one-step hydrothermal process was employed to synthesize the NiMoS catalyst on the CF, avoiding the negative impacts of the adhesive nafion which is commonly used. It is hypothesized that the catalyst can significantly extend the surface area of the original CF, provide new reaction sites, and improve the conductivity of the electrode [31]. The electrochemical reversibility of the species involved in the redox reactions. was assessed using CV tests. The following redox reactions are represented by one pair of redox peaks in the respective positive and negative electrolyte. As shown in Figure 1b,c, the reversibility of the reaction is usually judged by the symmetry of the curve [32], the redox couples $[Fe(CN)_6]^{4-/3-}$ and S_2^{2-}/S^{2-} show good reversibility and exhibit well-defined redox peak.

$$[Fe(CN)_6]^{3-} + e^- \leftrightarrow [Fe(CN)_6]^{4-} \quad E^0 = 0.46 \text{ V vs. } SHE$$

$$2S^{2-} \leftrightarrow S_2^{2-} + 2e^- \quad E^0 = -0.36 \text{ V vs. } SHE$$

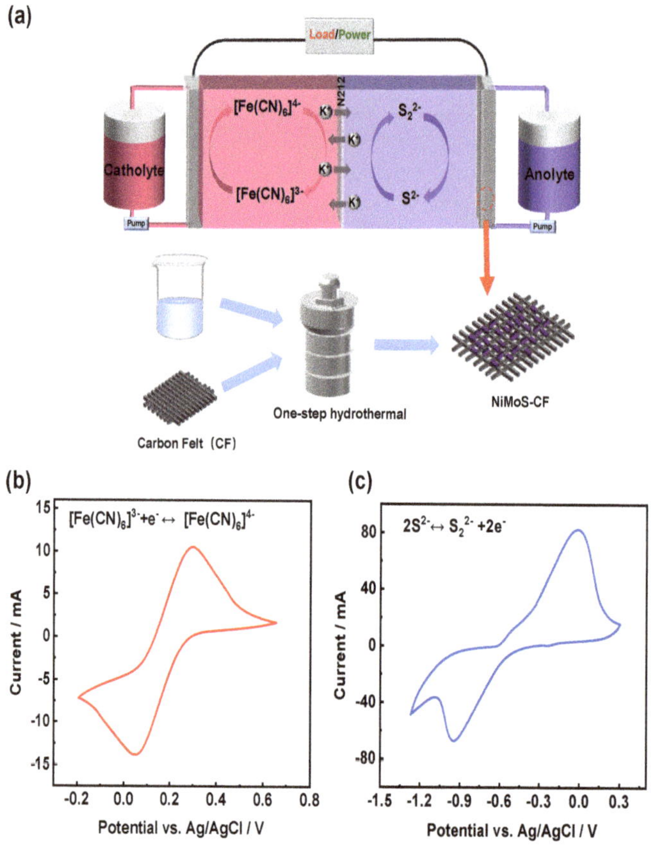

Figure 1. (**a**) Schematic illustration of the S/Fe redox flow battery system; CV curves of (**b**) ferricyanide/ferrocyanide and (**c**) sulfide/polysulfide in 1.0 M KCl electrolyte at 10 mV s^{-1}.

2.1. Structure and Morphology of NiMoS-CF Electrode

The SEM photographs reveal significant morphological distinctions between pristine CF and NiMoS-CF, indicating effective nucleation and growth of NiMoS on the surface. Specifically, the pristine CF (Figure 2a) exhibits a clean and smooth surface, whereas the NiMoS-CF (Figure 2b,c), after 12 h of hydrothermal treatment, displays a rough surface. Consequently, the electrode surface becomes relatively rough and this morphology not only provides additional active sites for redox reactions but promotes close contact with the electrolyte, thereby accelerating the kinetics of these reactions [33,34]. In the TEM image shown in Figure 2d, the distinct nanorods were observed. The spatially uniform distribution of Ni, Mo, and S elements can be revealed by HAADF-STEM analysis coupled with EDS mapping (Figure 2e–h). In Figure 2i, the HR-TEM image exhibits obvious lattice fringes with interplanar distances of 2.65 Å and 2.43 Å, aligning with the d-spacing of (269) and (140) planes of NiMoS. These crystalline planes are evident from the XRD patterns. Compared with the pristine CF and the standard NiMoS patterns, not only the diffraction peak from CF, but also the main peak of NiMoS can be obviously observed in the prepared NiMoS-CF, which confirms that the NiMoS catalyst is successfully grown on the CF (Figure 2j).

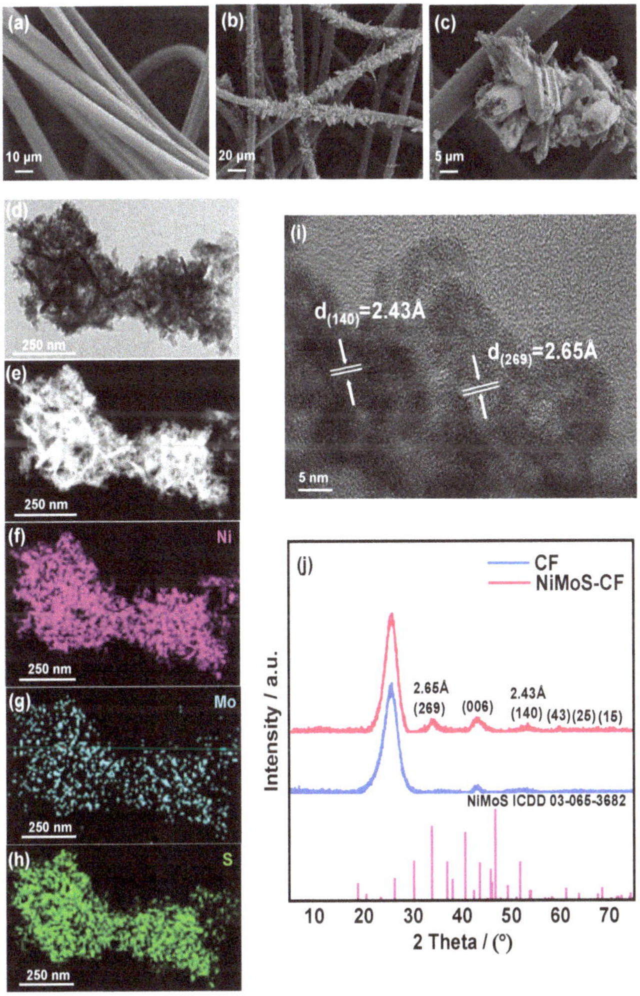

Figure 2. SEM images of the (**a**) pristine CF and (**b**,**c**) NiMoS-CF; (**d**) TEM image of NiMoS-CF; (**e**) HAADF-STEM image of NiMoS-CF; (**f–h**) EDS mapping of Ni, Mo and S; (**i**) HR-TEM image of NiMoS-CF; (**j**) XRD patterns of pristine CF and NiMoS-CF.

Also, significant differences in hydrophilicity were identified before and after the NiMoS modification on the CF. The contact angle of the pristine CF was 140° (Figure 3a), indicating that the surface of the material was lyophobic, while NiMoS-CF rapidly absorbed droplets during the test, showing hydrophilicity (Figure 3b). Specifically, the increased wettability facilitates better electrolyte accessibility and more efficient mass transport of reactants to the active sites, reduced the charge transfer resistance.

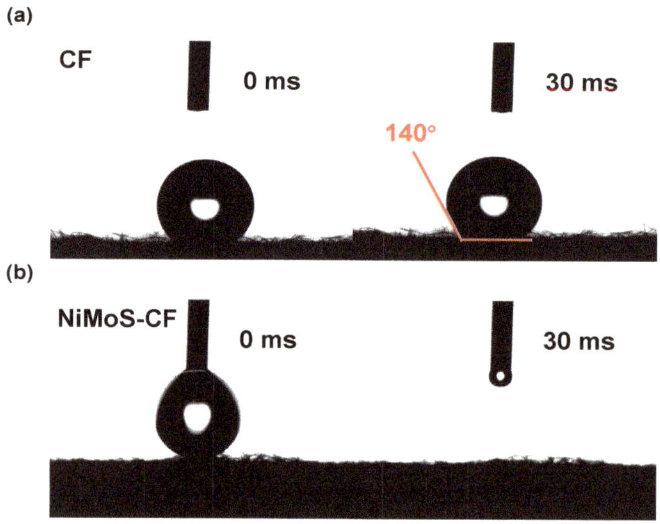

Figure 3. Contact angle measurements of (**a**) pristine CF and (**b**) NiMoS-CF.

The XRD patterns and XPS survey spectrum of NiMoS catalyst can be found in Figures S2 and S3. The unindexed peaks in the XRD pattern are attributed to the presence of minor secondary phases and impurities in the catalyst such as MoS_2 [35]. These phases likely arise from incomplete crystallization or residual precursors during the synthesis process. XPS analysis confirmed the presence of Ni, Mo, and S elements within the catalyst. The Ni 2p XPS spectra display two shakeup satellites (labeled Sat.) at 859.9 and 877.6 eV, along with two spin-orbit doublets, corresponding to Ni $2p_{3/2}$ (Ni^{2+} at 853.5 and Ni^{3+} at 855.2 eV) and Ni $2p_{1/2}$ (Ni^{2+} at 871.3 and Ni^{3+} at 873.3 eV) [36–38]. The Mo 3d XPS spectra are shown in Figure S3c, with peaks at 232.2 eV and 235.2 eV corresponding to Mo $3d_{5/2}$ and Mo $3d_{3/2}$, respectively [39]. The spectra of S 2p is illustrated in Figure S3d. Figure S3d illustrates the S 2p XPS spectra, where the peaks at 163.5 eV and 162.3 eV are assigned to $2p_{1/2}$ and $2p_{3/2}$ of S^{2-}, and the peak at 168.6 eV is indicative of sulfate species [40,41].

2.2. Catalytic Performance of NiMoS-CF Electrode for Sulfur Ions

2.2.1. Electrochemistry Tests

The electrochemical response of CF and NiMoS-CF to the redox reaction of K_2S was investigated in Figure 4a,b, detailed CV data are shown in Table S1, respectively. At the equilibrium potential of −0.455 V, the major redox reaction is attributed to the $S_2{}^{2-}/S^{2-}$ pairs [42]. At all test rates, the current densities of the main redox peaks on NiMoS-CF are higher than those on pristine CF. With a sweep rate of 10 mV s^{-1}, both pristine CF and NiMoS-CF present distinct redox peaks (Figure 4c). These findings suggest that the surface coverage of NiMoS enhances the electrochemical activity. The catalyst surface area is a key determinant of electrochemical performance, as indicated in previous studies [28,43–45]. The glassy carbon electrode (GCE) provides a flat, well-defined surface area that allows precise evaluation of the intrinsic catalytic activity of NiMoS. To minimize confounding factors such as substrate porosity, mass transport limitations, or uneven catalyst distribution that may arise when directly coating fibrous substrates like CF, replacing a glassy carbon electrode (GCE) for the working electrode (Figure S4), which is less susceptible to external factors and exhibits enhanced conductivity, multiple polysulfide redox peaks were observed (Figure S5). The NiMoS-GCE shows a significant peak-to-peak separation in the CV, this observation can be attributed to the intrinsic limitations of CV in capturing subtle kinetic improvements, especially in systems where substrate effects dominate. The GCE data and carbon felt-based results collectively provide a comprehensive understanding of both catalyst efficiency and scalability. According to the result, a more complex electrochemical reaction occurred among the polysulfides, and the valence states of S ions not only include S_2^{2-}/S^{2-} but may also involve additional polysulfide species, such as S_x^{2-} (x = 2–8). These redox reactions are directly linked to the energy conversion efficiency of the battery, as demonstrated in various sulfur-based batteries [26,31].

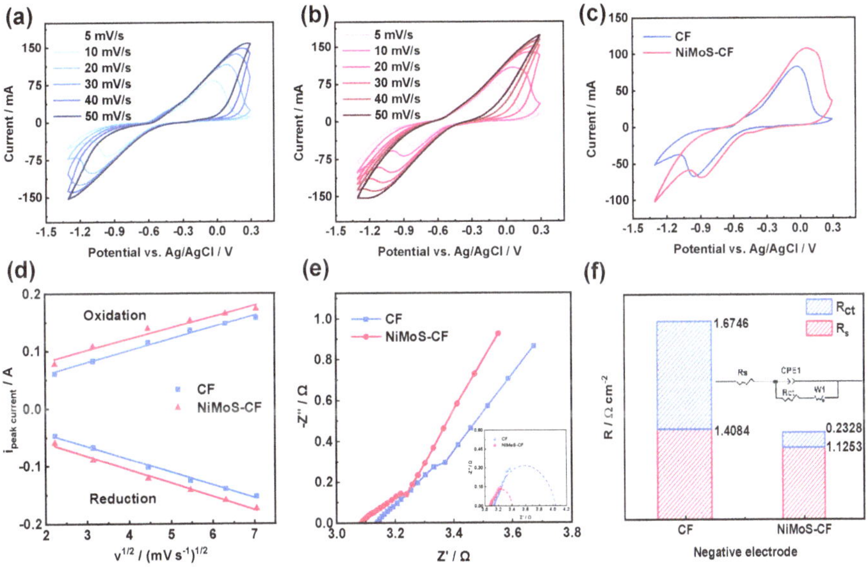

Figure 4. CV profiles for (**a**) pristine CF and (**b**) NiMoS−CF at various scan rates; (**c**) CV curves of pristine CF and NiMoS−CF at a scan rate of 10 mV s^{-1}; (**d**) peak currents fitting lines for the S_2^{2-}/S^{2-} redox pair on pristine CF and NiMoS−CF plotted against the square root of the scan rates; (**e**) Nyquist plots from the resulting EIS of pristine CF and NiMoS−CF with fitting curves; (**f**) comparison of resistance values derived from the Nyquist plots.

Since the diffusion constant of the active ion is proportional to the ratio of i and $v^{1/2}$, the linear lines developed for NiMoS-CF exhibit steeper slopes (Figure 4d), suggesting that the transfer process of active ions on CF and NiMoS-CF is controlled by diffusion, and the catalyst changes the diffusion rate rather than the mode of control. The resistance of both pristine CF and NiMoS-CF were characterized through electrochemical impedance spectroscopy (EIS) tests, as illustrated in Figure 4e. Furthermore, the measured charge transfer resistance (R_{ct}) of the NiMoS-CF (524 mΩ) was lower than that of the pristine CF (3768 mΩ), showing that the redox kinetics of the S_2^{2-}/S^{2-} pair are enhanced on the NiMoS-CF and significantly accelerates the redox reaction (Figure 4f, Table S2). In addition, in Figure 4e, the fitting curve of the high frequency region is added for better comparison.

ECSA refers to the Electrochemical Active Surface Area: a higher ECSA value implies that the effective area of the electrode is greater. Furthermore, the Tafel slope was quantitatively evaluated. The lower Tafel slope implies that the kinetics of the electrode are faster and its catalytic activity is higher. In the non-Faraday region, non-faraday capacitance of electrode is positively correlated with ESCA. If the measured capacitance is significantly lower than the theoretical value, it may indicate pore blockage, poor electrical conductivity, or low ion accessibility. In addition, the high double layer capacitance means that charge storage achieves a fast response through the adsorption/desorption of electrolyte ions on the electrode surface. CV curves were obtained by CV testing at different sweep speeds (Figure 5a,b), which were subsequently analyzed to determine the double-layer capacitance (C_{dl}) for the different electrodes (Figure 5c). The results of the calculations for (C_{dl}), ECSA, and Tafel slopeare summarized in Table 1. These findings demonstrate that NiMoS-CF possesses a larger double-layer capacitance, and the electrochemical area estimated from (C_{dl}) is greater than that of pristine CF. The decreased Tafel slope (Figure 5d) also indicates that NiMoS-CF exhibits faster reaction kinetics.

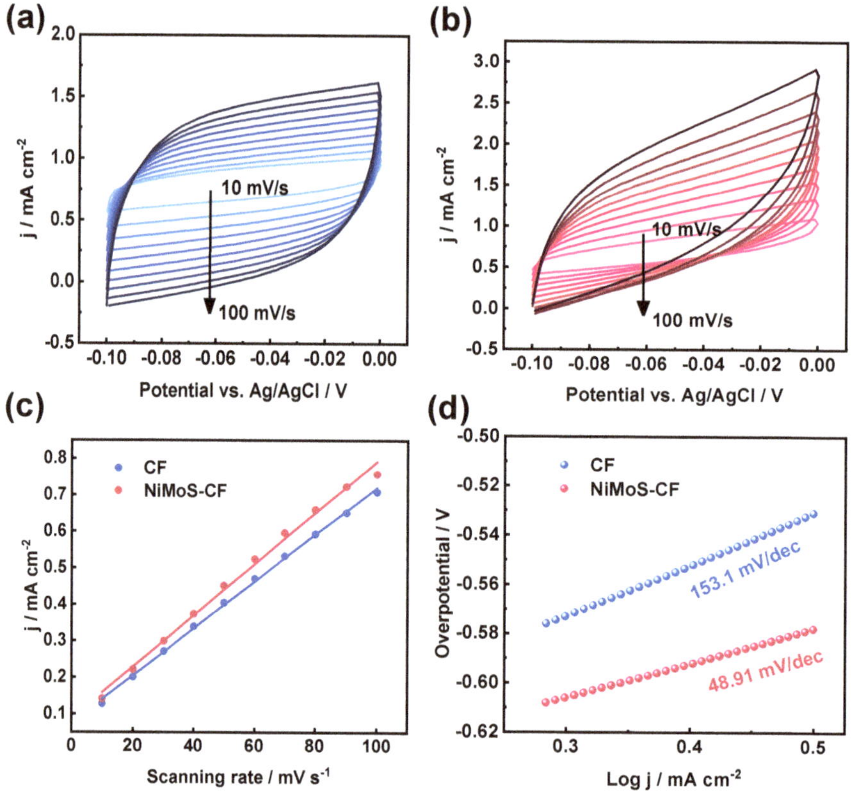

Figure 5. CV profiles for (**a**) pristine CF and (**b**) NiMoS−CF in non-Faraday intervals; (**c**) the corresponding C_{dl} comparison of different electrodes; (**d**) Tafel slopes for different electrodes.

Table 1. Diffusion rates of different electrodes.

Electrodes	Pristine CF	NiMoS-CF
D/($cm^2\ s^{-1}$)	9.42×10^{-5}	1.597×10^{-4}
k^0/(cm s^{-1})	1.3×10^{-3}	1.2×10^{-3}
C_{dl}/(mF cm^{-2})	6.44	7.02
ECSA/cm^2	362.25	394.875
Tafel Slope/(mV dec^{-1})	153.1	48.91

Furthermore, the structural properties of NiMoS catalyst, such as abundant edge active sites, provide excellent catalytic sites that enhance the adsorption and dissociation of sulfur ions, thereby accelerating the rate of redox reactions. During the reaction, sulfur is removed off the catalyst surface at a lower level, and the active metals Mo and Ni become unsaturated with ligands, generating vacant sites and enhancing the diffusion rate of sulfur ions [46].

To further investigate proliferation, the Randles–Sevcik equation was used to quantitatively calculate the diffusion coefficients of different electrodes. The data in Table 1 indicate that the NiMoS catalyst enhances transport capacity of ions, and the ions in the solution exhibit a greater diffusion capacity when utilizing the NiMoS-CF.

The standard rate constants can be used to determine the redox reaction's kinetics. Higher k^0 values indicate faster kinetics, reduced energy requirements, and lower overpotential, which are critical for the high-performance RFBs [47]. Reversible electrochemical reactions exhibit high charge transfer rates ($k^0 > 10^{-1}$ cm s^{-1}). Although Nicholson approach is the preferred method for calculating standard rate constants in reversible systems, the Lavagnini modification of Nicholson method in Equation (5) is more suitable for peak-to-peak separation over 200 mV in this experiment. Table 1 presents the results derived from the CV test. The higher redox kinetic parameters suggest that chemical reactions between the redox pairs are more likely to occur, allowing the system to reach equilibrium in a shorter amount of time with NiMoS-CF electrodes [47].

2.2.2. Flow Battery Tests

To assess the impact of electrodes on battery performance, PFRFBs were constructed with pristine CF and NiMoS-CF electrode, respectively. According to Figure 6a, NiMoS-CF cell exhibited significantly greater energy efficiency than pristine CF cell in the current density ranging from 20 to 100 mA cm^{-2}. At higher current densities, NiMoS-CF has a more pronounced positive effect compared to pristine CF serves as an anode. Specifically, the polarization of NiMoS-CF is lower than that of pristine CF RFBs. The energy efficiency is approximately equivalent to that of pristine CF cells when the current density above 100 mA cm^{-2}, and a similar trend is observed in the cells' voltage efficiency (Figure 6b). Consequently, the voltage loss in NiMoS-CF-based PFRFB is reduced as the polarization effect diminishes.

This assumption was supported by the polarization curve test conducted at 50% SOC (Figure 6c). The cell equipped with NiMoS-CF demonstrated lower charge voltage and higher discharge voltage in polarization tests, and the reduction of overpotential will lead to the improvement of EE.

The charging and discharging voltage curves in Figure 6d determine that the charging curve of the pristine CF electrode differs from that of the NiMoS-CF electrode, where the voltage initially decreases and then rises. It is believed that this behavior results from the electrochemical process that occurs when redox ions of various valence states diffuse to the electrode surface. A concentration gradient is created when the concentration difference of ions in the spatial domain decreases because the diffusion rate of ions on the electrode surface is less than the reaction rate. Concentration polarization and related losses are caused by this gradient [48]. At a current density of 100 mA cm^{-2}, battery based on NiMoS-CF exhibit lower charge and higher discharge plateaus compared to that based on pristine CF, verifying the battery with NiMoS-CF as the anode has reduced polarization. The curves confirm that NiMoS-CF electrode perform significantly better than pristine CF electrode; Specifically, the electrocatalytic activity of the pristine CF electrode is markedly improved following modification with NiMoS catalyst.

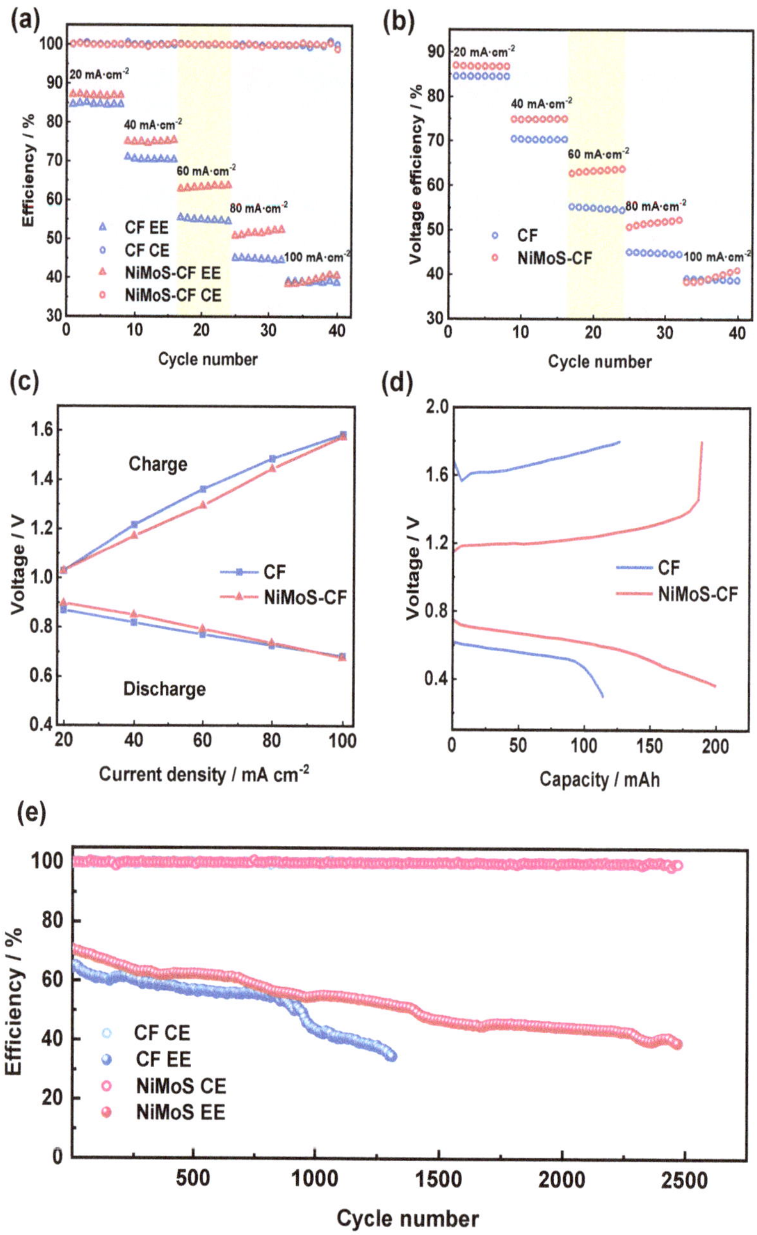

Figure 6. (**a**,**b**) The rate performances of PFRFBs with pristine CF and NiMoS−CF as anode, respectively; (**c**) polarization curves of PFRFBs with pristine CF and NiMoS−CF; (**d**) charge/discharge voltage curves of PFRFBs with pristine CF and NiMoS−CF; (**e**) cell performance of PFRFBs with pristine CF and NiMoS−CF.

To further assess and compared the characteristics of NiMoS-CF and pristine CF as negative electrodes, the CE and EE of the cells were measured at 40 mA cm^{-2} in Figure 6e. The data revealed that the battery equipped with NiMoS-CF electrode had an energy efficiency (EE) of 71% during the first cycle and 40% after 2500 cycles. That was higher than that of the battery with pristine CF electrode, which had an EE of 64% at the beginning of the cycle and 40% when it became inoperative due to severe capacity degradation after 1300 cycles. It is evident that the battery is capable of maintaining a stable voltage plateau and capacity output over numerous cycles of usage and is not prone to rapid performance degradation.

It is worth noting that the NiMoS-CF electrode's EE declined after 2500 cycles, owing primarily to CF degradation and electrolyte contamination. During the extended cycle, the CF may corrode or wear mechanically. As the cycle advances, the accumulation of chemical byproducts or impurities in the electrolyte can impede mass transfer, increase overpotential, and result in decreasing efficiency.

When compared to other representative batteries, the PFRFB with NiMoS-CF reported in this paper demonstrates exceptional performance, the capacity retention rate is as high as 99.9%, as illustrated in Table S3. In conclusion, following the modification of the NiMoS catalyst on the pristine CF electrode and subsequent battery testing, the cycle life, EE, and VE values of the battery are significantly enhanced.

3. Experimental

3.1. Materials

Nickel chloride hexahydrate ($NiCl_2{\cdot}6H_2O$, AR > 98.0%) was obtained from Shanghai McLean Biochemical Science and Technology Company (Shanghai, China). Sodium molybdate dihydrate ($Na_2MoO_4{\cdot}2H_2O$, AR > 98.0%), thiourea (CH_4N_2S, AR > 99.0%), ethanol absolute (C_2H_6O, AR > 99.5%), potassium ferricyanide ($K_3[Fe(CN)_6]$, AR > 99.5%), potassium polysulfide (K_2S, AR $\geq$ 40%), and potassium chloride (KCl, AR > 99.0%) were ordered from Sinopharm Chemical Reagent Company (Shanghai, China). The above chemicals were used directly for experiments without any purification. Nafion 212 membrane was provided by Jiangsu Kerun Membrane Materials Company (Suzhou, China). Prior to use, the commercial Nafion 212 was subjected to boiling in a 1.0 M KOH solution at 80 °C for 1 h. This process facilitated the transformation of the membrane from a proton (H^+) to a potassium (K^+) form of cation-exchange membrane. Before using, the treated Nafion 212 was soaked in deionized (DI) water.

3.2. Preparation of NiMoS Catalyst and NiMoS-CF

The one-step hydrothermal method was used to prepared the NiMoS catalyst and NiMoS-CF electrode. 0.3 mmol L^{-1} $NiCl_2{\cdot}6H_2O$, 0.3 mmol L^{-1} $Na_2MoO_4{\cdot}2H_2O$, and 0.25 mmol L^{-1} of CH_4N_2S were added to 100 mL of deionized water. The solution was magnetically agitated at room temperature (~25 °C) until dissolved. Different sizes of carbon felts (1.5 × 1.5 cm^2 and 2.0 × 2.0 cm^2) with the above solution were transferred to a 100-mL Teflon-lined autoclave. The reactor was kept in an oven at 180 °C for 24 h. Finally, the NiMoS catalyst and NiMoS-CF were collected, rinsed with DI water, and dried at 60 °C for 10 h before using.

3.3. Material Characterizations

The morphologies of pristine CF and NiMoS-CF electrodes were investigated by scanning electron microscopy (SEM, Zeiss GeminiSEM 300, Zeiss, Oberkochen, Germany). The structural and chemical information of the NiMoS-CF electrode were elucidated using high-resolution transmission electron microscopy (HR-TEM, model JEOL F200, JEOL, Tokyo, Japan) at an operating voltage of 200 kV. This analysis included high-angle annular dark-field scanning transmission electron microscopy (HAADF-STEM, JEOL) for imaging, companied by energy-dispersive X-ray spectroscopy (EDS) for elemental mapping. The characteristic diffraction peaks of pristine CF and NiMoS-CF were employed by X-ray diffraction (XRD, D8 ADVANCE and DAVINCI DESIGN, Karlsruhe, Germany) in the range of 3–90° with Cu Kα (λ = 1.5418 Å) radiation, an accelerating voltage of 40 kV and a current of 40 mA. The X-ray photoelectron spectrometer (XPS, ESCALAB250Xi, Thermo Fisher Scientific, Waltham, MA, USA) was performed for high-resolution XPS measurements equipped with Al Kα source. Hydrophilicity of pristine CF and NiMoS-CF were measured by contact angle test (CA, Dataphysics-OCA20, Stuttgart, Germany).

3.4. Electrochemical Measurements

Cyclic voltammetry (CV) experiments were conducted within a three-electrode configuration using an electrolyte solution composed of 50 mmol L^{-1} K_2S and 1.0 mol L^{-1} KCl. The setup included either pristine CF or NiMoS-CF (with dimensions of 1.5 × 1.5 cm^2) as the working electrode, pristine CF (2 × 2 cm^2) as the counter electrode, a Ag/AgCl electrode as the reference electrode, with test voltages in the range of −1.3 to 0.3 V and scanning rates between 5 and 50 mV s^{-1} on Chenhua Chi660e Electrochemical Workstation (CH Instrument, Shanghai, China). The electrochemical impedance spectroscopy (EIS) measurements were done in the amplitude of 15 mV and the frequency range of 10 MHz to 100 kHz. The glassy carbon electrode (GCE) was used in the same electrolyte system as working electrode, which was drop-coated with a catalyst ink of 3 mg NiMoS + 2 mg carbon powder + 80 μL Nafion + 900 μL anhydrous ethanol. S/Fe RFB cells (Figure S1) were fabricated using pristine CF or NiMoS-CF as the anode (2 × 2 cm^2) and pristine CF (2 × 2 cm^2) as the cathode, separated by a Nafion 212 ion-exchange membrane. The anolyte was prepared as a 40 mL aqueous solution containing 2.0 mol L^{-1} K_2S and 2.0 mol L^{-1} KCl, while the catholyte consisted of 40 mL of 1.0 mol L^{-1} $K_3[Fe(CN)_6]$ and 2.0 mol L^{-1} KCl. Both electrolytes were circulated through the cell at a constant flow rate of 70 mL min^{-1} during electrochemical testing. Tests of rate performance and long-term charge/discharge cycle were conducted on LAND battery test system (CT3002K, LAND Electronics, Wuhan, China).

At current densities ranging from 20 to 100 mA cm^{-2}, the rate performance at 50% State of Charge (SOC) was determined. For each specified current density, a series of eight charge/discharge cycles were executed to record the values of Coulombic Efficiency (CE), Voltage Efficiency (VE) and Energy Efficiency (EE). In constant current mode, long-term charge/discharge cycling experiments were conducted at 40 mA cm^{-2} within the voltage in the range of 0.3 to 1.8 V. Polarization curves were obtained on the same device.

3.5. Electrochemical Calculation

The diffusion coefficients for different electrodes were quantified according to the Randles-Sevcik equation [49]:

$$I_p = 0.4463nFAC(\frac{nFvD}{RT})^{\frac{1}{2}} \tag{1}$$

Or if the solution is at 25 °C [50]:

$$I_p = 269000n^{\frac{3}{2}}ACD^{\frac{1}{2}}V^{\frac{1}{2}} \tag{2}$$

where I_p represents the peak current in amperes, n denotes the number of electrons involved in the redox process (commonly 1), D is the diffusion coefficient (cm^2 s^{-1}), V is the scan rate (V s^{-1}), A is the electrode area (cm^2), and C is the concentration of the substance to be measured (mol cm^{-3}).

The Nicholson approach [51] was used to calculate standard rate constants k^0 for various electrodes.

$$\psi = \frac{(\frac{D_O}{D_R})^{\alpha/2}k^0}{(\frac{\pi D_O nFv}{RT})^{1/2}} \tag{3}$$

The dimensionless kinetic parameter, denoted as ψ, is derived from ΔEp [52], where R is the transfer coefficient, n is the number of electrons transferred in the redox process, F is the Faraday constant. R and T have their usual meanings. Typically, it is reasonable to assume that the diffusion coefficients for both the oxidized (D_O) and reduced (D_R) forms of the mediator are nearly identical, and the reduction and oxidation kinetics are relatively symmetrical ($R \sim 0.5$) [53].

From Equation (3), the slope of a plot of ψ versus $v^{1/2}$ can be used to determine k^0. The values of ψ at different peak separations can be estimated practically using Equation (4).

$$\psi = \frac{(-0.6288 + 0.0021\Delta Ep)}{1 - 0.017\Delta Ep} \tag{4}$$

where ΔEp is the redox peak current potential difference (V).

The Lavagnini method [54] is an improvement on the Nicholson method and can achieve peak-to-peak separation above 200 mV which is more suitable for this experiment:

$$\psi = 2.18(\beta/\pi)^{1/2}\exp\left[-(\beta^2 F/RT)\, n\Delta Ep\right] \tag{5}$$

where β is the transfer coefficient for the electrode process [55].

ECSA is the Electrochemical Active Surface Area, a higher ECSA value implies that the electrode's effective area is greater.

$$ESCA = \frac{C_{dl}}{C_s} A \tag{6}$$

where C_{dl} is the double capacitive electric layer, C_s is the differential capacitance of the smooth metal, which is taken at 0.04 mF cm^{-2} in this research [56], and A is the geometric surface area of the electrode.

In this experiment, all currents in non-faradic region were assumed to be non-faradic currents. The region for non-faradic is −0.1–0 V (vs. KCl saturated Ag/AgCl). The current i is obtained at the middle point of potential range from CV curves at 0.01–0.1 V s^{-1}. And of the system is taken as the average of absolute value of charging current. The electrolyte used in ECSA tests is 0.05 M K_2S with 1.0 M KCl.

4. Conclusions

In summary, we have successfully fabricated NiMoS catalyst on the carbon felt electrode via a one-step hydrothermal method. The electrochemical reaction kinetics and charge transfer behaviors of the obtained NiMoS-CF electrode were improved by the NiMoS modification, resulting in the accelerated kinetics of polysulfide redox reactions. The integrated PFRFB with NiMoS-modified carbon felt electrode as the anode, exhibited a significant improvement in the energy density and cycle stability, including an energy efficiency of 70%, voltage efficiency of 87%, outperforming the pristine CF-based battery with corresponding values of 64% and 84%, respectively. Furthermore, the cycle life of the PFRFB was extended to 2500 cycles with a coulombic efficiency consistently maintained above 99% at 40 mA cm^{-2}. This research provides a strategy for enhancing the performance of the S/Fe RFBs by employing sulfide bimetallic catalyst as the electrode booster, offering a promising direction for developing the cost-effective and high-performance energy storage systems.

Furthermore, while the three-electrode test described in this study can be utilized for high-precision electrochemical measurements, it is susceptible to effects from solution resistance, voltage window, and temperature. To avoid such impacts in future work, the test error will be addressed to the greatest extent possible.

Supplementary Materials: The following supporting information can be downloaded at: https://www.mdpi.com/article/10.3390/molecules30061219/s1; Figure S1: Photograph of the NiMoS-S/Fe cell stack; Figure S2: XRD pattern of NiMoS catalyst; Figure S3: (a) XPS survey spectrum of NiMoS catalyst; high-resolution spectra for (b) Ni 2p, (c) Mo 3d and (d) S 2p spectra of NiMoS catalyst; Figure S4: Glassy carbon electrode with NiMoS catalyst; Figure S5: CV curves of the original GCE and NiMoS modified GCE at 10 mV s^{-1}; Table S1: Kinetic parameters of S_2^{2-}/S^{2-} from CF and NiMoS-CF's CV data. Related to Figure 4a,b; Table S2: Fitting results for the as-prepared samples from the Nyquist plots; Table S3: An overall comparison with other representative RFBs. References [10,13,16,27,31,32,42,46,57–61] are cited in the Supplementary Materials.

Author Contributions: Conceptualization, D.M.; writing—original draft preparation, B.L.; writing—review and editing, H.M.; visualization, Z.Z.; visualization, F.W.; funding acquisition, project administration, Y.C.; funding acquisition, supervision, J.A.; resources, supervision, F.X.; funding acquisition, L.X. All authors have read and agreed to the published version of the manuscript.

Funding: This study received financial support from the National Natural Science Foundation of China (Grant No. W2432038), the special fund of Advantageous characteristic disciplines (Group) of Hubei Province, the Excellent Young and Middle-aged Scientific and Technological Innovation Team Project of Colleges and Universities in Hubei Province (Grant No. T2023041), China Postdoctoral Science Foundation (Grant No. 2023M730399), the Guangdong Foundation of Basic and Applied Foundation (2024A1515012826) and Guangdong Province Key Construction Discipline Research Ability Improvement Project (2024ZDJS027).

Institutional Review Board Statement: Not applicable.

Informed Consent Statement: Not applicable.

Data Availability Statement: Data will be made available on request.

Conflicts of Interest: The authors declare no known competing financial interests or personal relationships that could have appeared to influence the work reported in this paper.

References

1. Twitchell, J.; DeSomber, K.; Bhatnagar, D. Defining long duration energy storage. *J. Energy Storage* **2023**, *60*, 105787. [CrossRef]
2. Yuan, Z.; Liang, L.; Dai, Q.; Li, T.; Song, Q.; Zhang, H.; Hou, G.; Li, X. Low-cost hydrocarbon membrane enables commercial-scale flow batteries for long-duration energy storage. *Joule* **2022**, *6*, 884–905. [CrossRef]
3. Zuo, P.; Ye, C.; Jiao, Z.; Luo, J.; Fang, J.; Schubert, U.S.; McKeown, N.B.; Liu, T.L.; Yang, Z.; Xu, T. Near-frictionless ion transport within triazine framework membranes. *Nature* **2023**, *617*, 299–305. [CrossRef]
4. Yang, M.; Xu, Z.; Xiang, W.; Xu, H.; Ding, M.; Li, L.; Tang, A.; Gao, R.; Zhou, G.; Jia, C. High performance and long cycle life neutral zinc-iron flow batteries enabled by zinc-bromide complexation. *Energy Storage Mater.* **2022**, *44*, 433–440. [CrossRef]
5. Xing, F.; Liu, T.; Yin, Y.; Bi, R.; Zhang, Q.; Yin, L.; Li, X. Highly Active Hollow Porous Carbon Spheres@Graphite Felt Composite Electrode for High Power Density Vanadium Flow Batteries. *Adv. Funct. Mater.* **2022**, *32*, 2111267. [CrossRef]
6. Yu, Z.; Jia, X.; Cai, Y.; Su, R.; Zhu, Q.; Zhao, T.; Jiang, H. Electrolyte engineering for efficient and stable vanadium redox flow batteries. *Energy Storage Mater.* **2024**, *69*, 103404. [CrossRef]
7. Bui, T.T.; Shin, M.; Abbas, S.; Ikhsan, M.M.; Do, X.H.; Dayan, A.; Almind, M.R.; Park, S.; Aili, D.; Hjelm, J.; et al. Sulfonated para-Polybenzimidazole Membranes for Use in Vanadium Redox Flow Batteries. *Adv. Energy Mater.* 2024, *early view*. [CrossRef]
8. Jiang, Y.; Wang, Y.; Cheng, G.; Li, Y.; Dai, L.; Zhu, J.; Meng, W.; Xi, J.; Wang, L.; He, Z. Multiple-dimensioned defect engineering for graphite felt electrode of vanadium redox flow battery. *Carbon Energy* **2024**, *6*, e537. [CrossRef]
9. Li, X.; Yao, Y.; Liu, C.; Jia, X.; Jian, J.; Guo, B.; Lu, S.; Qin, W.; Wang, Q.; Wu, X. Lithium Ferrocyanide Catholyte for High-Energy and Low-cost Aqueous Redox Flow Batteries. *Angew. Chem.* **2023**, *135*, e202304667. [CrossRef]
10. Cheng, X.; Xuan, T.; Wang, L. A low-cost all-iron hybrid redox flow batteries enabled by deep eutectic solvents. *Chem. Eng. J.* **2024**, *491*, 151936. [CrossRef]
11. Gao, J.; Lee, K.; Amini, K.; Gordon, R.G.; Betley, T.A.; Aziz, M.J. A Highly Soluble Iron-Based Posolyte Species with High Redox Potential for Aqueous Redox Flow Batteries. *Adv. Funct. Mater.* **2024**, *34*, 2310140. [CrossRef]
12. Song, Y.; Zhang, K.; Li, X.; Yan, C.; Liu, Q.; Tang, A. Tuning the ferrous coordination structure enables a highly reversible Fe anode for long-life all-iron flow batteries. *J. Mater. Chem. A* **2021**, *9*, 26354–26361. [CrossRef]

13. Lan, J.; Li, K.; Yang, L.; Lin, Q.; Duan, J.; Zhang, S.; Wang, X.; Chen, J. Hierarchical Nano-Electrocatalytic Reactor for High Performance Polysulfides Redox Flow Batteries. *ACS Nano* **2023**, *17*, 20492–20501. [CrossRef]
14. Wang, B.; Zhou, W.; Zhang, Y.; Zhang, T.; Li, X.; Feng, Y.; Zhao, R.; Li, W.; Elzatahry, A.; Hassan, Y.; et al. An energetic K^+-S aqueous battery with 96% sulfur redox utilization. *Joule* **2024**, *8*, 2033–2048. [CrossRef]
15. Yan, S.; Huang, S.; Xu, H.; Li, L.; Zou, H.; Ding, M.; Jia, C.; Wang, Q. Redox Targeting-based Neutral Aqueous Flow Battery with High Energy Density and Low Cost. *ChemSusChem* **2023**, *16*, e202300710. [CrossRef]
16. Sreenath, S.; Nayanthara, P.S.; Pawar, C.M.; Ash, A.; Bhatt, B.; Verma, V.; Nagarale, R.K. An aqueous polysulfide redox flow battery with a semi-fluorinated cation exchange membrane. *Energy Adv.* **2024**, *3*, 203–214. [CrossRef]
17. Ortiz-Martínez, V.M.; Gómez-Coma, L.; Pérez, G.; Ortiz, A.; Ortiz, I. The roles of ionic liquids as new electrolytes in redox flow batteries. *Sep. Purif. Technol.* **2020**, *252*, 117436. [CrossRef]
18. Song, X.; Wang, C.; Shen, Z.; Guo, K.; Wu, J.; Guo, Z.; Liu, X.; Zhao, Y. Solvated metal complexes for balancing stability and activity of sulfur free radicals. *eScience* **2023**, *4*, 100225. [CrossRef]
19. Liang, H.; Kumar, P.; Ma, Z.; Zhao, F.; Cheng, H.; Xie, H.; Cao, Z.; Cavallo, L.; Li, Q.; Ming, J. Electrolyte Intermolecular Interaction Mediated Nonflammable Potassium-Ion Sulfur Batteries. *ACS Energy Lett.* **2024**, *9*, 3536–3546. [CrossRef]
20. Shin, M.; Noh, C.; Kwon, Y. Electrolyte optimization of alkaline aqueous redox flow battery using iron-2,2-bis(hydroxymethyl)-2,2′,2′-nitrilotriethanol complex as active material for anolyte. *Chem. Eng. J.* **2023**, *453*, 139738. [CrossRef]
21. Rahimi, M.; Molaei Dehkordi, A.; Roberts, E.P.L. Magnetic nanofluidic electrolyte for enhancing the performance of polysulfide/iodide redox flow batteries. *Electrochim. Acta* **2021**, *369*, 137687. [CrossRef]
22. Ahmad, A.; Aldawood, T.A.; Mansha, M.; Ali, S.; Tahir, M.N.; Khan, M.; Khan, I.A.; Khan, S.A. Optimized and cost-effective elemental-sulfur sodium polysulfide/sodium bromide aqueous electrolytes for redox flow batteries. *J. Power Sources* **2024**, *614*, 235013. [CrossRef]
23. Ding, M.; Fu, H.; Lou, X.; He, M.; Chen, B.; Han, Z.; Chu, S.; Lu, B.; Zhou, G.; Jia, C. A Stable and Energy-Dense Polysulfide/Permanganate Flow Battery. *ACS Nano* **2023**, *17*, 16252–16263. [CrossRef] [PubMed]
24. Amin, H.M.A.; Apfel, U.-P. Metal-Rich Chalcogenides as Sustainable Electrocatalysts for Oxygen Evolution and Reduction: State of the Art and Future Perspectives. *Eur. J. Inorg. Chem.* **2020**, *2020*, 2679–2690. [CrossRef]
25. Yang, Z.; Wang, B.; Chen, Y.; Zhou, W.; Li, H.; Zhao, R.; Li, X.; Zhang, T.; Bu, F.; Zhao, Z.; et al. Activating sulfur oxidation reaction via six-electron redox mesocrystal NiS_2 for sulfur-based aqueous batteries. *Natl. Sci. Rev.* **2023**, *10*, nwac268. [CrossRef]
26. Ma, D.; Hu, B.; Wu, W.; Liu, X.; Zai, J.; Shu, C.; Tadesse Tsega, T.; Chen, L.; Qian, X.; Liu, T.L. Highly active nanostructured CoS_2/CoS heterojunction electrocatalysts for aqueous polysulfide/iodide redox flow batteries. *Nat. Commun.* **2019**, *10*, 3367. [CrossRef]
27. Wei, X.; Xia, G.-G.; Kirby, B.; Thomsen, E.; Li, B.; Nie, Z.; Graff, G.G.; Liu, J.; Sprenkle, V.; Wang, W. An Aqueous Redox Flow Battery Based on Neutral Alkali Metal Ferri/ferrocyanide and Polysulfide Electrolytes. *J. Electrochem. Soc.* **2016**, *163*, A5150–A5153. [CrossRef]
28. Gao, M.; Huang, S.; Zhang, F.; Lee, Y.M.; Huang, S.; Wang, Q. Successive ionic layer adsorption and reaction–deposited copper sulfide electrocatalyst for high-power polysulfide-based aqueous flow batteries. *Mater. Today Energy* **2020**, *18*, 100540. [CrossRef]
29. Li, Z.; Lu, Y.-C. Polysulfide-based redox flow batteries with long life and low levelized cost enabled by charge-reinforced ion-selective membranes. *Nat. Energy* **2021**, *6*, 517–528. [CrossRef]
30. Radich, E.J.; Dwyer, R.; Kamat, P.V. Cu_2S Reduced Graphene Oxide Composite for High-Efficiency Quantum Dot Solar Cells. Overcoming the Redox Limitations of S_2^-/S_n^{2-} at the Counter Electrode. *J. Phys. Chem. Lett.* **2011**, *2*, 2453–2460. [CrossRef]
31. Astruc, D. Inorganic Electrochemistry. Theory, Practice and Application. By Piero Zanello. *Angew. Chem. Int. Ed.* **2004**, *43*, 3752–3753. [CrossRef]
32. Bard, A.J.; Faulkner, L.R. *Electrochemical Methods: Fundamentals and Applications*, 2nd ed.; Wiley: New York, NY, USA, 2001; ISBN 978-0-471-04372-0.
33. Nicholson, R.S. Theory and Application of Cyclic Voltammetry for Measurement of Electrode Reaction Kinetics. *Anal. Chem.* **1965**, *37*, 1351–1355. [CrossRef]
34. Amin, H.M.A.; Uchida, Y.; Kätelhön, E.; Compton, R.G. Determination of standard electrochemical rate constants from semicircular sweep voltammetry: A combined theoretical and experimental study. *J. Electroanal. Chem.* **2021**, *880*, 114891. [CrossRef]
35. Velický, M.; Bradley, D.F.; Cooper, A.J.; Hill, E.W.; Kinloch, I.A.; Mishchenko, A.; Novoselov, K.S.; Patten, H.V.; Toth, P.S.; Valota, A.T.; et al. Electron Transfer Kinetics on Mono- and Multilayer Graphene. *ACS Nano* **2014**, *8*, 10089–10100. [CrossRef] [PubMed]
36. Lavagnini, I.; Antiochia, R.; Magno, F. An Extended Method for the Practical Evaluation of the Standard Rate Constant from Cyclic Voltammetric Data. *Electroanalysis* **2004**, *16*, 505–506. [CrossRef]
37. Klingler, R.J.; Kochi, J.K. Electron-transfer kinetics from cyclic voltammetry. Quantitative description of electrochemical reversibility. *J. Phys. Chem.* **1981**, *85*, 1731–1741. [CrossRef]
38. Yang, Y.; Shao, Z. Boron and nitrogen co-doped carbon nanospheres for supercapacitor electrode with excellent specific capacitance. *Nanotechnology* **2022**, *33*, 185403. [CrossRef]

39. Lou, X.; Fu, H.; Xu, J.; Long, Y.; Yan, S.; Zou, H.; Lu, B.; He, M.; Ding, M.; Zhu, X.; et al. Cost-Effective Membrane and Advanced Electrode for Stable Polysulfide-Ferricyanide Flow Battery. *Energy Mater. Adv.* **2022**, *2022*, 9865618. [CrossRef]
40. Long, Y.; Xu, Z.; Wang, G.; Xu, H.; Yang, M.; Ding, M.; Yuan, D.; Yan, C.; Sun, Q.; Liu, M.; et al. A neutral polysulfide/ferricyanide redox flow battery. *iScience* **2021**, *24*, 103157. [CrossRef]
41. Zhou, J.; Yu, L.; Zhu, Q.; Huang, C.; Yu, Y. Defective and ultrathin NiFe LDH nanosheets decorated on V-doped Ni_3S_2 nanorod arrays: A 3D core–shell electrocatalyst for efficient water oxidation. *J. Mater. Chem. A* **2019**, *7*, 18118–18125. [CrossRef]
42. Yu, L.; Zhou, H.; Sun, J.; Qin, F.; Yu, F.; Bao, J.; Yu, Y.; Chen, S.; Ren, Z. Cu nanowires shelled with NiFe layered double hydroxide nanosheets as bifunctional electrocatalysts for overall water splitting. *Energy Environ. Sci.* **2017**, *10*, 1820–1827. [CrossRef]
43. Xie, J.; Zhu, K.; Min, J.; Yang, L.; Luo, J.; Liu, J.; Lei, M.; Zhang, R.; Ren, L.; Wang, Z. In-situ grown ultrathin MoS_2 nanosheets on MoO_2 hollow nanospheres to synthesize hierarchical nanostructures and its application in lithium-ion batteries. *Ionics* **2019**, *25*, 1487–1494. [CrossRef]
44. Yang, D.; Cao, L.; Feng, L.; Huang, J.; Kajiyoshi, K.; Feng, Y.; Liu, Q.; Li, W.; Feng, L.; Hai, G. Formation of hierarchical Ni_3S_2 nanohorn arrays driven by in-situ generation of VS_4 nanocrystals for boosting alkaline water splitting. *Appl. Catal. B Environ.* **2019**, *257*, 117911. [CrossRef]
45. Liu, M.; Kong, L.; Wang, X.; He, J.; Bu, X.-H. Engineering Bimetal Synergistic Electrocatalysts Based on Metal–Organic Frameworks for Efficient Oxygen Evolution. *Small* **2019**, *15*, 1903410. [CrossRef] [PubMed]
46. Amin, H.M.A.; Attia, M.; Tetzlaff, D.; Apfel, U.-P. Tailoring the Electrocatalytic Activity of Pentlandite $Fe_xNi_{9-x}S_8$ Nanoparticles via Variation of the Fe:Ni Ratio for Enhanced Water Oxidation. *ChemElectroChem* **2021**, *8*, 3863–3874. [CrossRef]
47. Chen, H.; Zou, Y.; Li, J.; Zhang, K.; Xia, Y.; Hui, B.; Yang, D. Wood aerogel-derived sandwich-like layered nanoelectrodes for alkaline overall seawater electrosplitting. *Appl. Catal. B Environ.* **2021**, *293*, 120215. [CrossRef]
48. Liu, C.; Jia, D.; Hao, Q.; Zheng, X.; Li, Y.; Tang, C.; Liu, H.; Zhang, J.; Zheng, X. P-Doped Iron–Nickel Sulfide Nanosheet Arrays for Highly Efficient Overall Water Splitting. *ACS Appl. Mater. Interfaces* **2019**, *11*, 27667–27676. [CrossRef]
49. Han, W.; Liu, Z.; Pan, Y.; Guo, G.; Zou, J.; Xia, Y.; Peng, Z.; Li, W.; Dong, A. Designing Champion Nanostructures of Tungsten Dichalcogenides for Electrocatalytic Hydrogen Evolution. *Adv. Mater.* **2020**, *32*, 2002584. [CrossRef]
50. Li, L.-Y.; Yan, S.; Huang, Y.-J.; Zhong, F.-F.; Cao, J.-C.; Ding, M.; Jia, C.-K. Carbon felt electrode coated with WS2 enables a high-performance polysulfide/ferricyanide flow battery. *Rare Met.* **2024**, *43*, 5039–5047. [CrossRef]
51. Khan, N.A.; Rashid, N.; Junaid, M.; Zafar, M.N.; Faheem, M.; Ahmad, I. NiO/NiS Heterostructures: An Efficient and Stable Electrocatalyst for Oxygen Evolution Reaction. *ACS Appl. Energy Mater.* **2019**, *2*, 3587–3594. [CrossRef]
52. Xu, J.; Zhang, W.; Fan, H.; Cheng, F.; Su, D.; Wang, G. Promoting lithium polysulfide/sulfide redox kinetics by the catalyzing of zinc sulfide for high performance lithium-sulfur battery. *Nano Energy* **2018**, *51*, 73–82. [CrossRef]
53. Luo, P.; Sun, F.; Deng, J.; Xu, H.; Zhang, H.; Wang, Y. China Tree-Like NiS-Ni_3S_2/NF Heterostructure Array and Its Application in Oxygen Evolution Reaction. *Acta Phys.-Chim. Sin.* **2018**, *34*, 1397–1404. [CrossRef]
54. Yuan, S.; Li, W.; Wei, J.; Zheng, J.; Fang, W.; Yi, X.; Lai, W. Preparation of nanosheets-assembled flower-like NiMoS catalyst and its hydrodesulfurization performance. *Acta Pet. Sin. Pet. Process. Sect.* **2023**, *39*, 1349–1360.
55. Wang, H.; Sayed, S.Y.; Luber, E.J.; Olsen, B.C.; Shirurkar, S.M.; Venkatakrishnan, S.; Tefashe, U.M.; Farquhar, A.K.; Smotkin, E.S.; McCreery, R.L.; et al. Redox Flow Batteries: How to Determine Electrochemical Kinetic Parameters. *ACS Nano* **2020**, *14*, 2575–2584. [CrossRef] [PubMed]
56. Ni, J.; Li, M. Optimized operation method of vanadium redox flow batteries based on microgrid load response. *J. Glob. Energy Interconnect.* **2019**, *2*, 608–616. [CrossRef]
57. Ma, Q. A Double-Layer Electrode for the Negative Side of Deep Eutectic Solvent Electrolyte-Based Vanadium-Iron Redox Flow Battery. *Energy* **2023**, *265*, 126291. [CrossRef]
58. Mushtaq, K. In-Situ Crossover Diagnostics to Assess Membrane Efficacy for Non-Aqueous Redox Flow Battery. *J. Energy Storage* **2021**, *40*, 102713. [CrossRef]
59. Jung, J. Highly Selective Composite Membranes Using Ladder-like Structured Polysilsesquioxane for a Non-Aqueous Redox Flow Battery. *J. Membr. Sci.* **2020**, *595*, 117520. [CrossRef]
60. Li, Y.; Geysens, P.; Zhang, X.; Sniekers, J.; Fransaer, J.; Binnemans, K.; Vankelecom, I.F.J. Cerium-Containing Complexes for Low-Cost, Non-Aqueous Redox Flow Batteries (RFBs). *J. Power Sources* **2020**, *450*, 227634. [CrossRef]
61. Zou, H.; Xu, Z.; Xiong, L.; Wang, J.; Fu, H.; Cao, J.; Ding, M.; Wang, X.; Jia, C. An Alkaline S/Fe Redox Flow Battery Endowed with High Volumetric-Capacity and Long Cycle-Life. *J. Power Sources* **2024**, *591*, 233856. [CrossRef]

molecules

MDPI

Review

Progress in MOFs and MOFs-Integrated MXenes as Electrode Modifiers for Energy Storage and Electrochemical Sensing Applications

Sanjeevamuthu Suganthi, Khursheed Ahmad * and Tae Hwan Oh *

School of Chemical Engineering, Yeungnam University, 280 Daehak-Ro, Gyeongsan 38541, Republic of Korea
* Correspondence: khursheed@yu.ac.kr (K.A.); taehwanoh@ynu.ac.kr (T.H.O.)

Abstract: The global energy demand and environmental pollution are the two major challenges of the present scenario. Recently, researchers focused on the preparation and investigation of catalysts for their capacitive properties for energy storage devices. Thus, supercapacitors have received extensive interest from researchers due to their promising energy storage features and decent cyclic stability/performance. The performance of the supercapacitors are significantly influenced by the physicochemical properties of the electrocatalyst. In this review article, we have compiled the previous reports on the fabrication of MOFs-based composite materials with MXenes for energy storage and electrochemical sensing applications. The metallic and bimetallic MOFs and MOFs/MXenes materials for supercapacitor applications are reviewed. In addition, MOFs/MXenes-based hybrid composites are also compiled towards the determination of various toxic/hazardous materials, such as metal ions like copper ions, mercury ions, and picric acid. We believe that present review article may benefit the researchers working on the preparation of MOFs-based catalysts for supercapacitor and electrochemical sensing applications.

Keywords: MOFs; MXenes; composites; electrochemistry; energy storage; sensors

1. Introduction

In recent years, it has been observed that energy crisis and environmental pollution are two major problems for today's world [1]. It is believed that they may worsen in the future if significant attention is not given to such challenges. The global energy crisis is a pressing issue which may significantly influence people's lifestyle [2,3]. It is necessary to focus on the design and utilization of energy storage or energy conversions devices [4]. It is expected that developing highly efficient energy storage and conversion systems may emerge as a key solution to address such energy-related issues [5,6]. In connection to this, supercapacitors have received tremendous attention from scientific scholars [7–9]. The performance of supercapacitors are significantly influenced by the physiochemical features of electrocatalysts (also known as electrode materials) [10]. Thus, it is very important to design and explore the highly efficient electrode materials for energy storage devices and applications. Previously, various electrode materials, such as MOFs [11,12], metal sulfides [13,14], MXenes [15,16], quantum dots [17,18], covalent organic frameworks [19,20], and two-dimensional (2D) materials [21,22] have been reported for the construction of supercapacitors. The MOFs have demonstrated significant potential as candidates for supercapacitors due to high specific surface areas and organized porous structures [23–25]. These crystalline materials are composed of metal ions or clusters coordinated with organic ligands, forming porous frameworks with expansive surface areas [11]. The MOFs can be prepared by utilizing various protocols, such as microwave-assisted, solvo-thermal, hydrothermal, electrochemical, sono-chemical, and mechano-chemical methods. Each method may offer distinct advantages for fine-tuning the physicochemical properties of the prepared MOFs. Ongoing research in MOF synthesis and applications holds great potential for advancements across multiple technological fields [26,27]. The MOFs may be well-suited for use as electrode modifiers for the construction of supercapacitors due to their impressive high specific surface area, abundance of pseudo-capacitive redox centers, and tunable pore size [28]. However, the poor conductive nature of the MOFs is a major challenge for the scientific community.

Similar to MOFs, MXenes ($M_{a+1}X_aT_n$, where a = 1, 2, or 3; M represents transition metals like Mo, Ti, Cr, Nb, Sc, and V; X denotes carbon, nitrogen, or both; and T_n includes surface terminations such as –Cl, –OH, –O, –F, etc.) have drawn substantial research interest in energy storage applications. This attention is due to their outstanding metallic conductivity, excellent hydrophilicity, distinctive layered structure, and rich surface-functional groups [29–31]. However, MXenes have relatively low thermodynamic and environmental stability compared to the MOFs [32]. Thus, researchers designed and combined the properties of the MOFs with MXenes to enhance the electrical conductivity of the MOFs and stability of the MXenes [33]. On the other hand, various toxic materials, metal ions, and compounds such as such as copper (Cu^{2+}), lead (Pb^{2+}), arsenic (As^{3+}), nitrite, benomyl, atrazine, phenol, chloramphenicol, picloram, methamidophos, tyrosine, mycophenolic acid, hygromycin B, and hydrogen peroxide (H_2O_2) are widely used in various applications [34–37]. Such compounds have negative impacts on the environment and human beings, and the monitoring of such materials is necessary [37]. In the previous decade, a large number of MOFs-based electrode materials were prepared and explored for energy storage and electrochemical sensing applications due to their excellent catalytic properties [38–44]. MOFs materials are promising electrode materials but suffer from low conductivity which need significant attentions [45–48]. Various methods have been developed for the synthesis of MOFs, but electrical conductivity of the MOFs remains a challenge [49–53]. MXene has excellent conductivity and decent catalytic behavior for electrochemical applications [54–58]. In this regard, MXene can be integrated with MOFs to improve the conductivity and electrochemical performance of the MOFs/MXene-based composite materials.

In this review, we have summarized recent advancements in the development of MOFs and MOF/MXene composites for the construction of supercapacitors and electrochemical sensing of various toxic substances and compounds. We believe that this article will serve as a valuable resource for researchers focused on advancing MOFs, MXene, and MOFs/MXenes-composite-based electrode materials for applications in supercapacitors and electrochemical sensing.

2. MOFs as Supercapacitor Material

The MOFs-based electrode materials have been utilized in various combinations. In this section, we have categorized MOFs as pristine MOFs and MOFs/MXenes.

2.1. Pristine MOFs Based Supercapacitors

In previous years, various metal-based MOFs, such as zinc (Zn), nickel (Ni), copper (Cu), cobalt (Co), and manganese (Mn), have been synthesized, and their potential for energy storage devices were evaluated by various research groups. In this connection, Farea et al. [59] reported the fabrication of a Zn-based MOF using trimesic as a ligand via a hydrothermal method. Authors characterized the synthesized Zn–MOF by employing Raman spectroscopy, ultraviolet–visible (UV–vis), and photo-luminescence (PL) spectroscopic studies. The synthesized Zn–MOF was further applied as an electrocatalyst for the construction of supercapacitors. The Zn–MOF catalyst exhibited specific capacitance of 58.6 F/g at 0.15 A/g. The Zn–MOFs-modified electrode also showed acceptable stability and retained 82.5% capacitance after 1000 charge–discharge cycles at 0.45 A/g. Hong et al. [60] synthesized Zn–MOF using novel strategies and ligands. In the study, the authors prepared hierarchical porous carbon (HPC) materials by using direct carbonization of Zn–MOFs in the absence of other carbon sources. The authors used Zn-containing MOFs precursors ($[Zn_4O(bdc)_3]$, $[Zn_2(bdc)(L\text{-}lac)(DMF)]$, $[Zn_2(bdc)_2(dabco)]$, and $[Zn_3(fumarate)_3(DMF)_2]$

(bdc = 1,4-benzenedicarboxylate; dabco = 1,4-diazabicyclo [2.2.2]octane; and L-lac = L-lactate)) to obtain the supercapacitor materials. Authors adopted CV and GCD techniques to evaluate the capacitive features of the proposed electrocatalyst and reported the specific capacitances between 164 and 203 F/g. Salehi et al. [61] reported the synthesis of Zn–MOF using an electrodeposition method. In brief, the authors used zinc (II) nitrate hexahydrate, which was dissolved in N, N-dimethyl-formamide (DMF) and stirred for 20 min. Furthermore, 1,4 dicarboxybenzene (tetraphthalic acid) was separately mixed in ethanol and stored for deposition process. The Zn–MOF film was deposited onto a nickel (Ni) substrate through an electro-deposition process in a two-electrode assembly connected to a DC power source, applying a current density of 30 mA/cm^2 for 20 min. The resulting Zn–MOF was then analyzed using techniques, including scanning electron microscopy (SEM), X-ray diffraction (XRD), and Fourier-transform infrared spectroscopy (FTIR). Results revealed that Zn–MOF on Ni foam exhibits a layered cuboid structure with uniform dimensions and thickness. The authors expected that such morphology may provide large contact area for electrochemical reactions and shorten the electrolyte pathway. The electrochemical properties of the Zn–MOF-modified Ni foam were evaluated using GCD, CV, and electrochemical impedance spectroscopy (EIS) methods. The obtained results demonstrated a specific capacitance of 288 F/g at 2 A/g and suggested the presence of pseudo-capacitive behavior with a retention cycle of 56% of its initial performance. Cao et al. [62] reported the synthesis of Co-Ni–MOF-derived LDH materials as shown in Figure 1.

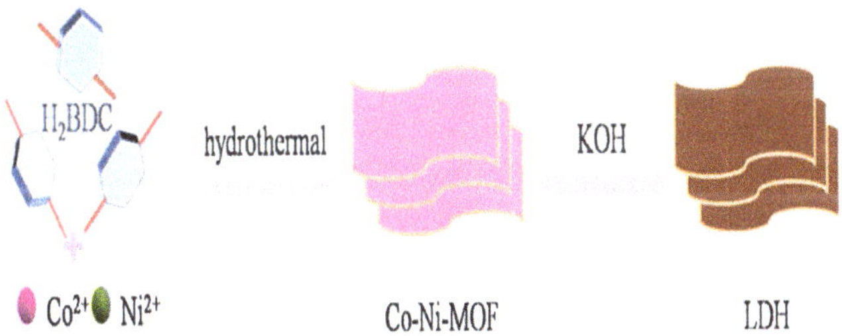

Figure 1. Schematic diagram of the preparation of Co-Ni-MOF-derived LDH. Reproduced with permission [62].

Furthermore, it was observed that LDH has relatively higher capacitance as observed by CV (Figure 2a). The GCD results also showed that LDH has higher specific capacitance compared to Co-Ni-MOF (Figure 2b,c). The EIS results also showed that LDH has higher conductivity and better charge transfer resistance compared to the Co-Ni-MOF (Figure 2d). This may be ascribed to the better electrocatalytic and conductive properties of the LDH. Du et al. [63] proposed the fabrication of a Cu-based MOF (Cu-MOF) material using a dual-temperature-zone chemical vapor deposition (CVD)-assisted approach. The $Cu_3(HHTP)_2$ (HHTP = 2,3,6,7,10,11-hexahydroxytriphenylene) nanowires arrays were prepared using the proposed CVD method, and the synthesized Cu-MOF was characterized by SEM, XRD, TEM and energy-dispersive X-ray spectroscopic (EDX) techniques.

The SEM results demonstrated that the obtained Cu-MOF consists of an oriented nanowire-like structure. The obtained results demonstrated a decent specific capacitance of 41.1 mF/cm^2 at 0.5 A/g. This work expands the synthesis approach from traditional liquid- or gas-phase reactions to a solid–solid reactions, which is anticipated to enable the fabrication of a wide range of conductive MOFs in supercapacitor applications. Liu et al. [64] synthesized Cu-based layered coordination polymer ([Cu(hmt)(tfbdc)(H_2O)] (where hmt = hexamethylenetetramine, and tfbdc = 2,3,5,6-tetrafluoroterephthalate; Cu-LCP) using a solvothermal method. The Cu-LCP showed higher specific capacitance of 1274 F/g at 1 A/g in a 1 M lithium hydroxide (LiOH)-based system. The Cu-LCP also retained about 88% specific capacitance after 2000 cycles. This may be ascribed to the structural surface properties of the prepared Cu-MOF. The SEM results showed that Cu-MOF comprised of a nanorods-like surface, which may provide better electro transportation. It was believed that the relatively small diameter of the synthesized Cu-LCP nanorods may reduce the ion diffusion length and increase the number of active sites available for electrochemical reactions. Liu et al. [65] proposed the preparation of a conjugated copper catecholate linker (dibenzo-[g,p]chrysene-2,3,6,7,10,11,14,15-octaol (8OH-DBC)=Cu-DBC)-based Cu-MOF using a D_2-symmetric redox-active ligand in a copper bis(dihydroxy) coordination geometry. The prepared Cu-DBC has decent conductive nature and can be used as an electrocatalyst for energy storage applications. The Cu-DBC was further used as an electrode material, and the obtained results demonstrated a specific capacitance of 479 F/g at 0.2 A/g. The authors also reported an acceptable retention-specific capacitance of 80% after 2000 cycles. Hou et al. [66] proposed the fabrication of intrinsically conductive Cu-MOF nanowire arrays on the self-supported poly-pyrrole membrane for the development of binder-free flexible supercapacitors. The Cu-MOF nanowire arrays may afford decent conductivity with enough active surface area for the accessibility of the liquid electrolyte, while the poly-pyrrole membrane may provide acceptable mechanical flexibility, extra capacitance, and improved charge transfer skeleton. The authors achieved the specific capacitance of 252.1 mF/cm^2. This research work not only introduces a durable and flexible electrode for supercapacitors which can operate across a wide temperature range but also provides important insights for designing Cu-MOF-based hybrid materials for the construction of supercapacitors. The development of new MOFs are the most promising and important task to further enhance the performance of the supercapacitor electrodes. Ma et al. [67] proposed new Cu-based MOF materials for energy storage applications. The synthesized Cu-MOF (Cu-atrz-BDC) has decent specific surface area of 13.96 m^2/g and average pore size of 14.6 nm. The Cu-atrz-BDC electrode demonstrated a high specific capacitance of 5525 F/g at 1 A/g. The proposed electrode materials exhibited decent mechanical stability and cyclic stability. Ramachandran et al. [68] reported the synthesis of copper benzene@1, 3, 5-tricarboxylate (Cu@BTC) using a hydrothermal method. The authors found that synthesized Cu@BTC comprises two mixed phases ($Cu_3(BTC)_2 \cdot 3H_2O$ + $\{[Cu(BTC\text{-}H_2)_2 \cdot (H_2O_2)] \cdot 3H_2O\}$). The authors also proposed the mechanism for the formation of two phases in the prepared Cu@BTC sample as shown in Figure 3.

The CV responses of the different electrodes (Cu@BTC-120, Cu@BTC-90, Cu@BTC-150, and Cu@BTC-180) were examined (Figure 4a). It was observed that the Cu@BTC-120-based electrode has better CV response compared to the other electrodes. The Cu@BTC-120-based electrode also demonstrates relatively better capacitance (Figure 4b). The GCD curves of the Cu@BTC-90, Cu@BTC-120, Cu@BTC-150, and Cu@BTC-180 electrodes were also obtained as shown in Figure 4c. The specific capacitance versus current density of the different electrodes are shown in Figure 4d. The Cu@BTC-120 electrode demonstrated a significant improvement in specific capacitance compared to other Cu@BTC electrodes, achieving a maximum specific capacitance of 228 F/g at a current density of 1.5 A/g.

The Cu@BTC-120 electrode also demonstrates excellent long-term stability, retaining approximately 89.7% of its capacitance after 3000 cycles. The improved performance may be ascribed to the efficient transport of electrolyte ions through the substantial void space between octahedral micro-crystals ($Cu_3(BTC)_2 \cdot 3H_2O$) and micro-rods ($\{[Cu(BTC\text{-}H_2)_2 \cdot (H_2O_2)] \cdot 3H_2O\}$) in the Cu@BTC-120 structure. In other studies, Shaikh and his team introduced Cu-MOF-based electrode materials for energy storage applications [69–71]. These materials demonstrated reasonable specific capacitance and power density. While the performance of the Cu-MOF-based materials was satisfactory, the authors incorporated supporting materials like carbon nanotubes, copper oxide, and graphene. Therefore, it is suggested that the improved performance may result from synergistic interactions between the MOF and these supporting materials. Liu et al. [72] synthesized a layered Co-based MOF ($\{[Co(Hmt)(tfbdc)(H_2O)_2] \cdot (H_2O)_2\}_n$, Co-LMOF, and Hmt = hexamethylenetetramine; H_2tfbdc = 2,3,5,6-tetrafluoroterephthalic acid) and its (Co-LMOF) physicochemical properties were characterized by using SEM, XRD, and FTIR techniques. The Co-LMOF has particles in nanometer diameter within the range of 35 to 250 nm. The Co-LMOF electrode demonstrated a specific capacitance of 2474 F/g at 1 A/g and retained about 94.3% of its initial capacitance after 2000 cycles. The outstanding electrochemical performance of the proposed electrode material may be attributed to the intrinsic properties of Co−LMOF, ample space for electrolyte storage and diffusion, and the nanoscale size of the particles. Ramachandran et al. [73] described the formation of Co-MOF in different solvents/mixtures by employing solvothermal method. The formation of Co-MOFs under different conditions is described in Figure 5. The Co-MOF (prepared in DMF/ethanol) was further used for supercapacitor application and obtained results showing the specific capacitance of 958.1 F/g at 2 A/g. The reported electrode also retained around 92% capacitance after 3000 cycles.

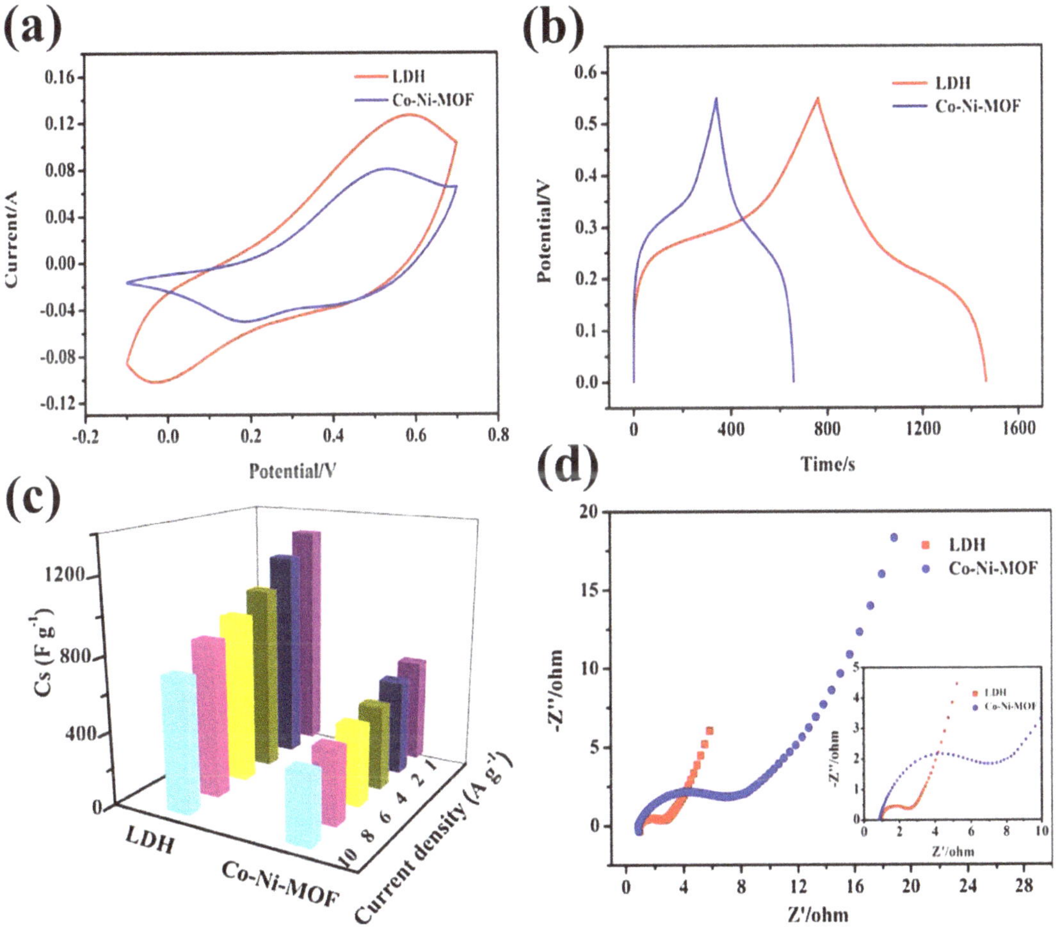

Figure 2. (**a**) CVs, (**b**) GCDs, (**c**) specific capacitance, and (**d**) EIS curves of Co-Ni-MOF and LDH. Reproduced with permission [62].

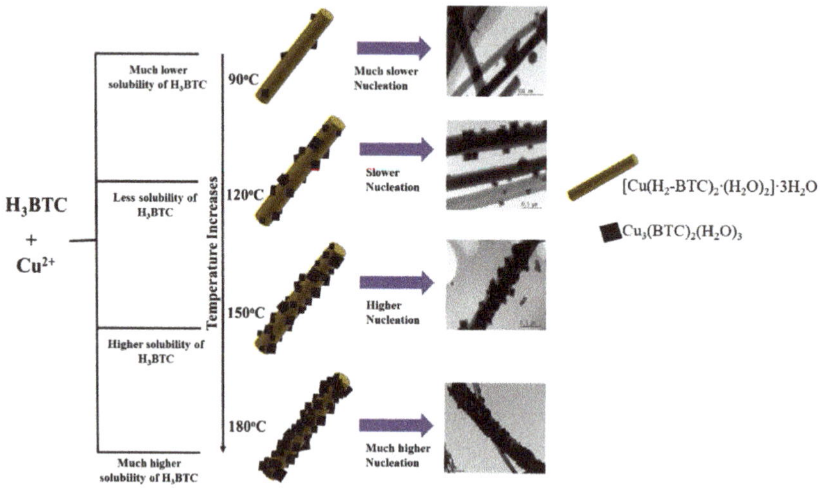

Figure 3. Schematic description for the synthesis of Cu-MOF. Reproduced with permission [68].

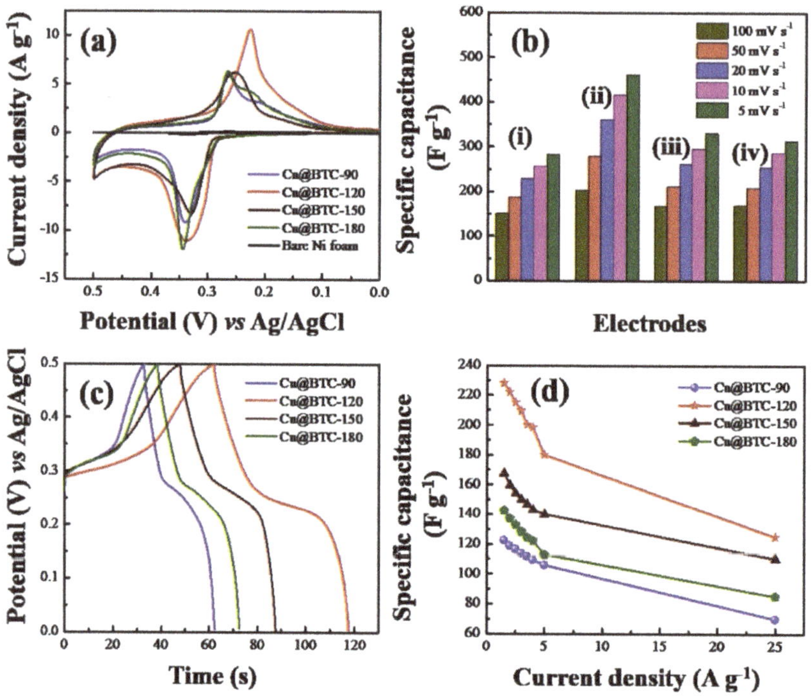

Figure 4. (**a**) CV responses and (**b**) specific capacitance of five different electrodes (i–iv) at different applied scan rates (5–100 mV/s). (**c**) GCD curves at 2 A/g and (**d**) specific capacitance versus current density of different electrodes. Reproduced with permission [68].

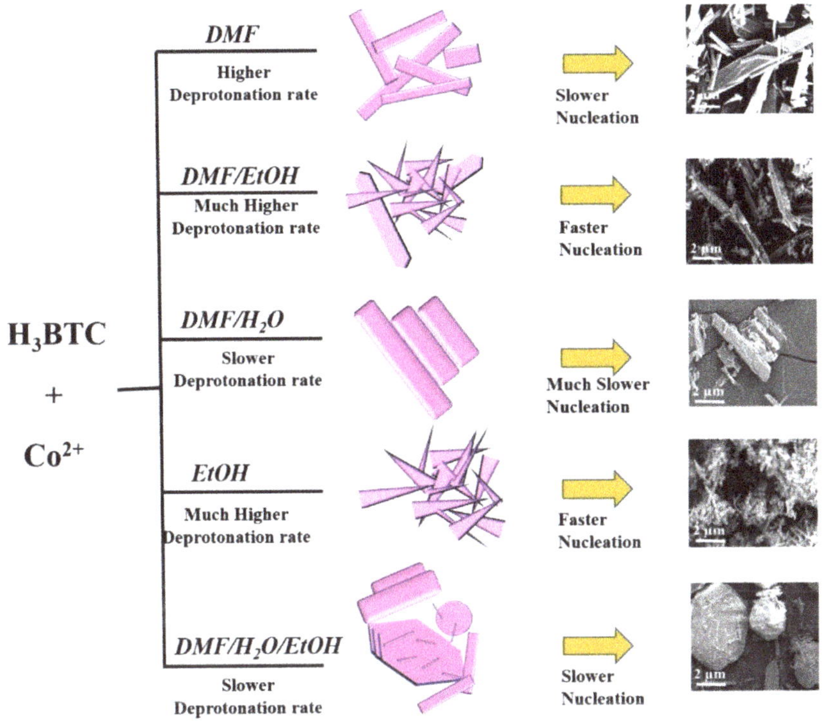

Figure 5. Schematic representation for the preparation of Co-MOFs under different conditions. Reproduced with permission [73].

The study proposed that Co-MOF with a large specific surface area and micro-pore volume may be synthesized by using a DMF/ethanol solvent mixture. It was also proposed that a hybrid structure featuring nano-needles and sharp-edged nanorods may provide more active sites for electrochemical reactions and result in a higher-charge storage capacity compared to other Co-MOFs synthesized in different solvents. In another study, Lee et al. [74] used three dicarboxylic acids with different molecular lengths as organic linkers to tune the surface area and pore-size of the Co-MOFs. It was observed that longer linkers generate larger pores and a greater surface area, while shorter linkers have the opposite effect. In addition, varying molecular sizes of the organic linkers may result in distinct surface architectures of the micro-structure films on the electrode surface. The Co-MOF-based electrodes displayed pseudo-capacitive properties, with the film featuring larger-pores, greater surface-area, and a continuously inter-connected leaflet-like micro-structure with fewer structural interfaces showing superior supercapacitive performance. Thus, of the three Co-MOFs studied, the one with the longer organic linker had larger pores, a greater surface area, and a continuously interconnected leaflet-like microstructure on the electrode surface with fewer structural interfaces, allowing for efficient charge transfer. This Co-MOF electrode exhibited the best specific capacitance of 179.2 F/g. Yang et al. [75] reported the synthesis of the layered Co-MOF, which has a nanosheet-like surface morphology, as revealed by SEM analysis. The authors adopted a simple oil bath synthesis method for the preparation of the Co-MOF. The authors adopted it as an electrode modifier for supercapacitor application and reported an interesting specific capacitance of 2564 F/g at 1 A/g, with acceptable stability of 3000 cycles with 95.9% retention cycles. The authors found that the presence of intrinsic properties, nanosheet-like surface, layered structure, and decent conductivity, of the Co-MOF are responsible for the improved electrochemical performance of the Co-MOF-based supercapacitor. The study suggested that Co-MOF may be a promising electrode material for supercapacitor applications. Lee et al. [76] investigated the pseudo-capacitor behavior of the Co-MOF-based film which demonstrated a specific capacitance of 206.76 F/g. The proposed electrode material exhibited decent stability of 1000 cycles with loss of only 1.5% of its initial performance. Thus, it is clear that Co-MOF are efficient electrode materials with excellent stability. Wang et al. [77] proposed a synthesis of new Co-ordination polymer based Co-MOF material which exhibited high surface area and high nitrogen content. The presence of nitrogen and high surface area may boost the electrochemical performance of the Co-MOF for energy storage applications. The authors used 3 molar potassium hydroxide (KOH) as electrolyte system and examined the supercapacitive properties of the Co-MOF. The Co-MOF-based electrode exhibited specific capacitance of 512 F/g at current density of 1 A/g. The proposed electrode material demonstrated the retention cycles of 97.4% of its initial performance after 40,000 cycles. The excellent performance of the electrode material may be attributed to the electrical properties, large specific surface area, and interconnected pores, which enhance side reactions with the electrolyte and facilitate rapid ion transport between the electrode and the electrolyte interface.

Punde et al. [78] have proposed the solvothermal and mechanochemical grinding methods for the preparation of cobalt–benzene-tricarboxylic-acid-based MOF (CoBTC MOF). The comparison between the performance of solvothermal- and mechanochemical-grinding-methods-based CoBTC MOF were examined. The mechanochemical-method-based electrode materials exhibited high specific surface area and improved porosity/diffusion process compared to the solvothermal method. The authors also integrated this MOF with graphene, and the proposed electrode showed specific capacitance of 608.2 F/g at current density of 0.25 A/g in 1 M KOH electrolyte system with capacitance retention of 94.9% after 2000 cycles. Wang et al. [79] reported the benign synthesis of a manganese (Mn)-based MOF for supercapacitor applications. The synthesized Mn-LMOF was further explored as an electrode material, and electrochemical investigations suggested that a high specific capacitance of 1098 F/g and 1178 F/g can be obtained in a 1 M KOH and 1 M LiOH electrolyte system, respectively, at the applied current density of 1 A/g. The authors found that Mn-LMOF has relatively good stability in the LiOH electrolyte compared to the KOH electrolyte and retained 92.6% specific capacitance of its initial performance after 2000 cycles. This acceptable stability and reasonable electrochemical performance of the Mn-LMOF was due to its layered structure and presence of nano-sized particles. Cheng et al. [80] also proposed a novel Mn-based MOF and evaluated its electrochemical activity for supercapacitor applications. In brief, a Mn-MOF (Mn(II)-porphyrins polycondensation polymer (THPP-PA-Mn)) was prepared via esterification of meso-tetra(p-hydroxyphenyl) porphyrin (THPP) with phthalic anhydride (PA) followed by metallizing it with manganese acetate, as shown in Figure 6a. The authors synthesized Mn-MOF at different temperature conditions, and their structural surface properties were examined by using SEM and transmission electron microscopy (TEM). The SEM images of the Mn-MOF 100, (d and e) Mn-MOF 120, (f and g) Mn-MOF 140, and (h and i) Mn-MOF 160 are shown in Figure 6b–i. The TEM and high-resolution TEM images of the Mn-MOF 140 are shown in Figure 6j and 6k,l, respectively. Furthermore, the prepared Mn-MOF was coated on the nickel foam, and its electrochemical performance was determined by using CV and GCD. The observations showed that a specific capacitance of 90.9 F/g at 2.5 A/g with a decay of 13.7% in the specific capacitance after 3000 cycles, which also suggested its good long-term cyclic stability.

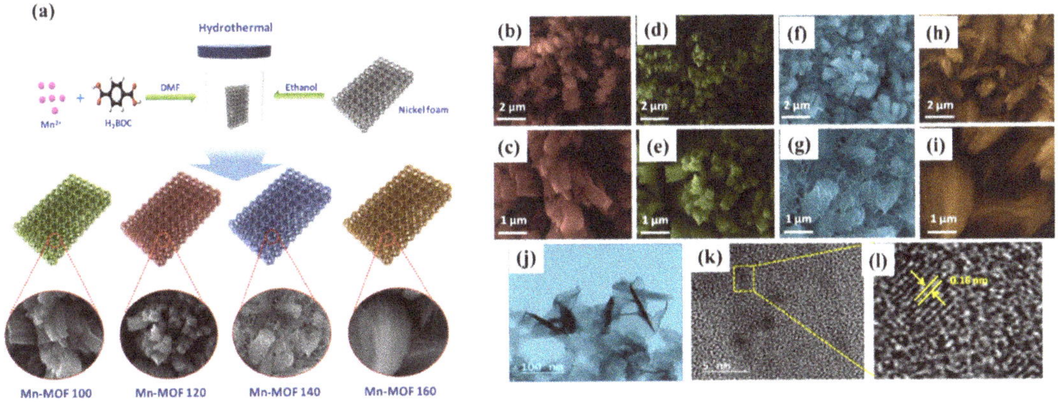

Figure 6. (**a**) Schematic graph for the synthesis of Mn-MOF. SEM images of Mn-MOF synthesized under different temperature conditions: (**b**,**c**) Mn-MOF 100, (**d**,**e**) Mn-MOF 120, (**f**,**g**) Mn-MOF 140, and (**h**,**i**) Mn-MOF 160. TEM image (**j**) and HRTEM images (**k**,**l**) of Mn-MOF 140. Reproduced with permission [81].

Shinde et al. [81] reported the hydrothermal synthesis of layered manganese-1,4-benzenedicarboxylic acid-based MOFs [Mn(BDC).nDMF]n (Mn-MOF) for energy storage applications. The Mn-MOF-based electrode exhibited the specific capacitance of 567.5 F/g at 1 A/g. The proposed electrode also retained its initial performance up to 81.80% after 10,000 cycles, which suggested its excellent stability for long-term energy storage applications. Sundriyal et al. [82] synthesized layered manganese-1,4-benzene dicarboxylate (Mn-BDC) (Mn-MOF) using simple strategies and used it as electrode material. The proposed electrode demonstrated specific capacitance of 1590 F/g at 3 A/g. The Mn-MOF-based electrode was also able to retain approximately 82% of its initial specific capacitance after 3000 consecutive GCD cycles. Gao et al. [83] synthesized Ni-MOF using a simple one-step hydrothermal method, with a modified mixed solution of DMF and water instead of pure DMF. The physicochemical characterization revealed that the Ni-MOF forms a loosely stacked layer-cuboid structure with abundant mesopores. Thus, it was believed that it may enhance charge transfer and ion transport. The authors used Ni-MOF as electrode material and obtained a specific capacitance of 804 F/g at 1 A/g with decent stability of 5000 cycles. Liu et al. [84] also explored Ni-MOF as suitable electrode material for energy storage applications. The authors recorded GCD graphs of the Ni-MOF (0 mL), Ni-MOF (0.5 mL), Ni-MOF (1 mL), and Ni-MOF (1.5 mL) at different current densities, as shown in Figure 7a, 7b, 7c, and 7d, respectively.

The Ni-MOF (1 mL); Ni-MOF (0.5 mL); Ni-MOF (0 mL); and Ni-MOF (1.5 mL) exhibited the specific capacitance of 2567.23 F/g; 2291.1 F/g; 1581.1 F/g; and 58.2651 F/g, respectively. The obtained showed that Ni-MOF (1 mL) demonstrated a high specific capacitance of 2567.23 F/g, which was higher than other materials. Sheberla et al. [85] proposed a Ni-based MOF (Ni_3(2,3,6,7,10,11-hexaiminotriphenylene)$_2$ ($Ni_3(HITP)_2$) high electrical conductivity that can serve as the sole electrode material in supercapacitor applications. The proposed Ni-MOF can be used as a binder-free electrode material, and electrochemical investigations revealed that Ni-MOF-based electrodes can bear stability up to 10,000 cycles. Zhang et al. [86] designed and synthesized Ni-based MOF (Ni-BTC (BTC = trimesic acid). The schematic diagram of the synthesis of Ni-MOF has been presented in Figure 8a. The surface morphological features of the Ni-BTC and Ni-BTC/IPA-3 were checked by using SEM, TEM, HRTEM, and SAED patterns, as shown in Figure 8b–j. The SEM and TEM images showed that Ni-BTC/IPA-3 consists of a nanosheets-like surface morphology.

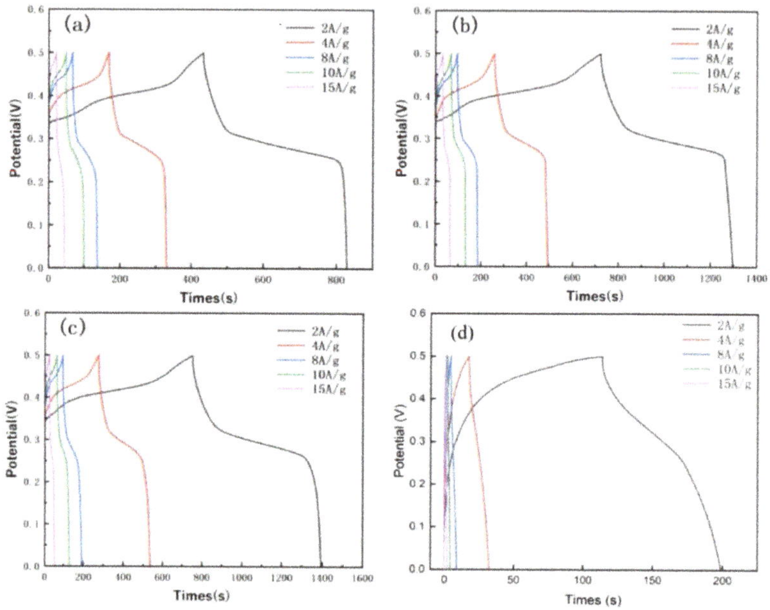

Figure 7. GCD graphs of the (**a**) Ni-MOF (0 mL), (**b**) Ni-MOF (0.5 mL), (**c**) Ni-MOF (1 mL), and (**d**) Ni-MOF (1.5 mL). Reproduced with permission [84].

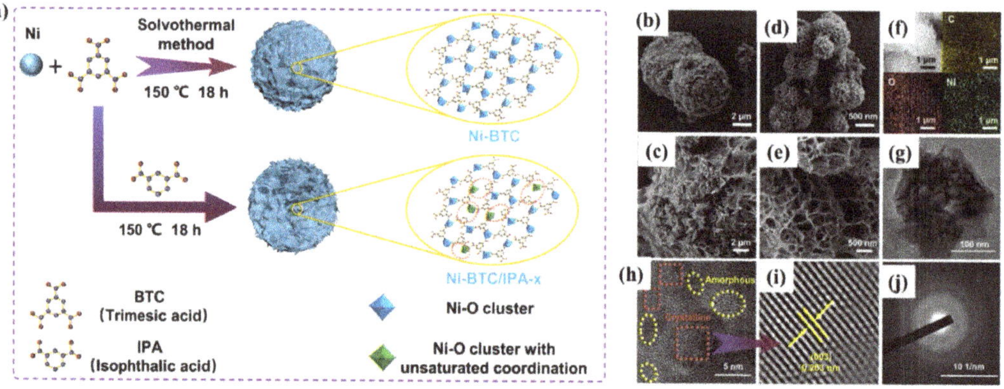

Figure 8. (**a**) Schematic diagram shows the synthesis of Ni-MOF. (**b**) SEM pictures of (**b**,**c**) Ni-BTC and (**d**,**e**) Ni-BTC/IPA-3. (**f**) EDX mapping for C, O and Ni elements, (**g**) TEM, (**h**) HRTEM image, (**i**) crystal faces intensity profile, and (**j**) SAED patterns for Ni-BTC/IPA-3. Reproduced with permission [86].

The electrical conductivity and active sites of the Ni-BTC were improved by embedding isophthalic acid, i.e., IPA as missing linkers, resulting in defective Ni-BTC/IPA$_{-x}$. The findings suggest that defects may play a key role in tuning the electronic structure of the metal center, as well as the pore structures and specific surface areas of MOFs, which helps to enhance the performance of supercapacitors. The CV curves of the Ni-BTC/IPA-3 and AC electrode at 50 mV/s are shown in Figure 9a. It can be seen that Ni-BTC/IPA-3 has better electrochemical activity.

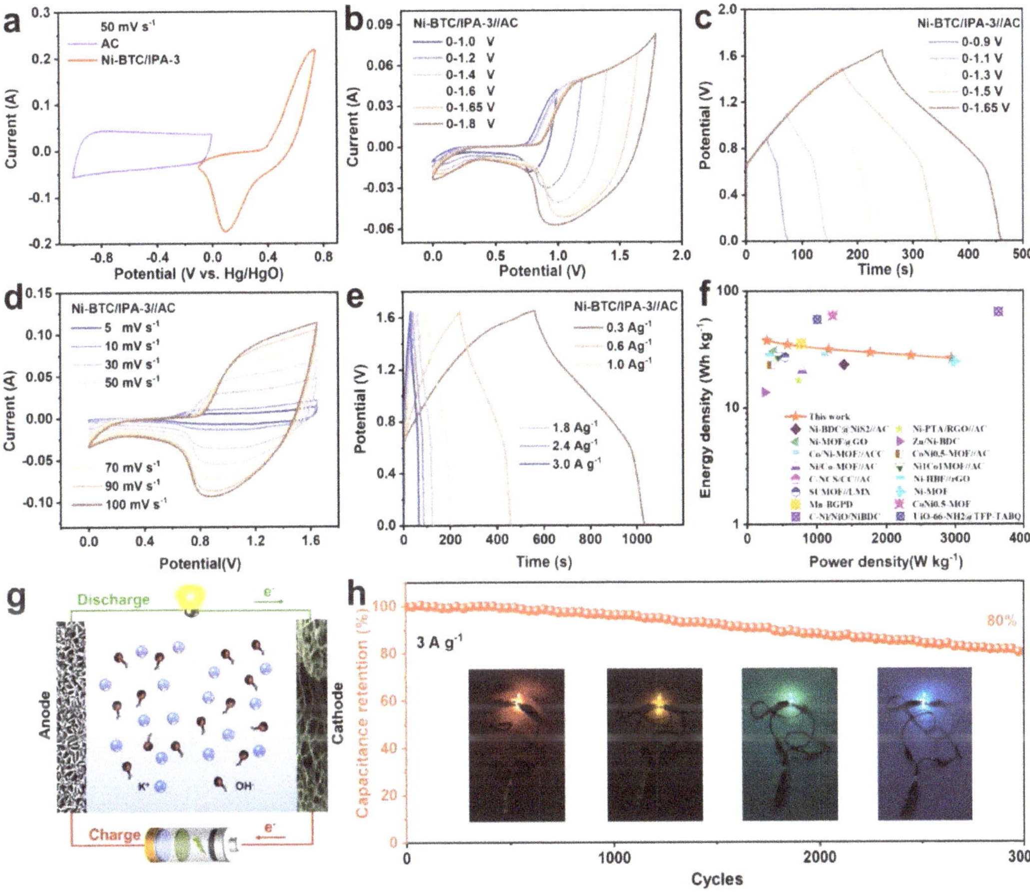

Figure 9. (**a**) CV curves of Ni-BTC/IPA-3 and AC electrode at 50 mV/s. (**b**) CVs at 50 mV/s. (**c**) GCD results at 0.6 A/g at different potential windows, (**d**) CV curves, (**e**) GCD graphs, (**f**) Ragone plots, (**g**) schematic structure of working principle, and (**h**) cyclic performance at 3 A/g. Reproduced with permission [86].

The CVs of the Ni-BTC/IPA-3//AC are shown in Figure 9b, whereas the GCD graphs are displayed in Figure 9c under different potential windows. The Ni-BTC/IPA-3//AC exhibited decent electrochemical properties for energy storage devices. The CVs of the Ni-BTC/IPA-3//AC at different applied scan rates are shown in Figure 9d, and the GCD curves are summarized in Figure 9d. The Ragone plot has been presented in Figure 9f. The schematic has been described in Figure 9g. The cyclic performance of the proposed electrode material has been presented in Figure 9g. The Ni-based MOF (Ni-BTC/IPA-3) exhibited specific capacitance of 1209 F/g at 0.5 A/g with a rate performance of 70.8%. Yang et al. [87] also synthesized a layered Ni-based MOF material and characterized it using various sophisticated techniques. The proposed Ni-MOF-based electrode demonstrated decent specific capacitance of 1127 and 668 F/g at 0.5 and 10 A/g, respectively. The proposed electrode also retained good stability of 3000 cycles and held more than 90% capacitance of its initial performance after 3000 cycles. The excellent electrochemical performance of the proposed electrode material was attributed to the intrinsic properties of the Ni-based MOF, particularly its layered structure and favorable exposed facets. Kang et al. [88] reported a new electrode material for supercapacitor applications. The Ni-based

MOF ($Ni_3(btc)_2.12H_2O$) was synthesized by using simple strategies. The authors used KOH as an electrolyte system, and the proposed electrode material (Ni-MOF) demonstrated specific capacitance of 726 F/g and exhibited excellent stability of 1000 cycles with retention cycles of 94.6%. Xu et al. [89] also reported a novel one-dimensional (1D) Ni-based MOF ($Ni(HOC_6H_4COO)_{1.48}(OH)_{0.52}\_1.1H_2O$) using a hydrothermal synthetic method. Ni-MOF has a rods-like surface morphology, which may beneficial for supercapacitor application. The synthesized Ni-MOF nanorods have an average length of 900 nm and diameter of 60 nm. The authors found that Ni-MOF-based electrode demonstrated specific capacitance of 1698, 1385, 1278, 1068, 998, and 838 F/g at 1, 2, 3, 5, 7, and 10 A/g, respectively. The excellent electrochemical properties of the Ni-MOF indicated that the synthesized Ni-based MOF nanorods are highly efficient electrode materials for supercapacitor application and hold significant promise for use in the development of high-performance electrochemical energy storage devices. The electrochemical performance of the reported MOFs for energy storage applications is summarized in Table 1.

Table 1. Performance of the pristine MOFs-based supercapacitors [59–89].

Electrode Material	Synthesis Method	Electrolyte	Specific Capacitance (F/g)	Current Density (A/g)	Refs.
Zn-MOF	Hydrothermal	3 M KOH	58.6	0.45	[59]
Zn-MOF	High temperature method	1 M H_2SO_4	203	-	[60]
Zn-MOF	Electro-deposition	3 M KOH	288	2	[61]
Co/Ni-MOF	Hydrothermal	2 M KOH	570	1	[62]
Cu-MOF	High temperature assisted stirring	1 M KOH	215	0.5	[63]
Cu-MOF	Solvothermal	1 M LiOH	1274	1	[64]
Cu-MOF	Hydrothermal	1 M KCl	479	0.2	[65]
Cu-MOF	Hydrothermal	3 M NaCl	463 mF/cm^2	1.25 mA/cm^2	[66]
Cu-MOF	Hydrothermal	1 M KOH	5525	1	[67]
Cu-MOF	Hydrothermal	3 M KOH	122	1.5	[68]
Cu-MOF	Stirring/slow evaporation	1 M Na_2SO_4	244.2	0.8	[69]
Cu-MOF	Layering method	1 M Na_2SO_4	685.33	1.6	[70]
Cu-MOF	Slow evaporation	1 M Na_2SO_4	380	1.6	[71]
Co-MOF	Mixing method	1 M KOH	2474	1	[72]
Co-MOF	Solvothermal	3 M KOH	958.1	2	[73]
Co-MOF	Solvothermal	0.5 M LiOH	179.2	-	[74]
Co-MOF	Stirring at high temperature	5 M KOH	2564	1	[75]
Co-MOF	Solvothermal	1 M LiOH	206.76	0.61	[76]
Co-MOF	Solvothermal	3 M KOH	512	1	[77]
Co-MOF	Simple grinding method	1 M KOH	608.2	0.25	[78]
Mn-MOF	Solvothermal	1 M LiOH	1098	1	[79]
Mn-MOF	Solvothermal	1 M KOH	109.3	0.25	[80]
Mn-MOF	Hydrothermal	2 M KOH	211.37	1	[81]
Mn-MOF	Solvothermal	1 M Na_2SO_4	1590	3	[82]
Ni-MOF	Hydrothermal	2 M KOH	804	1	[83]

Table 1. *Cont.*

Electrode Material	Synthesis Method	Electrolyte	Specific Capacitance (F/g)	Current Density (A/g)	Refs.
Ni-MOF	Hydrothermal	6 M KOH	2567.2	1	[84]
Ni-MOF	Hydrothermal	1 M $TEABF_4$/ACN	111	0.05	[85]
Ni-MOF	Solvothermal	3 MKOH	1209	0.5	[86]
Ni-MOF	Hydrothermal	6 M KOH	1127	0.5	[87]
Ni-MOF	Hydrothermal	KOH	726	0.25	[88]
Ni-MOF	Hydrothermal	6 M KOH	1036	1	[89]

2.2. *MOFs/MXenes-Based Supercapacitors*

It is a well-known factor that MOFs suffer due to their poor electrical conductivity, which really needs significant attention to further enhance the performance of the MOFs-based supercapacitors. It has been observed from the reported literature that integration of MOFs with conductive supporting materials may improve the electrical conductivity of the MOFs materials. In the past few years, MXenes materials with molecular formula of $M_{n+1}X_nT_x$ (where T_x represents the surface terminations like O, OH, F, and/or Cl that is bonded to the outer M layers) have received extensive interest of the scientific community. The MXenes materials, such as carbon nitride and vanadium carbide, etc., possess decent physicochemical properties, such as high electrical conductivity, and it would be of great significance to integrate MOFs with MXenes as hybrid materials for supercapacitor applications. In connection to this, Yang et al. [90] proposed the synthesis of Ni-MOF integrated V_2CT_x (Tx = –O, –OH and –F surface groups) MXenes composite on Ni foam by using temperature-controlled annealing process. The Ni-MOF/V_2CT_x-modified electrode was further used as supercapacitor electrode, which demonstrated specific capacitance of 1103.9 C/g at 1 A/g. The proposed electrode also has acceptable cyclic stability of 15,000 cycles. This stability may be ascribed to the presence of synergism between the Ni-MOF and V_2CT_x material. Zhu et al. [91] also reported the synthesis of a typical Ni-MOF (Ni-HHTP, HHTP = 2, 3, 6, 7, 10, 11-hexahydroxytriphenylene) and prepared its composite with MXene. The synthesized MXene@Ni-HHTP-2 was further explored as an electrode material which showed the specific capacitance of 416.6 F/g at 0.5 A/g. Xie et al. [92] reported that the synthesis of MOF derived Co-Fe oxide porous nanorods integrated with MXene and used it as a supercapacitor material. The proposed material exhibited decent performance in terms of capacitance and stability of 10,000 cycles. Shingte et al. [93] synthesized MOF-derived rhombohedral-shaped nickel ferrite nanoparticles (NFO NPs) into the $Ti_3C_2T_x$ MXene layers. The 8% NFO NPs-loaded MXene (MXene/NFO-8)-based electrode showed the specific capacitance of 660 F/g at current density of 1 A/g. Qu et al. [94] proposed the synthesis of Ni-MOF/$Ti_3C_2T_x$ using a sonochemical method. The authors recorded SEM, TEM, and HRTEM images of the synthesized Ni-MOF/$Ti_3C_2T_x$ to study the morphological features (Figure 10a–d). The authors observed that $Ti_3C_2T_x$ are uniformly dispersed on the surface of the Ni-MOF, which improved the electronic conductivity and inhibited the aggregation of Ni-MOF nanosheets. The GCD curves of the MXene, Ni-MOF, and Ni-MOF/MXene are displayed in Figure 10e. The Ni-MOF/$Ti_3C_2T_x$-based electrode showed the specific capacitance of 867.3 F/g at 1 A/g. Compared to the pristine Ni-MOF, the incorporation of $Ti_3C_2T_X$ improves the electrical conductivity of Ni-MOF nanosheets and effectively prevents their agglomeration, leading to enhanced electrochemical performance.

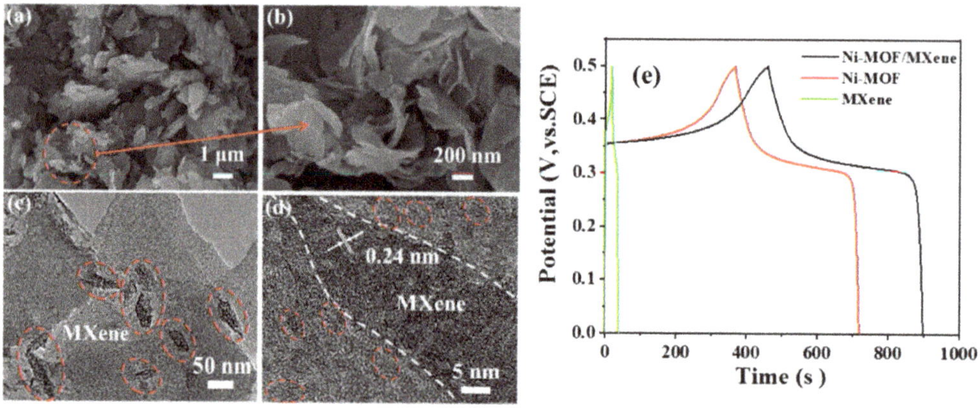

Figure 10. (**a**,**b**) SEM, (**c**) TEM, and (**d**) HRTEM image of Ni-MOF/MXene (red circle shows presence of NiO). (**e**) GCD curve of MXene, Ni-MOF, and Ni-MOF/MXene. Reproduced with permission [94].

Hassan et al. [95] prepared a novel MOF-5 using a simple hydrothermal approach (Figure 11a), where V_2CT_X MXene was subjected to standard HF etching, followed by the synthesis of a MOF-5/V_2CT_X composite in a one-step hydrothermal process. The MOF-5 and V_2CT_X, used as template or precursor sources, have some limitations, such as inconsistent pore size distribution in the carbonized product, durability issues, poor performance, and challenges in cost-effective synthesis. Therefore, authors used V_2CT_X MXene along with affordable conductive linkers for synthesizing MOF-5/V_2CT_X composites, which may offer an alternative approach to the preparation of versatile electrode materials. The synthesized MOF-5/V_2CT_X composite exhibited high porosity, numerous active sites, and uniform dispersion of the MOF-5 porous structure on the V_2CT_X MXene, which makes it a promising electrode materials for energy storage applications.

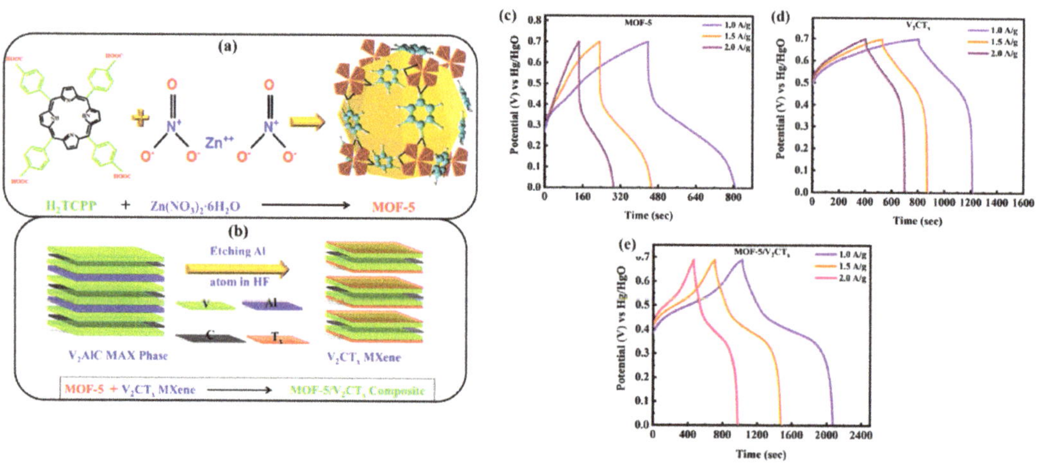

Figure 11. Schematic graph shows the synthesis of (**a**) MOF-5 and (**b**) V_2CT_X MXene and MOF-5/V_2CT_X MXene. GCD curves of the (**c**) MOF-5, (**d**) V_2CT_X MXene, and (**e**) MOF-5/V_2CT_X-MXene-based electrodes for supercapacitors. Reproduced with permission [95].

The CD curves of the MOF-5-, V_2CT_X MXene-, and MOF-5/V_2CT_X-MXene-based electrodes are shown in Figure 11c, 11d, and 11e, respectively. The synthesized MOF-5/V_2CT_X-based electrode demonstrated specific capacitance of 961 C/g in the three-electrode system and 235 C/g in the two-electrode system. Liu et al. [96] synthesized three-dimensional (3D) metal carbides, nitrides, and carbonitrides/MOFs composites ($Ti_3C_2T_x$/Cu-BTC, $Ti_3C_2T_x$/Fe,Co-PBA, $Ti_3C_2T_x$/ZIF-8, and $Ti_3C_2T_x$/ZIF-67) to improve the performance and stability of the supercapacitors. Further, a $Ti_3C_2T_x$/ZIF-67/CoV_2O_6 composite was prepared, which demonstrated specific capacitance of 285.5 F/g. In another work by Li et al. [97], MXene was directly utilized as a titanium (Ti) source to coordinate with an organic ligand, which formed Ti-MOF sheets. To enhance the electrochemical performance, mesopores were introduced during the preparation of 2D Ti-MOF, creating hierarchical porous Ti-MOF@$Ti_3C_2T_x$ hybrids. The Ti-MOF@$Ti_3C_2T_x$ exhibited decent electrochemical performance for energy storage applications. Wang et al. [98] adopted hydrothermal method for the synthesis of bimetallic Ni-CO-MOF on Ni foam and integrated with MXene as shown in Figure 12a. Further, MXene-Ni-Co@NiCo-MOF/NF (M-NC@NCM/NF) nanosheets was obtained using electrodeposition method. Due to the redistribution of metal cations between the MXene layers, nanosheets produced through electrodeposition in an electrolyte containing MXene display a distinctive hexagonal nanosheet morphology. The CV and GCD curves of the different electrodes are shown in Figure 12b and 12c, respectively. This unique structure improves electrochemical performance, maintaining a capacity retention rate of 75.3% after 5000 cycles, with a specific capacity reaching 2137.5 F g^{-1} at a current density of 1 A g^{-1}.

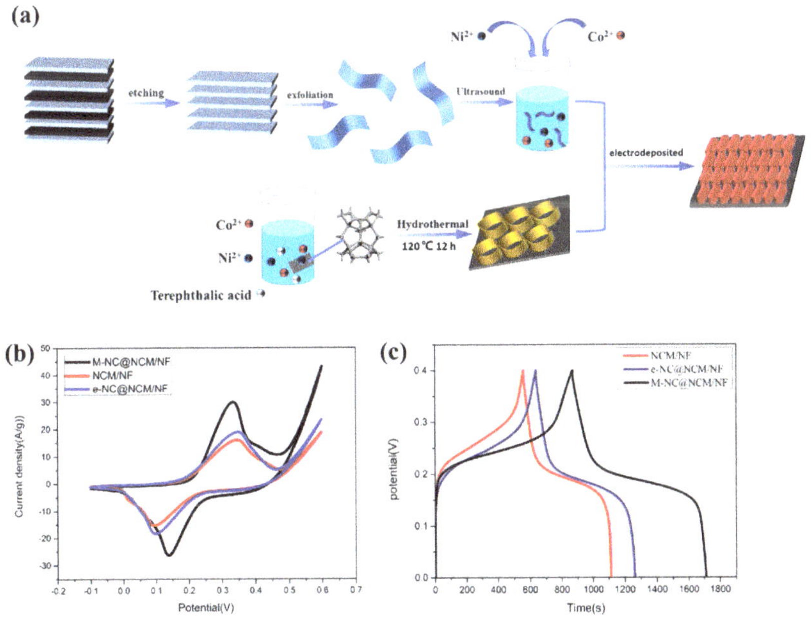

Figure 12. (**a**) Schematic graph shows the synthesis of M-NC@NCM/NF electrode. (**b**) CV and (**c**) GCD curves of the different electrodes. Reproduced with permission [98].

Shavita et al. [99] reported the synthesis of $Ti_3C_2T_x$/Ni-MOF composite using solvothermal method. It was believed that interaction between amino groups and MXene may facilitate the electron transfer pathway and enhance the electrochemical performance of the prepared $Ti_3C_2T_x$/Ni-MOF composite. The synthesized $Ti_3C_2T_x$/Ni-MOF composite showed an increased surface area, promoting efficient electrolyte ion diffusion, along with enhanced porosity, numerous redox-active sites, and improved electrical conductivity. The $Ti_3C_2T_x$/Ni-MOF-composite-based electrode demonstrated a specific capacitance of 536 F/g, which was higher than that of the pristine Ni-MOF (114 F/g) or $Ti_3C_2T_x$ (224 F/g). The enlarged surface area and expanded interlayer spacing between MXene sheets contributed to a higher capacitance of the $Ti_3C_2T_x$/Ni-MOF. Hassan et al. [100] also reported the synthesis of an Ag-MOF@V_2CT_x composite which showed a specific capacitance of 2180 C/g at 2 A/g. Ali et al. [101] used the hydrothermal method for the preparation of a NiMn-MOF/MXene composite and characterized it using various sophisticated techniques.

The synthesis of NiMn-MOF and hydrothermal treatment for MXene are shown in Figure 13a and 13b, respectively. The GCD graphs of the NiMn-MOF exhibited the high specific capacitance in a 1 M KOH electrolyte compared to the 2 M or 3 M KOH system as shown in Figure 13c. The GCD graphs of the NiMn-MOF at different applied current densities are summarized in Figure 13d. It can be observed that NiMn-MOF has high electrochemical performance at a current density of 1.3 A/g. The MXene-based electrode also exhibited improved specific capacitance compared to the NiMn-MOF at current density of 1.3 A/g (Figure 13e). Further, enhancement in the specific capacitance was also observed for the prepared NiMn-MOF/MXene composite compared to the pristine NiMn-MOF or MXene (Figure 13f–h). The proposed NiMn-MOF/MXene electrode material showed specific capacitance of 215.5 C/g with retention capacity of 90% after 5000 cycles. Ji et al. [102] investigated the synthesis of a range of bimetallic M_xN_y-MOF materials (such as Ni_1Co_1-MOF, Ni_1Co_3-MOF, Ni_3Co_1-MOF, Cu_1Co_1-MOF, and Ni_1Cu_1-MOF) and their composites with MXenes. The synthesis of the electrode materials are summarized in Figure 14a. The surface morphology of the synthesized Ni_1Co_1-MOF and (c) Ni_1Co_1-MOF@MXene. (d) EDX and (e) BET-BJH curves of Ni_1Co_1-MOF@MXene. SEM image of (f) Ni_1Co_3-MOF, (j) Ni_1Co_3-MOF@MXene, (g) Ni_3Co_1-MOF, (k) Ni_3Co_1-MOF@MXene, (h) Cu_1Co_1-MOF, (l) Cu_1Co_1-MOF@MXene, (i) Ni_1Cu_1-MOF, and (m) Ni_1Cu_1-MOF@MXene were checked by employing SEM. The obtained SEM, EDX, and BET results are displayed in Figure 14b–m. The Ni_1Cu_1-MOF/MXenes demonstrated a high specific capacitance of 1493.6 F/g at current density of 1 A/g with 52.7 retention after 5000 cycles (Figure 14n). Wei et al. [103] used hard template method for the preparation of MXene@NiCo-MOF and used it as an electrode material, which demonstrated a high specific capacitance of 2078.1 F/g at current density of 1 A/g.

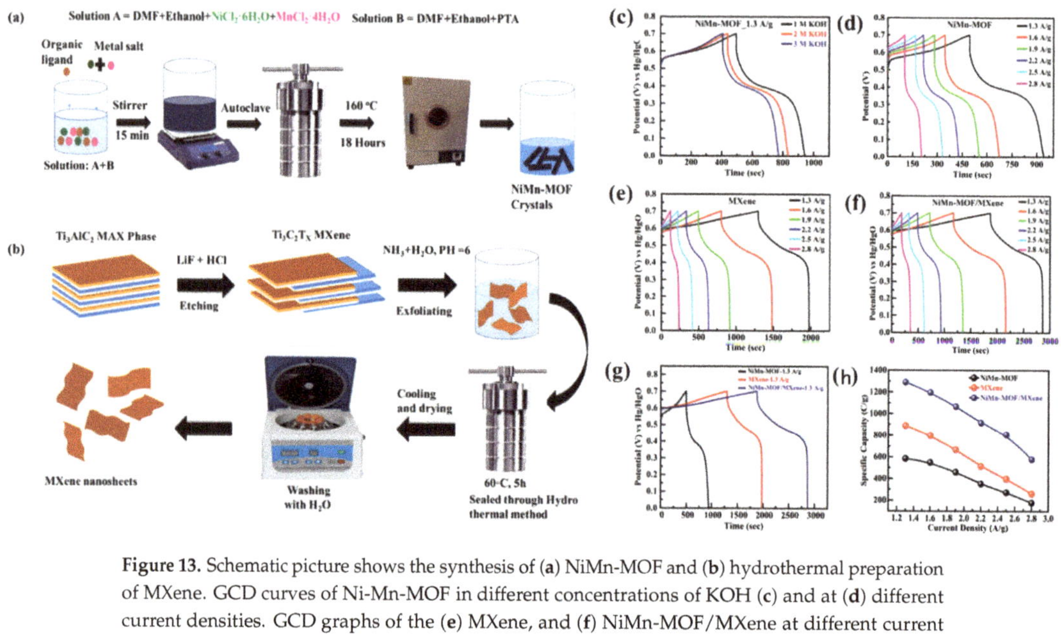

Figure 13. Schematic picture shows the synthesis of (**a**) NiMn-MOF and (**b**) hydrothermal preparation of MXene. GCD curves of Ni-Mn-MOF in different concentrations of KOH (**c**) and at (**d**) different current densities. GCD graphs of the (**e**) MXene, and (**f**) NiMn-MOF/MXene at different current density. GCD curves of the (**g**) NiMn-MOF, MXene, and NiMn-MOF/MXene. (**h**) Specific capacity of NiMn-MOF, MXene, and NiMn-MOF/MXene. Reproduced with permission [101].

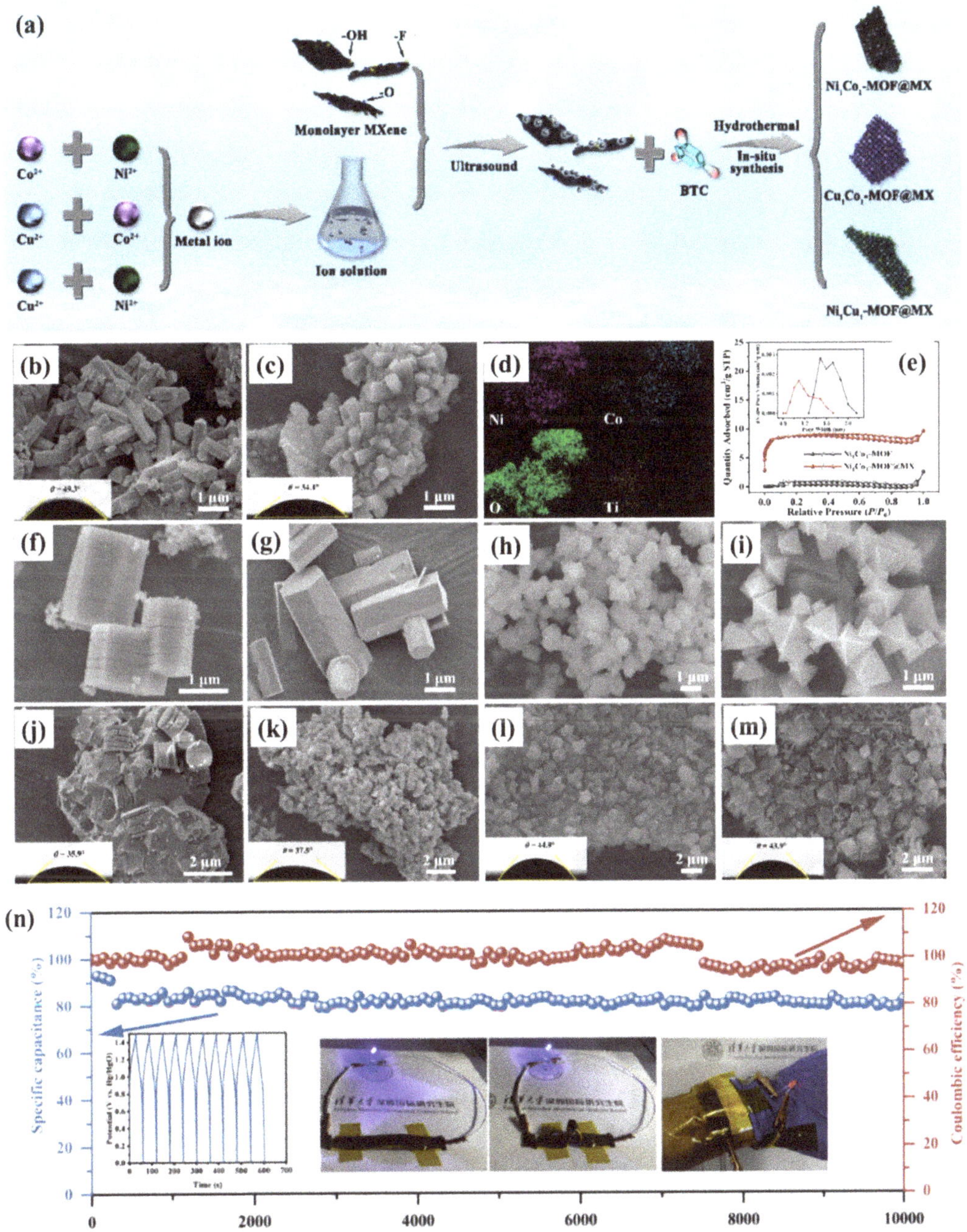

Figure 14. (a) Synthetic illustration for the preparation of electrode materials. SEM image and contact angle of (b) Ni_1Co_1-MOF and (c) Ni_1Co_1-MOF@MXene. (d) EDX and (e) BET-BJH curves of Ni_1Co_1-MOF@MXene. SEM image of (f) Ni_1Co_3-MOF, (j) Ni_1Co_3-MOF@MXene, (g) Ni_3Co_1-MOF, (k) Ni_3Co_1-MOF@MXene, (h) Cu_1Co_1-MOF, (l) Cu_1Co_1-MOF@MXene, (i) Ni_1Cu_1-MOF, and (m) Ni_1Cu_1-MOF@MXene. (n) Testing of Ni_1Co_1-MOF@MX//AC all-solid-state flexible supercapacitors. Reproduced with permission [102].

Zheng et al. [104] reported the synthesis of highly conductive and cyclically stable Ti_3C_2 MXene@pillared-layer [Ni(thiophene-2,5-dicarboxylate)(4,40-bipyridine)]$_n$ MOF composites (MXene@Ni-MOF). The synthesized material has good capacitive properties and reasonable high specific capacitance of 979 F/g was obtained at current density of 0.5 A/g. Chen et al. [105] reported a synthesis of conductive bismuth-catecholate metal−organic frameworks (Bi(HHTP)) with one-dimensional (1D) channels grown on the surfaces of $Ti_3C_2T_x$ nanosheets ($Ti_3C_2T_x$/Bi(HHTP)). The authors found that conductive Bi(HHTP) functions both as spacers to prevent the self-stacking of $Ti_3C_2T_x$ nanosheets and as an active component, delivering battery-type capacitance. Simultaneously, the $Ti_3C_2T_x$ serve as a framework for the conductive Bi(HHTP), further boosting the overall specific capacitance of the $Ti_3C_2T_x$/Bi(HHTP) composite. The specific capacitance of 326 F/g was obtained at current density of 0.5 A/g. Ramachandran et al. [106] described the synthesis of Co-MOF spheres and Co-MOF/$Ti_3C_2T_x$ on Ni foam which can be used as an innovative binder-free electrode material for supercapacitors. The Co-MOF@Ni and Co-MOF/$Ti_3C_2T_x$@Ni electrodes achieved maximum capacitance of 2872.5 F/g and 3741 F/g, respectively. It was expected that voids within the Co-MOF structure combined with the excellent conductivity of the layered $Ti_3C_2T_x$ increased active sites and facilitated rapid ion transport. Guo et al. [107] reported that zeolite-imidazole frameworks (ZIFs) may be the promising materials nut has some limitations such as agglomeration, poor conductivity and low stability. Thus, authors used MXene as substrate, and a Ni-doped ZIF-67 (NiCo-ZIF-7) was anchored. The MXene/NiCo-ZIF-67-based electrode exhibited the specific capacitance of 460 F/g. Yang et al. [108] used Co-BDC/$Ti_3C_2T_x$ as an electrode material and synthesized it by using thermal treatment. The Co-BDC/$Ti_3C_2T_x$ based supercapacitor electrode showed specific capacitance of 217.3 C/g. The strong interactions between Co-BDC and $Ti_3C_2T_x$ are responsible for the better specific capacitance of the proposed electrode. Yang et al. [109] reported the preparation of honeycomb-like MXene@$CoSe_2$/Ni_3Se_4, and the obtained results showed good electrochemical performance with specific capacity of 283 mAh/g at current density of 1 A/g. The proposed electrode also has excellent 80% retention of its initial specific capacitance after 5000 cycles. A self-powered smart system was developed by Xiang et al. [110] by incorporating flexible solid-state Zn-Co MOFs@MXene supercapacitors and polyacrylamide-$BaTiO_3$/NaCl (PAM-BTO/NaCl) organic ionic hydrogel sensors. The authors proposed that Zn-Co MOFs@MXene may be a promising electrode material for energy storage devices. Yue et al. [111] successfully prepared Ni/Co-MOF/thin-layer aminated $Ti_3C_2T_x$ based electrode material (Ni/Co-MOF@TCT-NH_2) using simple strategies. The Ni/Co-MOF@TCT-NH_2 showed a high specific capacitance of 1924 F/g at low current density of 0.5 A/g with excellent cyclic stability. Kshetri et al. [112] designed and synthesized Co-MOF on electroconductive MXene–carbon nanofiber mat (MX-CNF). The Co-MOF@MX-CNF was further used as starting material for the preparation of capacitive-type Co-PC@MX-CNF and battery-type MnO_2@Co_3O_4-PC@MX-CNF functional multi-component electrodes. The proposed work reported a specific capacitance of 426.7 F/g at a current density of 1 A/g. The performance of the reported supercapacitors are summarized in Table 2.

Table 2. Performance of the pristine MOFs/MXenes-based supercapacitors.

Electrode Material	Synthesis Method	Electrolyte	Specific Capacitance (F/g)	Current Density (A/g)	Refs.
Ni-BDC/V_2CT_x/NF-300	Solvothermal	1 M KOH	1103.9 C/g	1	[90]
MXene@Ni-HHTP	Sonochemical	3 M KOH	416.6	0.5	[91]
Co-Fe oxide/$Ti_3C_2T_x$	Hydrothermal	1 M LiCl	356.4 mF cm^2	-	[92]
MOF-NFO/MXene	Solvothermal	PVA/KOH	660	1	[93]

Table 2. *Cont.*

Electrode Material	Synthesis Method	Electrolyte	Specific Capacitance (F/g)	Current Density (A/g)	Refs.
Ni-MOF/MXene	Sonochemical	2 M KOH	867.3	1	[94]
MOF-5/V_2CT_x	Hydrothermal	3 M KOH	961 C/g	2	[95]
$Ti_3C_2T_x$/ZIF-67/CoV_2O_6	Versatile method	3 M KOH	285.5	1	[96]
HP-Ti-MOF@$Ti_3C_2T_X$	Soft template	1 M KOH	154	0.2	[97]
MXene@NiCo-MOF	Sonochemical	PVA/KOH	2078.1	1	[98]
$Ti_3C_2T_x$/Ni-MOF	Solvothermal	1 M H_2SO_4	139.4	1	[99]
Ag-MOF@V_2CT_x	Hydrothermal	1 M KOH	3114	2	[100]
NiMn-MOF@ MXene	Ultrasonication	1 M H_2SO_4	1289.7 C/g	1.3	[101]
Ni_1Co_{1-}MOF@MXene	Ultrasonication	PVA/KOH	1493.6	1	[102]
MXene@NiCo-MOF	Hydrothermal	PVA/KOH	2078.1	1	[103]
MXene@Ni-MOF	Stirring	3 M KOH	979	0.5	[104]
$Ti_3C_2T_x$Bi(HHTP)	Hydrothermal	3 M KOH	326	0.5	[105]
Co-MOF/$Ti_3C_2T_x$@NF	Hydrothermal	3 M KOH	3741	3 mA/cm^2	[106]
MXene/NiCoZDH	Hydrothermal	6 M KOH	877	1	[107]
Co-BDC/$Ti_3C_2T_x$-300	Solvothermal	1 M KOH	1453 C/g	2 A/m^2	[108]

3. MOFs and MOFs/MXenes in Electrochemical Sensors

It is well understood that MOFs and their hybrid composite materials have shown excellent performance for electrochemical applications. In the previous years, MOFs and MOFs/MXenes are also widely used as efficient electrode material for electrochemical sensing applications.

Zhao et al. [113] reported a novel 2D conductive MOF-based material using 2, 3, 7, 8, 12, 13-hexahydroxyl truxene and copper ions as starting materials. The synthesized c-MOF was further used for electrochemical sensing of paraquat and demonstrated a good limit of detection of 4.1×10^{-8} M and linear range of 0.2 to 5 μM. Elashery et al. [114] synthesized 2D Cu-MOF using simple strategies and characterized by sophisticated techniques. The synthetic process for the preparation of 2D Cu-MOF and electrode preparation has been illustrated in Figure 15. The synthesized Cu-MOF was further adopted as electrode material for the sensing of Cu (II) ions and reported decent LOD of 1×10^{-8} M.

Wen et al. [115] reported the synthesis of 2D bimetallic Ni- and Zn-based MOF nanosheets (NiZn–MOF NSs) with adjustable Ni/Zn ratios which were further utilized as electrode materials for the development of a tyrosinase biosensor (Figure 16a). The synthesized NiZn–MOF was characterized by employing XRD and SEM methods. The XRD pattern exhibited the presence of a decent crystalline nature in the prepared NiZn–MOF whereas SEM results demonstrated that SEM investigations showed that prepared material has flower-like cluster structure. The electrochemical sensing ability of the proposed electrode material was checked by using CV and amperometry methods. The fabricated sensor demonstrated excellent selectivity, long term stability, and high recovery in real samples. The working mechanism for the sensing of tyrosinase has been described in Figure 16b.

Wang et al. [116] synthesized 2D Yb-MOFs for the sensing of picric acid and obtained a LOD of 81.3 nM with linear range of 0.1 to 1 μM. Shi et al. [117] designed and prepared a two-dimensional leaf-like framework-L-embedded electrochemically reduced graphene oxide (ERGO@ZIF-L) and adopted it as an electrode modifier for the electrochemical sensing of benomyl. The authors optimized various parameters, such as material volume, solution pH, and accumulation time, to improve the performance of the proposed electrode for the sensing of benomyl. The Nyquist curves of the different electrodes (GCE (a), ZIF-

L/GCE (b), ERGO/GCE (c), and ERGO@ZIF-L/GCE (d)) in 5 mM $K_3Fe(CN)_6/K_4Fe(CN)_6$) are shown in Figure 17a. Kit can be seen that ERGO@ZIF-L/GCE has low resistance and high electrical conductivity compared to the other electrodes. The DPV results also showed that ERGO@ZIF-L/GCE has high electrochemical activity and better current response towards the sensing of 10 µM benomyl.

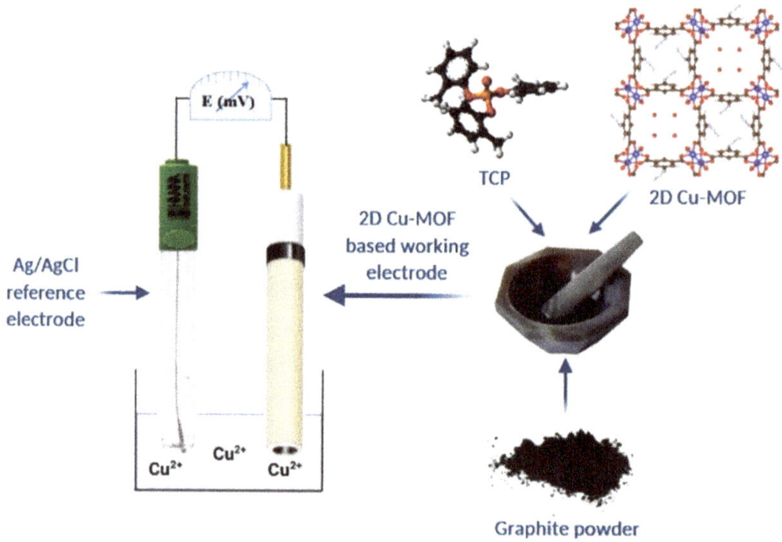

Figure 15. Schematic representation for the preparation of 2D Cu-MOF and working set up for the sensing of Cu^{2+} ions. Reproduced with permission [114].

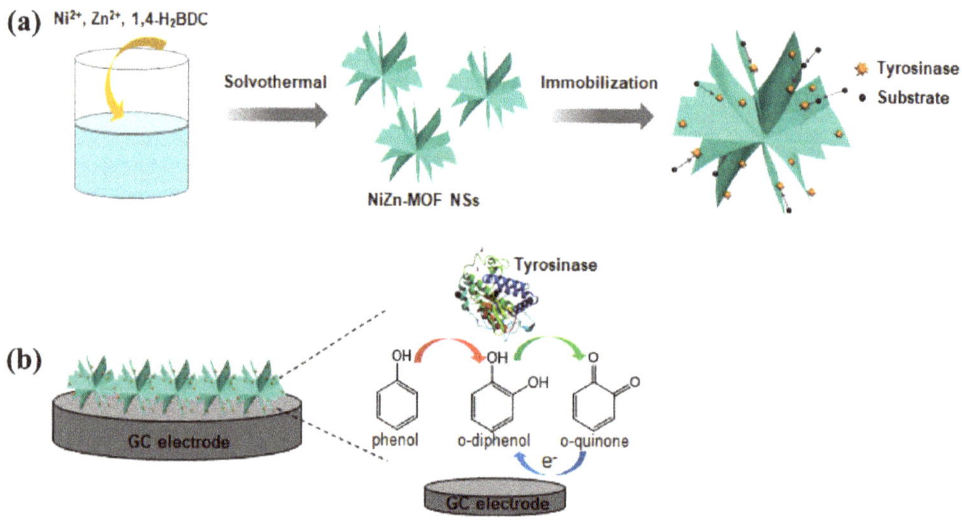

Figure 16. Schematic representation for the preparation of NiZn–MOF (**a**) and working mechanism for the sensing of tyrosinase (**b**). Reproduced with permission [115].

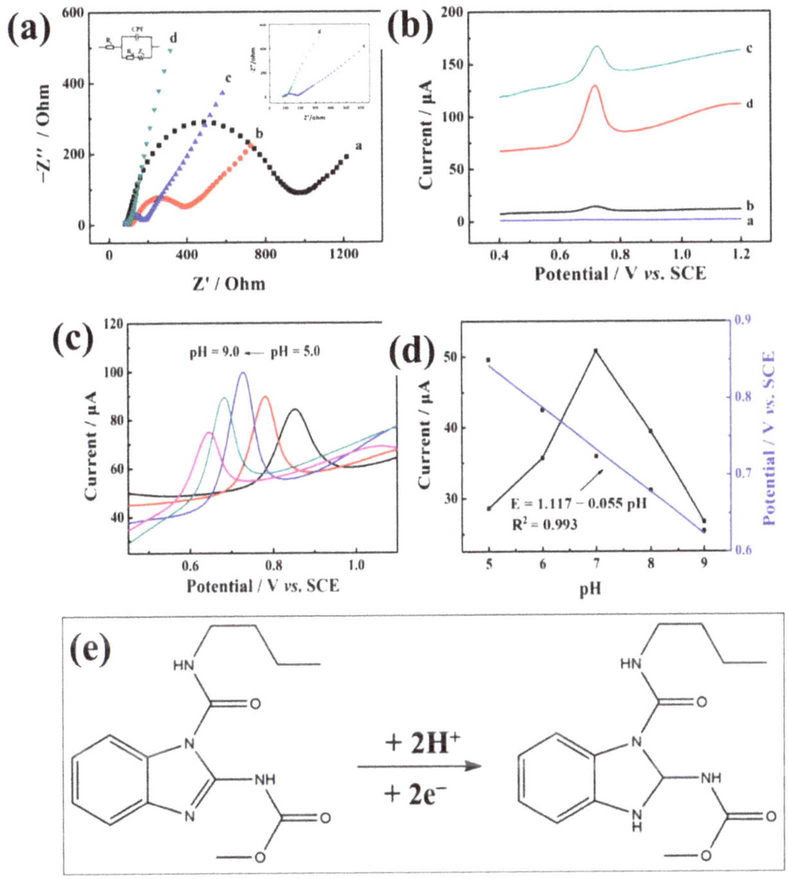

Figure 17. (**a**) EIS of different electrodes (GCE (a), ZIF-L/GCE (b), ERGO/GCE (c), and ERGO@ZIF-L/GCE (d)) in 5 mM $K_3Fe(CN)_6/K_4Fe(CN)_6$) system in 0.1 M KCl. (**b**) DPV curves of the different electrodes (GCE (a), ZIF-L/GCE (b), ERGO/GCE (c), and ERGO@ZIF-L/GCE (d)) in 10 μM benomyl. (**c**) DPV curves of the ERGO@ZIF-L/GCE in presence of 10 μM benomyl at different pH values. (**d**) Effects of pH values against current and potential peaks. (**e**) Working mechanism for the sensing of benomyl. Reproduced with permission [117].

This obtained results for the ERGO@ZIF-L/GCE was relatively higher compared to the other electrodes, as shown in Figure 17b. The pH of the solution was also optimized, and obtained results showed that ERGO@ZIF-L/GCE has high performance at pH of 7.0 (Figure 17c,d). The proposed sensors exhibited the LOD of 3 nM with linear range of 0.009 to 10.0 μM. In addition, the proposed sensor also demonstrated excellent repeatability, reproducibility, stability, and strong anti-interference performance. The sensing mechanism of benomyl has been described in Figure 17e. The authors believed that improved performance may be attributed to the excellent conductivity, high surface area, and active sites of the prepared electrode material. Du et al. [118] synthesized the glucose oxidase-like Co-MOF for the sensing of atrazine. The Co-MOF was coated on to the electrode surface, and its sensing activity was evaluated for the determination of atrazine. The authors obtained an LOD of 0.65 pM with good selectivity. Chen et al. [119] used Cu-MOF-anchored gold (Au) nanoparticles as electrode material for the determination of nitrite. The authors adopted simple wet chemical method for the synthesis of Cu-MOF, and Au NPs were deposited using an electro-deposition method. The synthesis of the Cu-MOF/Au has been illustrated in Figure 18a. The synthesized Cu-MOF/Au was coated on glassy carbon electrode (GCE) for electrochemical sensing purposes (Figure 18b). The sensing of the nitrite was carried out using an amperometry method. The amperometric results of the Cu-MOF/Au-based electrode for the sensing of nitrite are displayed in Figure 18c. The modified electrode showed an LOD of 82 nM with two linear range of 0.1–4000 and 4000–10,000 μM. The fabricated electrode also exhibited high selectivity (Figure 18d), sensitivity, and strong stability for nitrite detection. The excellent performance may be ascribed to the synergistic interactions, large surface area and porosity of Cu-MOF.

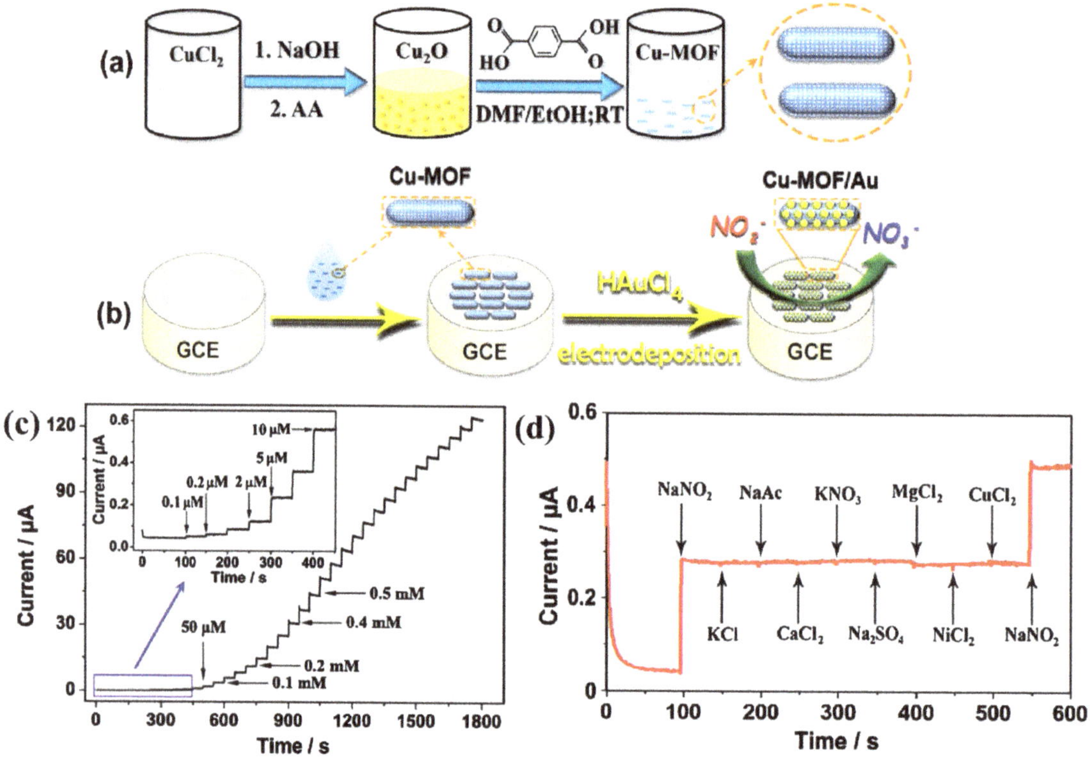

Figure 18. (**a**) Synthetic illustration for Cu-MOF. (**b**) Cu-MOF-based electrode modification illustration. (**c**) Amperometry results for the sensing of nitrite. (**d**) Selectivity results. Reproduced with permission [119].

Lu et al. [120] synthesized a 2D nickel phthalocyanine-based MOF nanosheet (2D NiPc-MOF) and used this synthesized MOF as electrode material for the determination of nitrite. The proposed electrode-material-based sensor demonstrated LOD of 2.3 μM with linear range of 0.01 to 11,500 mM. The authors believed that improved performance of the proposed sensor may be ascribed to the presence of high catalytic sites and larger electrochemical active surface area. Wang et al. [121] also proposed a cerium-based MOF (Ce-MOF) on carbon nanotubes (CNTs) substrates and applied it in electrochemical sensing applications. The obtained 2D Ce-MOF featured a distinctive layered structure with hierarchically connected channels, offering highly efficient and stable pathways for the transport of electrons, ions, and mass. The synthesized 2D layered CNTs@Ce-MOF nanosheet hybrid material utilizes Ce as the central ion and electrocatalytically active component for nitrite sensing. A decent LOD of 0.12 μM with linear ranges of 0.65 to 3.25 μM and 3.25 to 7000 μM were reported for the sensing of nitrite. In a report by Bai et al. [122], 2D

MOF M-TCPP (M = Cu, Co and Ni) nanofilms and the corresponding 2D MOF M-TCPP nanosheets were prepared in the presence and absence of PVP. Furthermore, Cu-TCPP nanofilm/CNT/GCE was fabricated and explored as the sensing material for the determination of hydrogen peroxide which demonstrated an interesting LOD of 5 nM with linear range of 0.01–3.75 μM and 3.75–377.75 μM. Saraf et al. [123] also investigated the bi-functional property of the Cu-MOF-integrated rGO as electrochemical electrode material for sensing and energy storage applications. The Cu-MOF/rGO-modified GCE demonstrated an excellent LOD of 33 nM with linear range of 3 to 40,000 μM for the determination of nitrite. The versatility of the Cu-MOF/rGO/GCE was evident from its reasonable selectivity in the presence of various interfering species, as well as its capability to detect the nitrite in real samples. Zhang et al. [124] reported a construction of a novel sensor based on a MIL-101(Cr)/XC-72-modified GCE for the sensing of chloramphenicol (CAP). The proposed sensor showed an LOD of 1.5×10^{-9} M with acceptable recoveries in real sample. Li et al. [125] fabricated a Zn/Ni-ZIF-8/XC-72/nafion as a sensing material and explored it potential for the determination of lead Pb(II) and Cu (II). The electrochemical performance of the MIL-101(Cr)/XC-72-modified GCE was determined by employing CV, electrochemical impedance spectroscopy (EIS), and differential pulse voltammetry (DPV). The proposed sensor exhibited LODs of 0.0150 and 0.0096 ppm for the determination of Pb (II) and Cu (II), respectively. Hadi et al. [126] described the fabrication and working of a glassy carbon electrode modified with a composite of multi-walled carbon nanotube and Cr-based metal–organic framework (MIL-101, Cr-BDC, BDC = 1,4-benzenedicarboxylate) for the sensing of picloram using square wave voltammetry (SWV) technique. The developed sensor showed an LOD of 0.06 μM and linear range of 0.1 to 12.5 μM. The proposed electrode was successfully applied to determine picloram in tap and river water samples spiked with picloram without requiring any purification step using the standard addition method. The obtained results showed high recovery rates ranging from 97.5% to 105.0%, demonstrating its reliability and accuracy. Song et al. [127] proposed a novel, reliable, and sensitive biosensor using ChE-Chit/MXene/Au NPs/MnO_2/Mn_3O_4/GCE as working electrode. The proposed sensor demonstrated LOD of 1.34×10^{-13} M for the sensing of methamidophos. The proposed sensor also exhibited good recoveries in real samples. Huang et al. [128] prepared [MIL-125(Ti)/$Fe_xTi_yO_z$]$_T$-X (MFTT-X) by using in situ solvothermal and calcination methods. The proposed material was used as a sensing material towards the detection of acetone. Chen et al. [129] proposed the fabrication of a tyrosine sensor using MXene/CNTs/Cu-MOF as an electrode material. The MXene/CNTs/Cu-MOF-based electrode showed an LOD of 0.19 μM with linear range of 0.53 to 232.46 μM. Wang et al. [130] used a molecularly imprinted electrochemical sensor (MIES) modified with Cu-MOF and $Ti_3C_2T_x$ as sensing platform for the detection of hygromycin B in food and obtained an LOD of 1.92×10^{-9} M.

Ge et al. [131] reported the fabrication of Zn-Co MOF/Ti_3C_2 MXene/Fe_3O_4-MGO/GCE for the sensing of mycophenolic acid and the developed sensor exhibited an LOD of 2.1×10^{-8} M for the determination of mycophenolic acid. Xiao et al. [132] synthesized an iron (Fe)-based MOF decorated with MXene (Fe-MOF/MXene) material by using a one-step hydrothermal method.

The Fe-MOF/MXene was applied as electrode material for the determination of arsenic (As (III)) using CV and square wave anodic stripping voltammetry techniques. The probable mechanism has been illustrated in Figure 19. The low detection limit of 0.58 ng/L was obtained for the sensing of As (III). The analysis of real water samples confirmed that the Fe-MOF/MXene-based sensor is suitable for the practical detection of As (III). The performance of the above-discussed electrochemical sensors are summarized in Table 3.

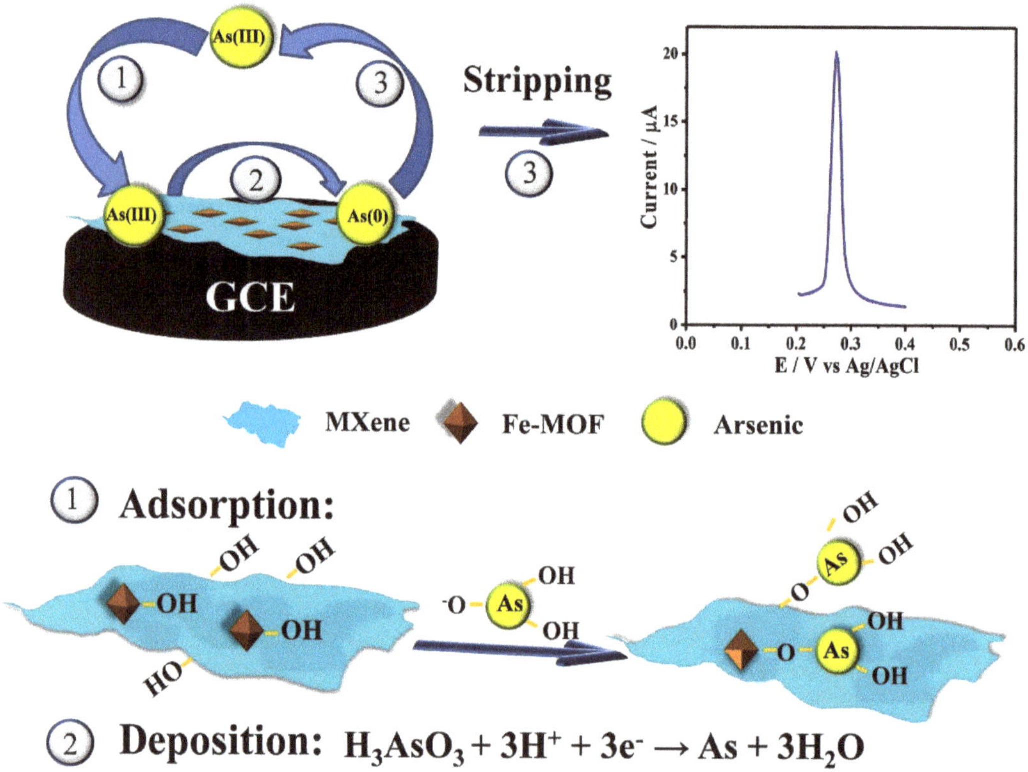

Figure 19. Schematic graph for the determination of As (III) using Fe-MOF/MXene/GCE. Reproduced with permission [132].

Table 3. Electrochemical sensing performance of the various reported sensors.

Electrode Material	Technique	LOD	Linear Range (µM)	Analyte	Real Sample	References
2D Cu-MOF	potentiometric	1×10^{-11} M	1.0×10^{-11} to 1.0×10^{-9} M	Cu^{2+}	Mate Tea, Tap water	[114]
NiZn–MOF NSs	Amperometry	6.5 nM	1 to 19	Phenol	Water sample	[115]
ERGO@ZIF-L	DPV	3 nM	0.009 to 10.0	Benomyl	Tomato	[117]
Co-MOF	DPV	0.65 pM	1 pM to 100 nM	Atrazine	-	[118]
Cu-MOF/Au	Amperometry	0.082	0.1–4000	Nitrite	River water	[119]
NiPc-MOFs	Amperometry	2.3	0.01 mM to 11,500 mM	Nitrite	Tap water	[120]
2D CNTs@Ce-MOF	DPV	0.12	3.25 to 7000 µM	Nitrite	-	[121]
Cu-TCPP nanofilm/CNT	Amperometry	5 nM	0.01 to 3.75 and 3.75 to 377.75	H_2O_2	-	[122]
Cu-MOF/rGO	Amperometry	33	3 to 40,000	Nitrite	Pond water	[123]
MIL-101(Cr)/XC-72	DPV	1.5 nM	1.0×10^{-8} to 2.0×10^{-5} M	Chloramphenicol	Honey, Eye drops	[124]
Zn/Ni-ZIF-8/XC-72/Nafion	DPV	0.0150 ppm	0.794 to 39.6 ppm	Pb^{2+}	Honey, Lake water	[125]
Zn/Ni-ZIF-8/XC-72/Nafion	DPV	0.0096	0.397 to 19.9 ppm	Cu^{2+}	-	[125]
Cr-MOF	Voltammetric	0.06	0.1–12.5	Picloram	River water	[126]
AChE-Chit/MXene/Au NPs/MnO_2/Mn_3O_4	DPV	1.34×10^{-13} M	10^{-12} to 10^{-6} M	Methamidophos	Fruit	[127]
MXene/CNTs/Cu-MOF	DPV	0.19	0.53 to 232.46	Tyrosine	Human serum	[129]
Cu-MOF/$Ti_3C_2T_x$	CV	1.92×10^{-9} M	5×10^{-9} to 5×10^{-6} M	Hygromycin B	Fish, Chicken	[130]

Table 3. *Cont.*

Electrode Material	Technique	LOD	Linear Range (μM)	Analyte	Real Sample	References
Zn-Co MOF/Ti_3C_2 MXene/Fe_3O_4-MGO	DPV	2.1×10^{-8} mol/L	3.21×10^{-8} mol/L to 8.9856×10^{-6} mol/L	Mycophenolic acid	Silage, maize, alfalfa, wheat	[131]
Fe-MOF/MXene	SWASV	0.58 ng/L	10 to 100 ng/L	As(III)	Lake water, River water	[132]

4. Conclusions and Future Perspectives

It is to be summarized that energy demand is increasing with increasing population globally. In this review, we have compiled the previously published articles on the use of MOFs and MOFs/MXenes as electrodes for energy storage and electrochemical sensing of toxic substances such as nitrite. The reported articles showed that MOFs/MXene-based electrode materials have superior performance compared to the pristine MOFs for energy storage and electrochemical sensing applications. However, several challenges remain to be addressed in the development of MOFs/MXenes composites for energy storage and electrochemical sensing applications. These challenges includes refining synthesis techniques for better control of composite morphology and size, exploring new MOF and MXene combinations to improve properties, studying the long-term stability and durability of the composites, and gaining a deeper understanding of the mechanisms driving their electrochemical behavior. Although, various efforts and approaches have been developed to prepare the MOFs/MXene composites. The MOFs/MXene composite demonstrated excellent performance, but still, several challenges, such as toxicity of the HF-etching agent, are the major concern. Thus, green synthetic procedures should be developed to prepare the MXenes and their composite materials.

Funding: This research received no external funding.

Institutional Review Board Statement: Not applicable.

Informed Consent Statement: Not applicable.

Data Availability Statement: No new data were created or analyzed in this study. Data sharing is not applicable to this article.

Conflicts of Interest: The authors declare no conflict of interest.

References

1. Corbos, R.-A.; Bunea, O.-I.; Jiroveanu, D.-C. The effects of the energy crisis on the energy-saving behavior of young people. *Energy Strategy Rev.* **2023**, *49*, 101184. [CrossRef]
2. Gajdzik, B.; Wolniak, R.; Nagaj, R.; Žuromskaitė-Nagaj, B.; Grebski, W.W. The Influence of the Global Energy Crisis on Energy Efficiency: A Comprehensive Analysis. *Energies* **2024**, *17*, 947. [CrossRef]
3. Guan, Y.; Yan, J.; Shan, Y.; Zhou, Y.; Hang, Y.; Li, R.; Liu, Y.; Liu, B.; Nie, Q.; Bruckner, B.; et al. Burden of the global energy price crisis on households. *Nat. Energy* **2023**, *8*, 304–316. [CrossRef]
4. Ye, W.; Jamshideasli, D.; Khodadadi, J.M. Improved Performance of Latent Heat Energy Storage Systems in Response to Utilization of High Thermal Conductivity Fins. *Energies* **2023**, *16*, 1277. [CrossRef]
5. Aghmadi, A.; Mohammed, O.A. Energy Storage Systems: Technologies and High-Power Applications. *Batteries* **2024**, *10*, 141. [CrossRef]
6. Chakraborty, M.R.; Dawn, S.; Saha, P.K.; Basu, J.B.; Ustun, T.S. A Comparative Review on Energy Storage Systems and Their Application in Deregulated Systems. *Batteries* **2022**, *8*, 124. [CrossRef]
7. Kumar, N.; Kim, S.-B.; Lee, S.-Y.; Park, S.-J. Recent Advanced Supercapacitor: A Review of Storage Mechanisms, Electrode Materials, Modification, and Perspectives. *Nanomaterials* **2022**, *12*, 3708. [CrossRef]
8. Rahman Khan, M.M.; Chakraborty, N. Conducting Polymer-Based Gel Materials: Synthesis, Morphology, Thermal Properties, and Applications in Supercapacitors. *Gels* **2024**, *10*, 553. [CrossRef]
9. Mensah-Darkwa, K.; Zequine, C.; Kahol, P.K.; Gupta, R.K. Supercapacitor Energy Storage Device Using Biowastes: A Sustainable Approach to Green Energy. *Sustainability* **2019**, *11*, 414. [CrossRef]
10. Forouzandeh, P.; Kumaravel, V.; Pillai, S.C. Electrode Materials for Supercapacitors: A Review of Recent Advances. *Catalysts* **2020**, *10*, 969. [CrossRef]

11. Qian, Y.; Zhang, F.; Pang, H. A review of MOFs and their composites-based photocatalysts: Synthesis and applications. *Adv. Funct. Mater.* **2021**, *31*, 2104231. [CrossRef]
12. Xiao, X.; Zou, L.; Pang, H.; Xu, Q. Synthesis of micro/nanoscaled metal–organic frameworks and their direct electrochemical applications. *Chem. Soc. Rev.* **2020**, *49*, 301–331. [CrossRef] [PubMed]
13. Lim, K.; Handoko, A.; Johnson, L.; Meng, X.; Lin, M.; Subramanian, G.; Anasori, B.; Gogotsi, Y.; Vojvodic, A.; Seh, Z. 2H-MoS2 on Mo_2CT_x MXene nanohybrid for efficient and durable electrocatalytic hydrogen evolution. *ACS Nano* **2020**, *14*, 16140–16155. [CrossRef] [PubMed]
14. Qian, Y.; Yu, J.; Zhang, Y.; Zhang, F.; Kang, Y.; Su, C.; Shi, H.; Kang, D.; Pang, H. Interfacial microenvironment modulation enhancing catalytic kinetics of binary metal sulfides heterostructures for advanced water splitting electrocatalysts. *Small Methods* **2022**, *6*, 2101186. [CrossRef]
15. Dong, K.; Liang, J.; Wang, Y.; Zhang, L.; Xu, Z.; Sun, S.; Luo, Y.; Li, T.; Liu, Q.; Li, N.; et al. Conductive two-dimensional magnesium metal-organic frameworks for high-efficiency O_2 electroreduction to H_2O_2. *ACS Catal.* **2022**, *12*, 6092–6099. [CrossRef]
16. Wang, Y.; Nian, Y.; Biswas, A.; Li, W.; Han, Y.; Chen, J. Challenges and opportunities in utilizing MXenes of carbides and nitrides as electrocatalysts. *Adv. Energy Mater.* **2020**, *11*, 2002967. [CrossRef]
17. Liu, Y.; Li, X.; Zhang, Q.; Li, W.; Xie, Y.; Liu, H.; Shang, L.; Liu, Z.; Chen, Z.; Gu, L.; et al. A general route to prepare low-ruthenium-content bimetallic electrocatalysts for pH-universal hydrogen evolution reaction by using carbon quantum dots. *Angew. Chem. Int. Ed.* **2020**, *59*, 1718–1726. [CrossRef]
18. Li, G.; Yin, Z.; Guo, H.; Wang, Z.; Yan, G.; Yang, Z.; Liu, Y.; Ji, X.; Wang, J. Metalorganic quantum dots and their graphene-like derivative porous graphitic carbon for advanced lithium-ion hybrid supercapacitor. *Adv. Energy Mater.* **2019**, *9*, 1802878. [CrossRef]
19. Cui, X.; Lei, S.; Wang, A.; Gao, L.; Zhang, Q.; Yang, Y.; Lin, Z. Emerging covalent organic frameworks tailored materials for electrocatalysis. *Nano Energy* **2020**, *70*, 104525. [CrossRef]
20. Peng, H.; Huang, S.; Tranca, D.; Richard, F.; Baaziz, W.; Zhuang, X.; Samori, P.; Ciesielski, A. Quantum capacitance through molecular infiltration of 7,7,8,8-tetracyanoquinodimethane in metal-organic framework/covalent organic framework hybrids. *ACS Nano* **2021**, *15*, 18580–18589. [CrossRef]
21. Zhang, X.; Chen, A.; Chen, L.; Zhou, Z. 2D materials bridging experiments and computations for electro/photocatalysis. *Adv. Energy Mater.* **2021**, *12*, 2003841. [CrossRef]
22. Hu, M.; Zhang, H.; Hu, T.; Fan, B.; Wang, X.; Li, Z. Emerging 2D MXenes for supercapacitors: Status, challenges and prospects. *Chem. Soc. Rev.* **2020**, *49*, 6666–6693. [CrossRef] [PubMed]
23. Li, W.; Guo, X.; Geng, P.; Du, M.; Jing, Q.; Chen, X.; Zhang, G.; Li, H.; Xu, Q.; Braunstein, P.; et al. Rational design and general synthesis of multimetallic metal-organic framework nano-octahedra for enhanced Li-S battery. *Adv. Mater.* **2021**, *33*, 2105163. [CrossRef] [PubMed]
24. Qian, Y.; Zhang, F.; Zhao, S.; Bian, C.; Mao, H.; Kang, D.; Pang, H. Recent progress of metal-organic framework-derived composites: Synthesis and their energy conversion applications. *Nano Energy* **2023**, *111*, 108415. [CrossRef]
25. Bai, Y.; Liu, C.; Shan, Y.; Chen, T.; Zhao, Y.; Yu, C.; Pang, H. Metal-organic frameworks nanocomposites with different dimensionalities for energy conversion and storage. *Adv. Energy Mater.* **2021**, *12*, 2100346. [CrossRef]
26. Naser, S.A.E.; Badmus, K.O.; Khotseng, L. Synthesis, Properties, and Applications of Metal Organic Frameworks Supported on Graphene Oxide. *Coatings* **2023**, *13*, 1456. [CrossRef]
27. Kidanemariam, A.; Cho, S. Recent Advances in the Application of Metal–Organic Frameworks and Coordination Polymers in Electrochemical Biosensors. *Chemosensors* **2024**, *12*, 135. [CrossRef]
28. Cao, Z.; Momen, R.; Tao, S.; Xiong, D.; Song, Z.; Xiao, X.; Deng, W.; Hou, H.; Yasar, S.; Altin, S.; et al. Metal–Organic Framework Materials for Electrochemical Supercapacitors. *Nano-Micro Lett.* **2022**, *14*, 181. [CrossRef]
29. Xie, X.; Zhang, N. Positioning MXenes in the photocatalysis landscape: Competitiveness, challenges, and future perspectives. *Adv. Funct. Mater.* **2020**, *30*, 2002528. [CrossRef]
30. Wu, C.; Unnikrishnan, B.; Chen, I.; Harroun, S.; Chang, H.; Huang, C. Excellent oxidation resistive MXene aqueous ink for micro-supercapacitor application. *Energy Storage Mater.* **2020**, *25*, 563–571. [CrossRef]
31. Morales-García, A.; Calle-Vallejo, F.; Illas, F. MXenes: New horizons in catalysis. *ACS Catal.* **2020**, *10*, 13487–13503. [CrossRef]
32. Do, H.H.; Cho, J.H.; Han, S.M.; Ahn, S.H.; Kim, S.Y. Metal–Organic-Framework- and MXene-Based Taste Sensors and Glucose Detection. *Sensors* **2021**, *21*, 7423. [CrossRef] [PubMed]
33. Wan, M.; Shi, C.; Qian, X.; Qin, Y.; Jing, J.; Che, H. Metal-organic Framework ZIF-67 Functionalized MXene for Enhancing the Fire Safety of Thermoplastic Polyurethanes. *Nanomaterials* **2022**, *12*, 1142. [CrossRef]
34. Gulcin, İ.; Alwasel, S.H. Metal Ions, Metal Chelators and Metal Chelating Assay as Antioxidant Method. *Processes* **2022**, *10*, 132. [CrossRef]
35. Das, T.K.; Poater, A. Review on the Use of Heavy Metal Deposits from Water Treatment Waste towards Catalytic Chemical Syntheses. *Int. J. Mol. Sci.* **2021**, *22*, 13383. [CrossRef]
36. Ayach, J.; El Malti, W.; Duma, L.; Lalevée, J.; Al Ajami, M.; Hamad, H.; Hijazi, A. Comparing Conventional and Advanced Approaches for Heavy Metal Removal in Wastewater Treatment: An In-Depth Review Emphasizing Filter-Based Strategies. *Polymers* **2024**, *16*, 1959. [CrossRef]

37. Simon, F.-G.; Kalbe, U. Measurement of the Environmental Impact of Materials. *Materials* **2022**, *15*, 2208. [CrossRef]
38. Zhang, S.; Wang, M.; Wang, X.; Song, J.; Yang, X. Electrocatalysis in MOF Films for Flexible Electrochemical Sensing: A Comprehensive Review. *Biosensors* **2024**, *14*, 420. [CrossRef]
39. Klinowski, J.; Paz, F.A.A.; Silva, P.; Rocha, J. Microwave-Assisted Synthesis of Metal–Organic Frameworks. *Dalton Trans.* **2011**, *40*, 321–330. [CrossRef]
40. Zhao, Z.; Li, H.; Zhao, K.; Wang, L.; Gao, X. Microwave-assisted synthesis of MOFs: Rational design via numerical simulation. *Chem. Eng. J.* **2022**, *428*, 131006. [CrossRef]
41. Phan, P.T.; Hong, J.; Tran, N.; Le, T.H. The Properties of Microwave-Assisted Synthesis of Metal–Organic Frameworks and Their Applications. *Nanomaterials* **2023**, *13*, 352. [CrossRef] [PubMed]
42. Taylor-Pashow, K.M.L.; Della Rocca, J.; Xie, Z.; Tran, S.; Lin, W. Postsynthetic Modifications of Iron-Carboxylate Nanoscale Metal-Organic Frameworks for Imaging and Drug Delivery. *J. Am. Chem. Soc.* **2009**, *131*, 14261–14263. [CrossRef] [PubMed]
43. Khan, N.A.; Kang, I.J.; Seok, H.Y.; Jhung, S.H. Facile synthesis of nano-sized metal- organic frameworks, chromium-benzenedicarboxylate, MIL-101. *Chem. Eng. J.* **2011**, *166*, 1152–1157. [CrossRef]
44. Zhu, Y.-J.; Chen, F. Microwave-Assisted Preparation of Inorganic Nanostructures in Liquid Phase. *Chem. Rev.* **2014**, *114*, 6462–6555. [CrossRef]
45. Tran, V.A.; Tran, N.T.; Doan, V.D.; Nguyen, T.-Q.; Pham Thi, H.H.; Vo, G.N.L. Application Prospects of MXenes Materials Modifications for Sensors. *Micromachines* **2023**, *14*, 247. [CrossRef]
46. Ali, A.; Majhi, S.M.; Siddig, L.A.; Deshmukh, A.H.; Wen, H.; Qamhieh, N.N.; Greish, Y.E.; Mahmoud, S.T. Recent Advancements in MXene-Based Biosensors for Health and Environmental Applications—A Review. *Biosensors* **2024**, *14*, 497. [CrossRef]
47. Yahia, M.; Lozano, L.A.; Zamaro, J.M.; Téllez, C.; Coronas, J. Microwave-assisted synthesis of metal–organic frameworks UiO-66 and MOF-808 for enhanced CO_2/CH_4 separation in PIM-1 mixed matrix membranes. *Sep. Purif. Technol.* **2024**, *330*, 125558. [CrossRef]
48. Skoda, D.; Kazda, T.; Munster, L.; Hanulikova, B.; Styskalik, A.; Eloy, P.; Debecker, D.P.; Vilcakova, J.; Cech, O.; Simonikova, L.; et al. Microwave-assisted synthesis of platelet-like cobalt metal-organic framework, its transformation to porous layered cobalt-carbon nanocomposite discs and their utilization as anode materials in sodium-ion batteries. *J. Energy Storage* **2020**, *27*, 101113. [CrossRef]
49. Moradi, E.; Rahimi, R.; Safarifard, V.; Azari, S. A Sonochemically-Synthesized Microporous Metal-Organic Framework for the Rapid and Efficient Ultrasonic-Assisted Removal of Mercury (II) Ions in a Water Solution and a Study of the Antibacterial Activity. *Proceedings* **2019**, *41*, 31. [CrossRef]
50. Pokhrel, N.; Vabbina, P.K.; Pala, N. Sonochemistry: Science and Engineering. *Ultrason. Sonochem.* **2016**, *29*, 104–128. [CrossRef]
51. Wang, X.; Li, Q.; Yang, N.; Yang, Y.; He, F.; Chu, J.; Gong, M.; Wu, B.; Zhang, R.; Xiong, S. Hydrothermal synthesis of NiCo-based bimetal-organic frameworks as electrode materials for supercapacitors. *J. Solid State Chem.* **2019**, *270*, 370–378. [CrossRef]
52. Qi, Q.; Liu, Z.; Chen, X.; Yu, J.; Li, X.; Wang, R.; Liu, Y.; Chen, J. Promoted electrochemical performance by MOF on MOF composite catalyst of microbial fuel cell: CuCo-MOF@ZIF-8 and the comparison between two-step hydrothermal method and dual-solution method. *Biosens. Bioelectron.* **2024**, *264*, 116693. [CrossRef] [PubMed]
53. Wu, L.-Z.; Zhou, X.-Y.; Zeng, P.-C.; Huang, J.-Y.; Zhang, M.-D.; Qin, L. Hydrothermal synthesis of Ni(II) or Co(II)-based MOF for electrocatalytic hydrogen evolution. *Polyhedron* **2022**, *225*, 116035. [CrossRef]
54. Su, H.; Jin, C.; Zhang, X.; Yu, Z.; Zeng, X. Recent progress in the synthesis and electrocatalytic application of MXene-based metal phosphide composites. *Carbon Neutralization* **2024**, *3*, 1009–1035. [CrossRef]
55. Zeng, X.; Zhang, Q.; Jin, C.; Huang, H.; Gao, Y. Fe-Induced Electronic Transfer and Structural Evolution of Lotus Pod-like $CoNiFeP_x$@P,N-C Heterostructure for Sustainable Oxygen Evolution. *Energy Environ. Mater.* **2024**, *7*, e12628. [CrossRef]
56. Zeng, X.; Ye, Y.; Wang, Y.; Yu, R.; Moskovits, M.; Stucky, G.D. Honeycomb-like MXene/$NiFeP_x$–NC with "continuous" single-crystal enabling high activity and robust durability in electrocatalytic oxygen evolution reactions. *J. Adv. Ceram.* **2023**, *12*, 553–564. [CrossRef]
57. Zeng, X.; Duan, D.; Zhang, X.; Li, X.; Li, K.; Yu, R.; Moskovits, M. Doping and interface engineering in a sandwich $Ti_3C_2T_x/MoS_{2-x}P_x$ heterostructure for efficient hydrogen evolution. *J. Mater. Chem. C* **2022**, *10*, 4140–4147. [CrossRef]
58. Zhang, Z.; Liang, T.; Jin, C.; Zhang, S.; Cui, Y.; Chen, J.; Zeng, X. Synergistically coupling CoS/FeS_2 heterojunction nanosheets on a MXene via a dual molten salt etching strategy for efficient oxygen evolution reaction. *J. Mater. Chem. A* **2024**, *12*, 14517–14530. [CrossRef]
59. Farea, M.A.; Mohammed, H.Y.; Murshed, M.N.; Narwade, V.N.; Deore, K.B.; Shirsat, M.D. A zinc-metal organic framework as a stable and efficient electrode for high-performance supercapacitor. *Mater. Today Proc.* **2023**, *92*, 689–693. [CrossRef]
60. Hong, S.; Kim, Y.; Kim, Y.; Suh, K.; Yoon, M.; Kim, K. Hierarchical Porous Carbon Materials Prepared by Direct Carbonization of Metal–Organic Frameworks as an Electrode Material for Supercapacitors. *Bull. Korean Chem. Soc.* **2021**, *42*, 309–314. [CrossRef]
61. Salehi, S.; Aghazadeh, M.; Karimzadeh, I. Zn-MOF electrode material for supercapacitor applications. *PPAM* **2022**, *2*, 77–82.
62. Cao, F.; Gan, M.; Ma, L.; Li, X.; Yan, F.; Ye, M.; Zhai, Y.; Zhou, Y. Hierarchical sheet-like Ni–Co layered double hydroxide derived from a MOF template for high-performance supercapacitors. *Synth. Met.* **2017**, *234*, 154–160. [CrossRef]
63. Du, X.; Zhang, J.; Wang, H.; Huang, Z.; Guo, A.; Zhao, L.; Niu, Y.; Li, X.; Wu, B.; Liu, Y. Solid–solid interface growth of conductive metal–organic framework nanowire arrays and their supercapacitor application. *Mater. Chem. Front.* **2020**, *4*, 243–251. [CrossRef]

64. Liu, Q.; Liu, X.; Shi, C.; Zhang, Y.; Feng, X.; Cheng, M.-L.; Su, S.; Gu, J. A copper-based layered coordination polymer: Synthesis, magnetic properties and electrochemical performance in supercapacitors. *Dalton Trans.* **2015**, *44*, 19175–19184. [CrossRef]
65. Liu, J.; Zhou, Y.; Xie, Z.; Li, Y.; Liu, Y.; Sun, J.; Ma, Y.; Terasaki, O.; Chen, L. Conjugated copper–catecholate framework electrodes for efficient energy storage. *Angew. Chem. Int. Ed.* **2020**, *59*, 1081–1086. [CrossRef]
66. Hou, R.; Miao, M.; Wang, Q.; Yue, T.; Liu, H.; Park, H.S.; Qi, K.; Xia, B.Y. Integrated Conductive Hybrid Architecture of Metal–Organic Framework Nanowire Array on Polypyrrole Membrane for All-Solid-State Flexible Supercapacitors. *Adv. Energy Mater.* **2020**, *10*, 1901892. [CrossRef]
67. Ma, Y.; Gao, G.; Su, H.; Rong, H.; Lai, L.; Liu, Q. A Cu 4 cluster-based MOF as a supercapacitor electrode material with ultrahigh capacitance. *Ionics* **2021**, *27*, 1699–1707. [CrossRef]
68. Ramachandran, R.; Zhao, C.; Luo, D.; Wang, K.; Wang, F. Synthesis of copper benzene-1, 3, 5-tricarboxylate metal organic frameworks with mixed phases as the electrode material for supercapacitor applications. *Appl. Surf. Sci.* **2018**, *460*, 33–39. [CrossRef]
69. Rajak, R.; Saraf, M.; Kumar, P.; Natarajan, K.; Mobin, S.M. Construction of a Cu-Based Metal–Organic Framework by Employing a Mixed-Ligand Strategy and Its Facile Conversion into Nanofibrous CuO for Electrochemical Energy Storage Applications. *Inorg. Chem.* **2021**, *60*, 16986–16995. [CrossRef]
70. Hussain, N.; Abbas, Z.; Ansari, S.N.; Kedarnath, G.; Mobin, S.M. Phosphorization Engineering on a MOF-Derived Metal Phosphide Heterostructure (Cu/Cu_3P@NC) as an Electrode for Enhanced Supercapacitor Performance. *Inorg. Chem.* **2023**, *62*, 17083–17092. [CrossRef]
71. Ansari, S.N.; Saraf, M.; Gupta, A.K.; Mobin, S.M. Functionalized Cu-MOF@CNT Hybrid: Synthesis, Crystal Structure and Applicability in Supercapacitors. *Chem. Asian J.* **2019**, *14*, 3566–3571. [CrossRef] [PubMed]
72. Liu, X.; Shi, C.; Zhai, C.; Cheng, M.; Liu, Q.; Wang, G. Cobalt-Based Layered Metal–Organic Framework as an Ultrahigh Capacity Supercapacitor Electrode Material. *ACS Appl. Mater. Interfaces* **2016**, *8*, 4585–4591. [CrossRef] [PubMed]
73. Ramachandran, R.; Zhao, C.; Luo, D.; Wang, K.; Wang, F. Morphology-dependent electrochemical properties of cobalt-based metal organic frameworks for supercapacitor electrode materials. *Electrochim. Acta* **2018**, *267*, 170–180. [CrossRef]
74. Lee, D.Y.; Shinde, D.V.; Kim, E.-K.; Lee, W.; Oh, I.-W.; Shrestha, N.K.; Lee, J.K.; Han, S.-H. Supercapacitive property of metal–organic-frameworks with different pore dimensions and morphology. *Microporous Mesoporous Mater.* **2013**, *171*, 53–57. [CrossRef]
75. Yang, J.; Ma, Z.; Gao, W.; Wei, M. Layered Structural Co-Based MOF with Conductive Network Frames as a New Supercapacitor Electrode. *Chem. Eur. J.* **2017**, *23*, 631–636. [CrossRef]
76. Lee, D.Y.; Yoon, S.J.; Shrestha, N.K.; Lee, S.-H.; Ahn, H.; Han, S.-H. Unusual energy storage and charge retention in Co-based metal–organic-frameworks. *Microporous Mesoporous Mater.* **2012**, *153*, 163–165. [CrossRef]
77. Wang, L.; Shao, D.; Guo, J.; Zhang, S.; Lu, Y. Superstable porous Co-coordination polymer as the electrode material for supercapacitor. *J. Solid State Chem.* **2019**, *277*, 630–635. [CrossRef]
78. Punde, N.; Rawool, C.; Rajpurohit, A.; Karna, S.; Srivastava, A. Hybrid Composite Based on Porous Cobalt-Benzenetricarboxylic Acid Metal Organic Framework and Graphene Nanosheets as High Performance Supercapacitor Electrode. *ChemistrySelect* **2018**, *3*, 11368–11380. [CrossRef]
79. Wang, X.; Liu, X.; Rong, H.; Song, Y.; Wen, H.; Liu, Q. Layered manganese-based metal–organic framework as a high capacity electrode material for supercapacitors. *RSC Adv.* **2017**, *7*, 29611–29617. [CrossRef]
80. Cheng, Z.; Qiu, Y.; Tan, G.; Chang, X.; Luo, Q.; Cui, L. Synthesis of a Novel Mn(II)-porphyrins polycondensation polymer and its application as pseudo-capacitor electrode material. *J. Organomet. Chem.* **2019**, *900*, 120940. [CrossRef]
81. Shinde, P.A.; Seo, Y.; Lee, S.; Kim, H.; Pham, Q.N.; Won, Y.; Jun, S.C. Layered manganese metal-organic framework with high specific and areal capacitance for hybrid supercapacitors. *Chem. Eng. J.* **2020**, *387*, 122982. [CrossRef]
82. Sundriyal, S.; Shrivastav, V.; Sharma, M.; Mishra, S.; Deep, A. Redox Additive Electrolyte Study of Mn–MOF Electrode for Supercapacitor Applications. *ChemistrySelect* **2019**, *4*, 2585–2592. [CrossRef]
83. Gao, S.; Sui, Y.; Wei, F.; Qi, J.; Meng, Q.; He, Y. Facile synthesis of cuboid Ni-MOF for high-performance supercapacitors. *J. Mater. Sci.* **2018**, *53*, 6807–6818. [CrossRef]
84. Liu, R.; Shang, M.Y.; Liu, C.; Hao, Y.; Yang, F.; Shi, J.Y.; Chen, Y.; Wang, Y.F.; Feng, J.Q.; Yang, P.F.; et al. Effect of hydrochloric acid on the properties of NiMOF nanostructures as supercapacitor electrode materials. *Chem. Phys. Lett.* **2024**, *850*, 141474. [CrossRef]
85. Sheberla, D.; Bachman, J.C.; Elias, J.S.; Sun, C.-J.; Shao-Horn, Y.; Dincă, M. Conductive MOF electrodes for stable supercapacitors with high areal capacitance. *Nat. Mater.* **2017**, *16*, 220–224. [CrossRef]
86. Zhang, H.; Yang, Y.; Deng, Y.; Guo, B.; Jia, X.; Zhang, G.; Li, G.; Wang, K. Construction of 2D MOF, nanosheets Missing-Link. Defects Enhanc. supercapacitor Perform. *J. Alloys Compd.* **2024**, *999*, 175049. [CrossRef]
87. Yang, J.; Xiong, P.; Zheng, C.; Qiu, H.; Wei, M. Metal–organic frameworks: A new promising class of materials for a high performance supercapacitor electrode. *J. Mater. Chem. A* **2014**, *2*, 16640–16644. [CrossRef]
88. Kang, L.; Sun, S.-X.; Kong, L.-B.; Lang, J.-W.; Luo, Y.-C. Investigating metal-organic framework as a new pseudo-capacitive material for supercapacitors. *Chin. Chem. Lett.* **2014**, *25*, 957–961. [CrossRef]
89. Xu, J.; Yang, C.; Xue, Y.; Wang, C.; Cao, J.; Chen, Z. Facile synthesis of novel metalorganic nickel hydroxide nanorods for high performance supercapacitor. *Electrochim. Acta* **2016**, *211*, 595–602. [CrossRef]

90. Yang, X.; Tian, Y.; Li, S.; Wu, Y.-P.; Zhang, Q.; Li, D.-S.; Zhang, S. Heterogeneous NiMOF/V_2CTx–MXene hierarchically-porous nanorods for robust and high energy density hybrid supercapacitors. *J. Mater. Chem. A* **2022**, *10*, 12225–12234. [CrossRef]
91. Zhu, R.; Gu, Y.; Liu, L.; Lu, J.; Pang, H. Integration of conductive MOF and MXene for high-performance supercapacitor. *N. J. Chem.* **2024**, *48*, 10593–10598. [CrossRef]
92. Xie, W.; Wang, Y.; Zhou, J.; Zhang, M.; Yu, J.; Zhu, C.; Xu, J. MOF-derived $CoFe_2O_4$ nanorods anchored in MXene nanosheets for all pseudocapacitive flexible supercapacitors with superior energy storage. *Appl. Surf. Sci.* **2020**, *534*, 147584. [CrossRef]
93. Shingte, S.R.; Chavan, V.D.; Dhavale, R.P.; Kim, D.-K.; Park, H.-H.; Dongale, T.D.; Patil, P.B. Optimizing supercapacitive performance of MXene through MOF-derived nickel ferrite nanoparticle integration. *J. Energy Storage* **2024**, *92*, 112169. [CrossRef]
94. Qu, Y.; Shi, C.; Cao, H.; Wang, Y. Synthesis of Ni-MOF/$Ti_3C_2T_x$ hybrid nanosheets via ultrasonific method for supercapacitor electrodes. *Mater. Lett.* **2020**, *280*, 128526. [CrossRef]
95. Hassan, H.; Umar, E.; Iqbal, M.W.; Alrobei, H.; Afzal, A.M.; Alqorashi, A.K.; Yaseen, T.; Sunny, M.A. Innovative MOF-5/V_2CT_x composite for high-performance, ultra-fast supercapacitors and hydrogen evolution reaction. *Electrochim. Acta* **2024**, *489*, 144277. [CrossRef]
96. Liu, C.; Bai, Y.; Li, W.; Yang, F.; Zhang, G.; Pang, H. In Situ Growth of ThreeDimensional MXene/Metal–Organic Framework Composites for HighPerformance Supercapacitors. *Angew. Chem. Int. Ed.* **2022**, *61*, e202116282. [CrossRef]
97. Li, S.; Cheng, Q.; Ye, P.; Zhang, Y.; Zhang, L.; Liu, F.; Qiu, H.; Qu, X.; Nie, Y. Hierarchical two-dimensional Ti-MOF derived from MXene for hybrid supercapacitor electrodes. *Appl. Organomet. Chem.* **2024**, *38*, e7547. [CrossRef]
98. Wang, J.; Gong, J.; Zhang, H.; Lv, L.; Liu, Y.; Dai, Y. Construction of hexagonal nickel-cobalt oxide nanosheets on metal-organic frameworks based on MXene interlayer ion effect for hybrid supercapacitors. *J. Alloys Compd.* **2021**, *870*, 159466. [CrossRef]
99. Shavita, K.K.; Thakur, A.L.; Sharma, S.S. Exploring MXene-MOF composite for supercapacitor application. *Mater. Chem. Phys.* **2024**, *322*, 129463. [CrossRef]
100. Hassan, H.; Shoaib, M.; Iqbal, M.W.; Alqorashi, A.K.; Sunny, M.A.; Yaseen, T.; Alrobei, H.; Afzal, A.M. Synergistic performance of Ag-MOF@V_2CTx composite for asymmetric supercapacitors and hydrogen evolution reaction. *J. Phys. Chem. Solids* **2024**, *193*, 112151. [CrossRef]
101. Ali, M.; Afzal, A.M.; Mumtaz, S.; Yusuf, K.; Deifalla, A. Designing of NiMn-MOF@ MXene nanorods@AC@NF electrode material for high performance electrochemical based energy storage devices. *J. Energy Storage* **2024**, *92*, 111972. [CrossRef]
102. Ji, Y.; Li, W.; You, Y.; Xu, G. Architecting MxNy-MOF@MXene composites for excellent supercapacitor electrodes. *Colloids Surf. A Physicochem. Eng. Asp.* **2024**, *699*, 134680. [CrossRef]
103. Wei, J.; Hu, F.; Pan, Y.; Lv, C.; Bian, L.; Ouyang, Q. Design strategy for metal–organic framework assembled on modifications of MXene layers for advanced supercapacitor electrodes. *Chem. Eng. J.* **2024**, *481*, 148793. [CrossRef]
104. Zheng, S.; Zhou, H.; Xue, H.; Braunstein, P.; Pang, H. Pillared-layer Ni-MOF nanosheets anchored on Ti_3C_2 MXene for enhanced electrochemical energy storage. *J. Colloid Interface Sci.* **2022**, *614*, 130–137. [CrossRef]
105. Chen, S.; Zhang, H.; Liu, Y.; Guo, J.; Li, X.; Zhang, M.; Zhang, D.; Fang, P.; He, C. Conductive Bismuth-Catecholate Metal–Organic Frameworks Grown on $Ti_3C_2T_x$ Nanosheets for High-Performance Supercapacitors. *ACS Appl. Nano Mater.* **2024**, *7*, 14310–14320. [CrossRef]
106. Ramachandran, R.; Rajavel, K.; Xuan, W.; Lin, D.; Wang, F. Influence of $Ti_3C_2T_x$ (MXene) intercalation pseudocapacitance on electrochemical performance of CoMOF binder-free electrode. *Ceram. Int.* **2018**, *44*, 14425–14431. [CrossRef]
107. Guo, H.; Zhang, J.; Xu, M.; Wang, M.; Yang, F.; Wu, N.; Zhang, T.; Sun, L.; Yang, W. Zeolite-imidazole framework derived nickel-cobalt hydroxide on ultrathin MXene, nanosheets Long. life High. Perform. Supercapacitance. *J. Alloys Compd.* **2021**, *888*, 161250. [CrossRef]
108. Yang, X.; Xu, C.; Li, S.; Wu, Y.-P.; Wu, X.-Q.; Yin, Y.-M.; Li, D.-S. Thermal treatment for promoting interfacial interaction in Co-BDC/$Ti_3C_2T_x$ hybrid nanosheets for hybrid supercapacitors. *J. Colloid Interface Sci.* **2022**, *617*, 633–640. [CrossRef]
109. Yang, Y.; Huang, X.; Sheng, C.; Pan, Y.; Huang, Y.; Wang, X. In-situ formation of MOFs derivatives $CoSe_2$/Ni_3Se_4 nanosheets on MXene nanosheets for hybrid supercapacitor with enhanced electrochemical performance. *J. Alloys Compd.* **2022**, *920*, 165908. [CrossRef]
110. Xiang, G.-T.; Chen, N.; Lu, B.; Xu, J.-L.; Rodriguez, R.D.; Sheremet, E.; Hu, Y.-D.; Chen, J.-J. Flexible solid-state Zn-Co MOFs@MXene supercapacitors and organic ion hydrogel sensors for self-powered smart sensing applications. *Nano Energy* **2023**, *118*, 108936. [CrossRef]
111. Yue, L.; Chen, L.; Wang, X.; Lu, D.; Zhou, W.; Shen, D.; Yang, Q.; Xiao, S.; Li, Y. Ni/Co-MOF@aminated MXene hierarchical electrodes for high-stability supercapacitors. *Chem. Eng. J.* **2023**, *451*, 138687. [CrossRef]
112. Kshetri, T.; Khumujam, D.D.; Singh, T.I.; Lee, Y.S.; Kim, N.H.; Lee, J.H. Co-MOF@MXene-carbon nanofiber-based freestanding electrodes for a flexible and wearable quasi-solid-state supercapacitor. *Chem. Eng. J.* **2022**, *437*, 135338. [CrossRef]
113. Zhao, Q.; Li, S.-H.; Chai, R.-L.; Ren, X.; Zhang, C. Two-Dimensional Conductive Metal−Organic Frameworks Based on Truxene. *ACS Appl. Mater. Interfaces* **2020**, *12*, 7504–7509. [CrossRef] [PubMed]
114. Elashery, S.E.; Oh, H. Exploitation of 2D Cu-MOF nanosheets as a unique electroactive material for ultrasensitive Cu(II) ion estimation in various real samples. *Anal. Chim. Acta* **2021**, *1181*, 338924. [CrossRef]
115. Wen, Y.; Li, R.; Liu, J.; Zhang, X.; Wang, P.; Zhang, X.; Zhou, B.; Li, H.; Wang, J.; Li, Z.; et al. Promotion effect of Zn on 2D bimetallic NiZn metal organic framework nanosheets for tyrosinase immobilization and ultrasensitive detection of phenol. *Analytica Chimica Acta* **2020**, *1127*, 131–139. [CrossRef]

116. Wang, X.Y.; Xiao, S.Y.; Jiang, Z.W.; Zhen, S.J.; Huang, C.Z.; Liu, Q.Q.; Li, Y.F. An ultrathin 2D Yb(III) metal-organic frameworks with strong electrochemiluminescence as a "on-off-on" platform for detection of picric acid and berberine chloride form. *Talanta* **2021**, *234*, 122625. [CrossRef]
117. Shi, M.; Peng, G.; Xue, S.; Xu, J.; Gao, Y.; Liu, S.; Duan, X.; Lu, L. 2D Leaf-Like Structured ZIF-L Embedded Electrochemically Reduced Graphene Oxide Composite as an Electrochemical Sensing Platform for Sensitively Detecting Benomyl. *Molecules* **2022**, *27*, 6857. [CrossRef]
118. Du, B.; Lu, G.; Zhang, Z.; Feng, Y.; Liu, M. Glucose oxidase-like Co-MOF nanozyme-catalyzed self-powered sensor for sensitive detection of trace atrazine in complex environments. *Anal. Chim. Acta* **2023**, *1280*, 341817. [CrossRef]
119. Chen, H.; Yang, T.; Liu, F.; Li, W. Electrodeposition of gold nanoparticles on Cu-based metal-organic framework for the electrochemical detection of nitrite. *Sens. Actuators B Chem.* **2019**, *286*, 401–407. [CrossRef]
120. Lu, S.; Ji, H.; Hummel, M.; Wu, Y.; Wang, K.; Qi, X.; Gu, Z. Two-dimensional conductive phthalocyanine based metal–organic frameworks for electrochemical nitrite sensing. *RSC Adv.* **2021**, *11*, 4472–4477. [CrossRef]
121. Wang, S.; Xue, Y.; Yu, Z.; Huang, F.; Jin, Y. Layered 2D MOF nanosheets grown on CNTs substrates for efficient nitrite sensing. *Mater. Today Chem.* **2023**, *30*, 101490. [CrossRef]
122. Bai, W.; Li, S.; Ma, J.; Cao, W.; Zheng, J. Ultrathin 2D metal–organic framework (nanosheets and nanofilms)-based xD–2D hybrid nanostructures as biomimetic enzymes and supercapacitors. *J. Mater. Chem. A* **2019**, *7*, 9086–9098. [CrossRef]
123. Saraf, M.; Rajak, R.; Mobin, S.M. A fascinating multitasking Cu-MOF/rGO hybrid for high performance supercapacitors and highly sensitive and selective electrochemical nitrite sensors. *J. Mater. Chem. A* **2016**, *4*, 16432–16445. [CrossRef]
124. Zhang, W.; Zhang, Z.; Li, Y.; Chen, J.; Li, X.; Zhang, Y.; Zhang, Y. Novel nanostructured MIL-101(Cr)/XC-72 modified electrode sensor: A highly sensitive and selective determination of chloramphenicol. *Sens. Actuators B* **2017**, *247*, 756–764. [CrossRef]
125. Li, D.; Qiu, X.; Guo, H.; Duan, D.; Zhang, W.; Wang, J.; Ma, J.; Ding, Y.; Zhang, Z. A simple strategy for the detection of Pb(II) and Cu(II) by an electrochemical sensor based on Zn/Ni-ZIF-8/XC-72/Nafion hybrid materials. *Environ. Res.* **2021**, *202*, 111605. [CrossRef]
126. Hadi, M.; Bayat, M.; Mostaanzadeh, H.; Ehsani, A.; Yeganeh-Faal, A. Sensitive electrochemical detection of picloram utilising a multi-walled carbon nanotube/Cr-based metal-organic framework composite-modified glassy carbon electrode. *Int. J. Environ. Anal. Chem.* **2018**, *98*, 1441838. [CrossRef]
127. Song, D.; Jiang, X.; Li, Y.; Lu, X.; Luan, S.; Wang, Y.; Li, Y.; Gao, F. Metal−organic frameworks-derived MnO_2/Mn_3O_4 microcuboids with hierarchically ordered nanosheets and Ti_3C_2 MXene/Au NPs composites for electrochemical pesticide detection. *J. Hazard. Mater.* **2019**, *373*, 367–376. [CrossRef]
128. Huang, D.; Wang, Y.; Wang, X.; Li, H.; Tan, X.; Chen, Y.; Wang, W.; Cheng, Q.; Yi, M.; Han, G.; et al. Rational in situ construction of Fe-modified MXene-derived MOFs as high-performance acetone sensor. *Chem. Eng. J.* **2022**, *444*, 136526. [CrossRef]
129. Chen, J.; Chen, Y.; Li, S.; Yang, J.; Dong, J.; Lu, X. MXene/CNTs/Cu-MOF electrochemical probe for detecting tyrosine. *Carbon* **2022**, *199*, 110–118. [CrossRef]
130. Wang, H.; Cai, L.; Wang, Y.; Liu, C.; Fang, G.; Wang, S. Covalent molecularly imprinted electrochemical sensor modulated by borate ester bonds for hygromycin B detection based on the synergistic signal amplification of Cu-MOF and MXene. *Food Chem.* **2022**, *383*, 132382. [CrossRef]
131. Ge, Y.; Camarada, M.B.; Liu, P.; Qu, M.; Wen, Y.; Xu, L.; Liang, H.; Liu, E.; Zhang, X.; Hao, W.; et al. A portable smart detection and electrocatalytic mechanism of mycophenolic acid: A machine learning-based electrochemical nanosensor to adapt variable-pH silage microenvironment. *Sens. Actuators B Chem.* **2022**, *372*, 132627. [CrossRef]
132. Xiao, P.; Zhu, G.; Shang, X.; Hu, B.; Zhang, B.; Tang, Z.; Yang, J.; Liu, J. An Fe-MOF/MXene-based ultra-sensitive electrochemical sensor for arsenic (III) measurement. *J. Electroanal. Chem.* **2022**, *916*, 116382. [CrossRef]

molecules

MDPI

Article

Facile Growing of Ni-MOFs on Ni Foam by Self-Dissociation Strategy for Electrochemical Energy Storage

Hongmei Li [1], Yang Li [2,*], Shuxian Song [2], Yuhan Tian [2], Bo Feng [1], Boru Li [1], Zhiqing Liu [2] and Xu Zhang [2,*]

1 College of Material Science and Engineering, Shenyang Aerospace University, Shenyang 110136, China
2 School of Chemical Engineering, Ocean and Life Sciences, Dalian University of Technology, Panjin 124221, China
* Correspondence: chyangli@dlut.edu.cn (Y.L.); zhangxu@dlut.edu.cn (X.Z.)

Abstract: Metal–organic frameworks (MOFs) with redox metal centers have come into view as potential materials for electrochemical energy storage. However, the poor electrical conductivity largely impedes the potentiality of MOFs to construct high-performance electrodes in supercapacitors. In this work, a self-dissociation strategy has been applied to construct Ni-MOF microbelts on Ni foam (NF), where the NF is used as both a support and a Ni source. The transmission channels between the Ni-MOF and NF are favorable for the charge transport due to the in situ self-assembly of the TPA linkers with the dissociated Ni ions from the Ni foam. The grown Ni-MOF microbelt arrays can offer abundant active sites for redox reactions. The prepared Ni-MOF/NF-s electrode can yield a high capacitance of 1124 F g^{-1} at 1 A g^{-1} and retains 590 F g^{-1} at 10 A g^{-1}. This design may offer a controllable protocol for the construction of MOF microbelt arrays on various metal substrates.

Keywords: metal-organic frameworks; self-dissociation strategy; microbelts; supercapacitor

1. Introduction

With the great progress of electronic devices and electronic vehicles, advanced electrochemical energy storage technologies are urgently being developed [1,2]. Different from batteries with the limited charging–discharging currents due to diffusion-controlled redox reactions, supercapacitors—in which the ions are rapidly transported to the surface of the electrode—exhibit remarkable rate capabilities and a long cycle life [3,4]. Hence, they have received great attention. Apart from carbon-based double electric layer supercapacitors, asymmetric supercapacitors with faradaic pseudocapacitance electrode materials show the desired performance owing to reversible and rapid surface redox reactions [5–7].

Metal oxides [8,9], hydroxides [10,11], and sulfides [12–14] have been widely studied as electrode materials for supercapacitors. Other than inorganic solids, metal–organic frameworks (MOFs) with a porous structure and tunable functionality also show excellent potential applications for electrochemical energy storage [15–17]. The central metal ions of MOFs can perform as active sites for redox reactions and the controllable pore structure can provide open channels for the transmission of the electrolyte [18,19]. For instance, Wang et al. [20] synthesized a Co-based layered MOF, which showed a high specific capacitance, good rate capability, and long-term cycling life in 1M KOH. Guo et al. [21] prepared hierarchical porous Zr-MOFs as electrode materials for supercapacitors, showing a specific capacitance of the prepared HP-UiO-66 of 849 F g^{-1} at a current density of 0.2 A g^{-1}. Liu et al. [22] prepared a Mn-based MOF by a simple solution method; due to its 3D structure, it reached an admirable specific capacitance of 832.6 F g^{-1} at 1 A g^{-1}. The cuboid Ni-MOF fabricated by He et al. also exhibited a high specific capacitance of

804 Fg^{-1} at 1 Ag^{-1} [23]. However, poor electrical conductivity largely impedes the potentiality of MOFs to construct high-performance electrodes in supercapacitors. To improve the electrical conductivity, hybridizing conductive additives, such as carbon materials [24], conducting polymers [25], and metal foams [26] is an effective strategy. Zhou et al. [24] reported that capacitance performance can be successfully improved by the hybridization of Ni-MOFs with graphene nanosheets. The good electronic conductivity of graphene can enable excellent electron transportation for boosting the performance. Zhang et al. [27] have grown CoNi-MOF nanosheets on carbon fiber paper, which displayed a high specific capacitance of 2033 F g^{-1} at 1 A g^{-1}. Ni-MOF nanosheets were grown on Ni foam via a hydrothermal method reported by Wang et al. [26]. The as-prepared Ni-MOF/Ni foam exhibited a high specific capacitance with an outstanding cyclic stability. In addition, a Ni-doped Mn-MOF synthesized by Liu et al. [28] showed a high specific capacitance of 1676.6 F g^{-1} at 1 A g^{-1}. However, all these grown processes are conducted by mixing substrates with metal precursors, which may lead to the instability of growth in MOFs and make it difficult to meet the needs of structural control.

Ni foam (NF) is a suitable kind of support skeleton with a large pore structure and high conductivity for the growth of electrode materials [29,30]. It can dissociate Ni ions in an acidic solvent. In this work, we have used NF as a support and Ni as a source by taking advantage of its self-dissociation characteristic to grow Ni-MOF ($[Ni_3(OH)_2(C_8H_4O_2)_2(H_2O)_4]\cdot 2H_2O$) arrays directly. The terephthalic acid and metal Ni can undergo a replacement reaction in this acidic environment with a pH of around 5. In this reaction, the Ni can be oxidated into Ni^{2+} and H^+ can be reduced to H_2. The as-prepared Ni-MOF/NF-s exhibits a high specific capacitance of 1124 F g^{-1} at 1 A g^{-1}. and a good cyclic stability with a capacitance retention of 79% after 4000 cycles at 10 A g^{-1}. Additionally, the assembled hybrid supercapacitor demonstrates the maximum energy density of 38.3 Wh kg^{-1} at the power density of 725 W kg^{-1}, indicating the promising practical application in energy storage fields.

2. Results and Discussion

A schematic illustration for the synthesis of Ni-MOF/NF-s is shown in Figure 1. The NF was immersed in the DMF/H_2O mixed solvents with dissolved terephthalic acid (TPA) ligands. The pH of the solution measured was around 5. The TPA ligands can release H^+ in water, which can react with NF to dissociate Ni ions for the formation of Ni-MOF arrays by a one-step hydrothermal reaction.

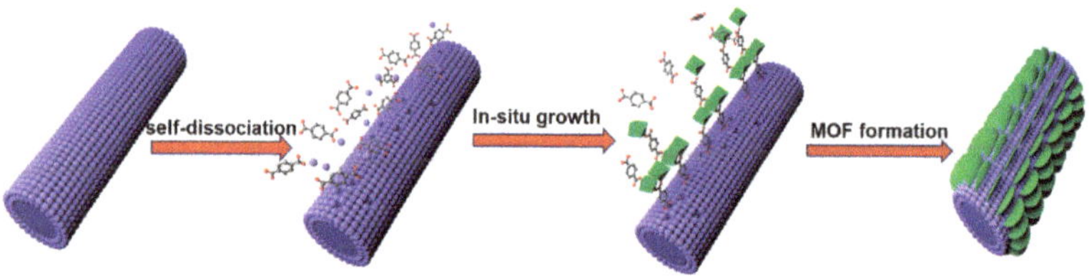

Figure 1. The schematic illustration for the synthesis of Ni-MOF/NF-s.

The morphology of Ni-MOF/NF-s and Ni-MOF/NF can be seen in Figure 2. It can be seen that the Ni-MOF was grown on the NF successfully. The formed microbelts are vertically based on the surface and the thickness of the nanosheet is about 170 nm. In addition, the microbelts are assembled into arrays on the NF, which can provide more available surface sites and mass transport channels to improve the electrochemical performance. The morphology of all the Ni-MOF/NF-s is different from the agglomerated nanosheets of Ni-MOF/NF. The difference may be based on whether a large amount of Ni^{2+} ions are present in the liquid phase. If there is a large amount of Ni^{2+} present in the liquid phase, and the MOFs at different locations may be formed simultaneously. But for Ni-MOF-NF-s, the Ni^{2+} ions released from the NF exist on the surface, and the growth of Ni-MOF may be based on the bottom up.

Figure 2. SEM images of (**a**,**b**) Ni-MOF/NF-s and (**c**,**d**) Ni-MOF/NF.

The morphology of Ni-MOF/NF-s is further observed by TEM and scanning TEM, as seen in Figure 3 below, the result of which confirms the 2D structure of the grown Ni-MOF with a smooth surface. Furthermore, the corresponding EDS test from the scanning TEM image shows that the of Ni, O, and C elements are uniformly dispersed through the Ni-MOF microbelts.

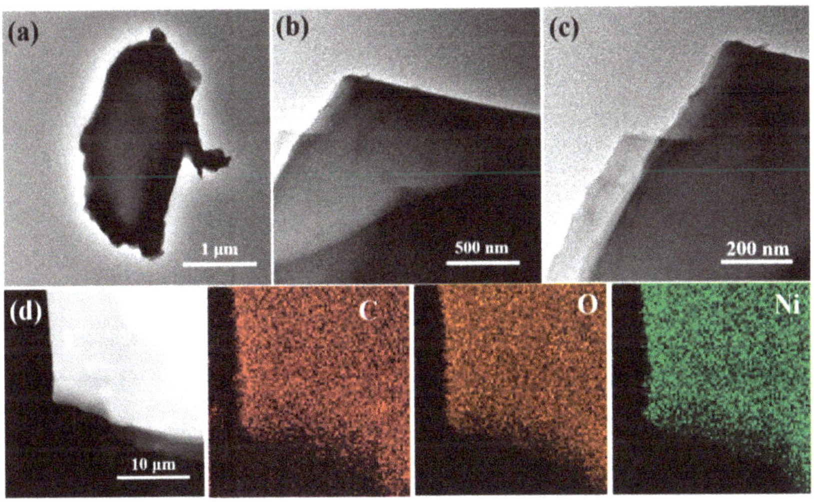

Figure 3. TEM images of (**a**–**c**) Ni-MOF/NF-s and (**d**) the elemental mapping images of Ni, C, and O.

The survey spectra of XPS (Figure 4a) show the coexistence of Ni, O, and C elements in Ni-MOF/NF-s and Ni-MOF/NF. Figure 4b shows similar high-resolution C 1s spectra for Ni-MOF/NF-s and Ni-MOF/NF, where the peaks come from the TPA linkers. The peak at 284.6 eV is related to the carbon of C-C configurations and the peak at 288.1 eV can be related to the carbon of carboxyl groups [31]. Moreover, the high-resolution XPS spectra of Ni 2p of Ni-MOF/NF-s and Ni-MOF/NF are shown in Figure 4c. Ni-MOF/NF-s displays two major peaks at 856.3 (Ni 2p 3/2) and 874.0 eV (Ni 2p 1/2) with their corresponding satellite peaks at 861.9 and 880.3 eV, indicating the divalent state of Ni [32]. Ni-MOF/NF also exhibits deconvoluted Ni 2p 3/2 and Ni 2p 1/2 peaks like those of Ni-MOF/NF-s. Furthermore, in the high-resolution O 1s spectra, both samples show the deconvoluted peaks of Ni-O bonds and O in carboxyl groups. The above results confirm that the composition of Ni-MOF synthesized by the self-dissociation method is consistent with that of the conventional method.

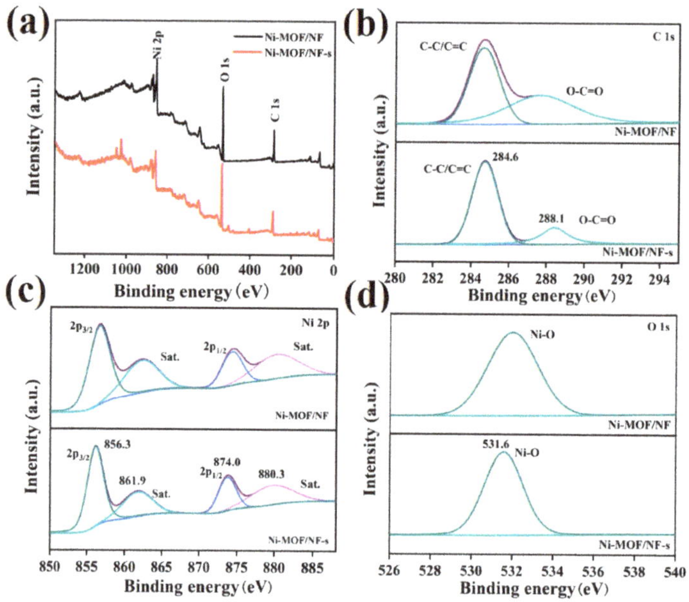

Figure 4. (**a**) XPS survey, (**b**) XPS C 1s, (**c**) XPS Ni 2p, and (**d**) XPS O 1s of Ni-MOF/NF-s and Ni-MOF/NF.

It can be seen from the XRD pattern that all samples have a high intensity peak at about 45°. Because of the high phase strength of nickel foam, the crystal structure of all samples is difficult to find. Therefore, Ni-MOF samples without nickel foam were prepared, and the corresponding XRD characterization is shown in Figure 5a. Significant peaks at 9.2°, 12.1°, 15.6°, 18.5°, 24.0°, and 29.0° (CCDC no. 638866) are similar to reported MOFs synthesized with a metal center Ni^{2+} and the organic ligand TPA [33,34], which can be related to $[Ni_3(OH)_2(C_8H_4O_2)_2(H_2O)_4]{\cdot}2H_2O$. An FT-IR spectrograph was used to further analyze the components of Ni-MOF/NF-s and Ni-MOF/NF. The peaks appearing at ca. 743, 812, 1090, and 1373 cm^{-1} are related to the deformation vibration of the aromatic C-H bonds of TPA linkers. Furthermore, the peaks present at ca. 1575 cm^{-1} correspond to the stretching vibration of the -COO^- in TPA. In addition, the broad peak at around 3400 cm^{-1} can be assigned to the adsorbed H_2O molecules [35].

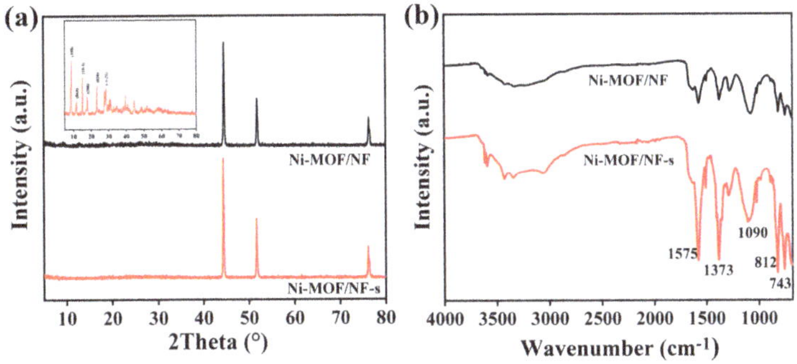

Figure 5. (**a**) XRD and (**b**) FT-IR spectra of Ni-MOF/NF-s and Ni-MOF/NF.

The electrochemical properties of the as-prepared Ni-MOF/NF-s are first evaluated in the three-electrode device compared with those of Ni-MOF/NF. The cyclic voltammograms (CVs) curves of various Ni-MOF/NFs at scan rates of 5 mV s^{-1} and 50 mV s^{-1} are displayed in Figure 6a,b. The couple of peaks in the CV curves are attributed to redox reactions of different valence states of Ni, indicating the pseudocapacitive characteristics [36]. Notably, the Ni-MOF/NF-s electrode exhibits larger enclosed curve areas both at 10 and 50 mV s^{-1}. This discrepancy indicates a higher reactivity and superior rate capability in the former electrode. The storage energy mechanism is further understood by the equation $i = av^b$, where i and v are the current and scan rate in the CV curves, and a and b are fitting values [37]. A value of $b = 0.5$ means a diffusion-controlled character, while $b = 1$ represents surface-induced behavior. In Figure 6c, the b value for Ni-MOF/NF-s is 0.67, higher than the 0.52 of Ni-MOF/NF, and closer to 1, suggesting better surface control behavior. The galvanostatic charge–discharge (GCD) profiles of Ni-MOF/NF-s and Ni-MOF/NF are shown in Figure 6d. The charge–discharge plateaus in the profiles further confirm the pseudocapacitive characteristics of the electrodes. Moreover, the Ni-MOF/NF-s electrode shows a longer discharge time than that of Ni-MOF/NF, implying a better electrochemical performance. In detail, the Ni-MOF/NF-s electrode delivers a higher capacitance of 1124 F g^{-1} at a current density of 1 A g^{-1} than 681 F g^{-1} of Ni-MOF/NF. When the current density is increased to 10 A g^{-1}, the Ni-MOF/NF-s electrode still shows a higher capacitance of 590 F g^{-1}, maintaining its 52% initial capacitance. The Nyquist plots tested by the electrochemical impedance spectroscopy are shown in Figure 6f, where the semicircle at the high frequency stands for the charge-transfer resistance and the slope at the low frequency reflects ion mobility [38,39]. The Ni-MOF/NF-s electrode has the better ion mobility (higher slope value) and lower charge-transfer resistance (smaller semicircle radius) compared with Ni-MOF/NF. This result is further confirmed by analysis of the ion diffusion kinetic using the equation of $Z' = R_s + R_{ct} + \sigma\omega^{-1/2}$. Z' is the real part, R_s is the internal resistance, R_{ct} is the charge-transfer resistance, and the ω is the angular frequency. The Ni-MOF/NF-s electrode shows a smaller σ (Warburg diffusion coefficient) value of 1.7 than the 2.4 of Ni-MOF/NF-s and Ni-MOF/NF, as seen in Figure 6g, implying enhanced ion diffusion kinetics. The OH-diffusion coefficient (D) in the electrolyte is calculated by the equation $D = \frac{R^2T^2}{2A^2n^4F^4C^2\sigma^2}$, where R (J mol^{-1} K^{-1}) relates to the gas constant, T (K) relates to the absolute temperature, A (cm^2) relates to the electrode area, n is the transfer F (C mol^{-1}) which relates to the Faraday constant, and C (mol L^{-1}) is the concentration of OH^- [40]. The D value of Ni-MOF/NF-s is 4.7×10^{-17} cm^2 s^{-1}, which is higher than that of the Ni-MOF/NF material (3.3×10^{-17} cm^2 s^{-1}), indicating that the Ni-MOF/NF-s ion diffusion rate is faster, exhibiting excellent ion diffusion/transport dynamics. Since the Ni^{2+} of the Ni-MOFs is coming from the NF, the enhanced structural stability and tight contact interfaces between the Ni-MOF and NF may be favorable for better charge storage and conductivity. In addition, Ni-MOF/NF-s also shows a better cycling stability with a capacitance retention of 79% after 4000 cycles at 10 A g^{-1} than the 65% of Ni-MOF/NF.

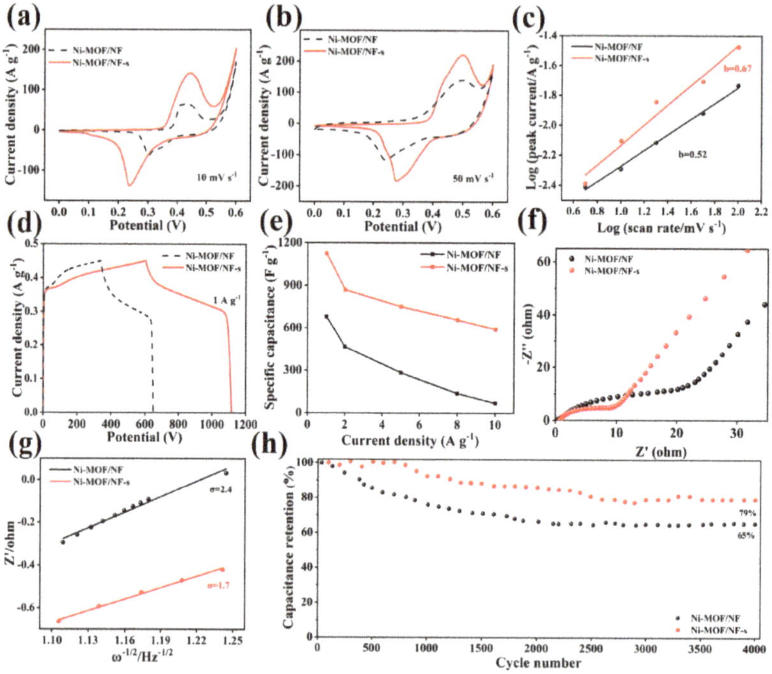

Figure 6. The electrochemical performance of Ni-MOF/NF-s and Ni-MOF/NF: (**a**) CV curves at 5 mV s^{-1}, (**b**) CV curves at 50 mV s^{-1}, (**c**) b-value determination of the peak currents of CV curves, (**d**) GCD curves at 1 A g^{-1}, (**e**) specific capacities at 1–10 A g^{-1}, (**f**) Nyquist plots, (**g**) linear relation of Z' vs. $\omega^{-1/2}$, and (**h**) cycling stability at 10 A g^{-1}.

To estimate the practical application of Ni-MOF/NF-s in the hybrid supercapacitor, Ni-MOF/NF-s is used as a positive electrode and active carbon is used as a negative electrode to assemble the hybrid supercapacitor (Ni-MOF/NF-s//AC). The performance of active carbon can be found in Figure 7a,b. The CV and GCD curves confirm the double layer energy storage characteristic of the active carbon. Figure 7c shows the CV curves of the hybrid supercapacitor at different scan rates (5–20 mV s^{-1}), which indicates pseudocapacitance and electric double-layer capacitance in the supercapacitor. The GCD curves at various current densities (1–10 A g^{-1}) for the hybrid supercapacitor are shown in Figure 7d, with the nonlinear feature proving the same result as found in CV analysis. The calculated capacitances of the hybrid supercapacitor are shown in Figure 7e. It yields a specific capacitance of 131 F g^{-1} at a current density of 1 A g^{-1} and maintains 77 F g^{-1} at 10 A g^{-1}. These results are compared with previously published Ni-MOF based electrodes [17,41–47], thereby illustrating the excellent performance of our supercapacitor. The Ragone plot shown in Figure 7f shows that Ni-MOF/NF-s//AC has the maximum energy density of 38.3 Wh kg^{-1} at a power density of 725 W kg^{-1}, and it can remain at the energy density of 22.5 Wh kg^{-1} when the power density is up to 7250 W kg^{-1}. As comparison, Ni-MOF/NF//AC shows a lower specific capacitance of 112 F g^{-1} at a current density of 1 A g^{-1} and maintains a poor performance of 24 F g^{-1} at 10 A g^{-1} (Figure S1).

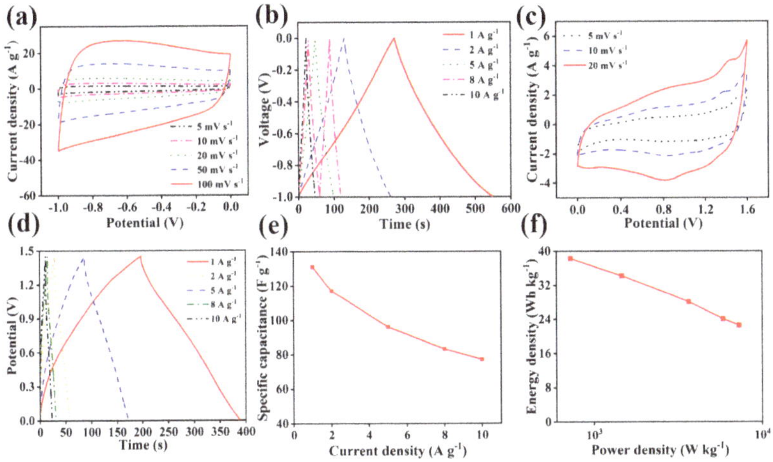

Figure 7. Electrochemical performance diagram of active carbon: (**a**) CV curves and (**b**) GCD curves; and Ni-MOF/NF-s//AC asymmetric supercapacitor: (**c**) CV curves, (**d**) GCD curves, (**e**) specific capacity diagram, and (**f**) Ragone plots.

The cycling test of the Ni-MOF/NF-s//AC can be tested at a current density of 10 A g^{-1}, where the 71% of its initial specific capacitance can be remained after 4000 cycles (Figure S2), implying a good cycling performance.

3. Materials and Methods

$Ni(NO_3)_2·6H_2O$ (Macklin, Shanghai, China), terephthalic acid (Macklin), ethanol (Fuyu Group, Tianjin, China), and HCl (Xihua Group, Guangzhou, China) were used as received.

NF was firstly cut into pieces with a size of 1 × 3.5 × 0.2 cm. The 0.1 M HCl solution and ethanol were used to wash NF to remove the oxides and impurities. To prepare the Ni-MOF/NF-s, the NF was soaked in 30 mL DMF + 5 mL water solution with 40 mg terephthalic acid (TPA). After sonicating for 30 min, the solution was then transferred into a 50 mL Teflon-lined autoclave and kept at 150 °C for 10 h. After reaction, the product was washed with water and ethanol, as well as dried at 60 °C for 6 h, to obtain Ni-MOF/NF-s.

To prepare the Ni-MOF/NF, the NF was soaked in 30 mL DMF + 5 mL water solution with 50 mg $Ni(NO_3)_2·6H_2O$ and 40 mg terephthalic acid (TPA). After sonicating for 30 min, the solution was then transferred into a 50 mL Teflon-lined autoclave and kept at 150 °C for 10 h. After reaction, the product was washed with water and ethanol to obtain Ni-MOF-s/NF.

4. Conclusions

In summary, Ni-MOFs/NF has been easily synthesized through the self-dissociation strategy by using NF as a support and Ni as a source. The in situ self-assembly of the TPA linkers with the dissociated Ni ions from the Ni foam guarantees the unblocked channels of charge transport. The Ni-MOF microbelt arrays can offer numerous active redox sites for improving the electrochemical performance. The as-made Ni-MOFs/NF-s electrode exhibits a high capacitance of 1124 F g^{-1} at 1 A g^{-1} and keeps 590 F g^{-1} at 10 A g^{-1}. In addition, the hybrid supercapacitor based on the Ni-MOFs/NF-s shows a high energy density of 38.3 Wh kg^{-1} at a power density of 725 W kg^{-1}.

Supplementary Materials: The following supporting information can be downloaded at https://www.mdpi.com/article/10.3390/molecules30030513/s1, Figure S1. Electrochemical performance Ni-MOF/NF//AC asymmetric supercapacitor: (a) CV curves, (b) GCD curves, (c) specific capacity diagram. Figure S2. Cyclic performance diagram of Ni-MOF/NF-s//AC. Table S1. Comparison of the electrochemical properties of Ni-MOF-based asymmetric supercapacitors.

Author Contributions: Conceptualization, H.L. and X.Z.; methodology, Y.T.; software, S.S.; validation, B.F. and B.L.; formal analysis, Z.L.; investigation, H.L.; resources, X.Z.; data curation, X.Z.; writing—original draft preparation, H.L.; writing—review and editing, Y.L. and X.Z.; All authors have read and agreed to the published version of the manuscript.

Funding: This research was funded by Shenyang Science and Technology Project, grant number 23-407-3-22 and 22-316-1-05.

Institutional Review Board Statement: Not applicable.

Informed Consent Statement: Not applicable.

Data Availability Statement: All data supporting the findings of this study are available within the paper and its Supplementary Materials, which are published online.

Conflicts of Interest: The authors declare no conflicts of interest.

References

1. Zhang, W.; Li, W.; Li, S. Self-template activated carbons for aqueous supercapacitors. *Sustain. Mater. Technol.* **2023**, *36*, 00582. [CrossRef]
2. Liang, Z.; Shen, J.; Xu, X.; Li, F.; Liu, J.; Yuan, B.; Yu, Y.; Zhu, M. Advances in the development of single-atom catalysts for high-energy-density lithium-sulfur batteries. *Adv. Mater.* **2022**, *34*, 2200102. [CrossRef]
3. Zhang, L.; Qin, X.; Zhao, S.; Wang, A.; Luo, J.; Wang, Z.L.; Kang, F.; Lin, Z.; Li, B. Advanced matrixes for binder-free nanostructured electrodes in lithium-ion batteries. *Adv. Mater.* **2020**, *32*, 1908445. [CrossRef] [PubMed]
4. Huang, J.; Yuan, K.; Chen, Y. Wide voltage aqueous asymmetric supercapacitors: Advances, strategies, and challenges. *Adv. Funct. Mater.* **2022**, *32*, 2108107. [CrossRef]
5. Shi, X.; Zheng, S.; Wu, Z.S.; Bao, X. Recent advances of graphene-based materials for high-performance and new-concept supercapacitors. *J. Energy Chem.* **2018**, *27*, 25–42. [CrossRef]
6. Jiang, L.; Sheng, L.; Fan, Z. Biomass-derived carbon materials with structural diversities and their applications in energy storage. *Sci. China Mater.* **2018**, *61*, 133–158. [CrossRef]
7. Chen, X.; Liu, Q.; Bai, T.; Wang, W.; He, F.; Ye, M. Nickel and cobalt sulfide-based nanostructured materials for electrochemical energy storage devices. *Chem. Eng. J.* **2021**, *409*, 127237. [CrossRef]
8. Sreejesh, M.; Dhanush, S.; Rossignol, F.; Nagaraja, H.S. Microwave assisted synthesis of rGO/ZnO composites for non-enzymatic glucose sensing and supercapacitor applications. *Ceram. Int.* **2017**, *43*, 4895–4903. [CrossRef]
9. Zhai, T.; Wan, L.; Sun, S.; Chen, Q.; Sun, J.; Xia, Q.; Xia, H. Phosphate ion functionalized Co_3O_4 ultrathin nanosheets with greatly improved surface reactivity for high performance pseudocapacitors. *Adv. Mater.* **2017**, *29*, 1604167. [CrossRef] [PubMed]
10. Li, H.; Musharavati, F.; Zalenezhad, E.; Chen, X.; Hui, K.N.; Hui, K.S. Electrodeposited Ni-Co layered double hydroxides on titanium carbide as a binder-free electrode for supercapacitors. *Electrochim. Acta.* **2018**, *261*, 178–187. [CrossRef]
11. Lee, D.; Mathur, S.; Kim, K.H. Bilayered $NiZn(CO_3)(OH)_2$-$Ni_2(CO_3)(OH)_2$ nanocomposites as positive electrode for supercapacitors. *Nano Energy* **2021**, *86*, 106076. [CrossRef]
12. Liu, Y.; Jiang, G.; Sun, S.; Xu, B.; Zhou, J.; Zhang, Y.; Yao, J. Decoration of carbon nanofibers with $NiCo_2S_4$ nanoparticles for flexible asymmetric supercapacitors. *J. Alloys Compd.* **2018**, *731*, 560–568. [CrossRef]
13. Zhao, W.; Zheng, Y.; Cui, L.; Jia, D.; Wei, D.; Zheng, R.; Barrow, C.; Yang, W.; Liu, J. MOF derived Ni-Co-S nanosheets on electrochemically activated carbon cloth via an etching/ion exchange method for wearable hybrid supercapacitors. *Chem. Eng. J.* **2019**, *371*, 461–469. [CrossRef]
14. Zhang, X.; Tian, Y.; Lu, W.; Yang, S.; Qu, N.; Zhang, Q.; Lei, D.; Liu, A. Design of oxygen-doped Co_3S_4 hollow nanosheets by suppressed sulfurization for supercapacitors. *Chemelectrochem* **2021**, *8*, 3629–3636. [CrossRef]
15. Yang, B.; Li, B.; Xiang, Z. Advanced MOF-based electrode materials for supercapacitors and electrocatalytic oxygen reduction. *Nano Res.* **2022**, *16*, 1338–1361. [CrossRef]
16. Zheng, S.; Zhou, H.; Xue, H.; Braunstein, P.; Pang, H. Pillared-layer Ni-MOF nanosheets anchored on Ti_3C_2 MXene for enhanced electrochemical energy storage. *J. Colloid Interface Sci.* **2022**, *614*, 130–137. [CrossRef] [PubMed]
17. Zhang, X.; Yang, S.; Lu, W.; Lei, D.; Tian, Y.; Guo, M.; Mi, P.; Qu, N.; Zhao, Y. MXenes induced formation of Ni-MOF microbelts for high-performance supercapacitors. *J. Colloid Interface Sci.* **2021**, *592*, 95–102. [CrossRef] [PubMed]
18. Du, W.; Bai, Y.L.; Xu, J.; Zhao, H.; Zhang, L.; Li, X.; Zhang, J. Advanced metal-organic frameworks (MOFs) and their derived electrode materials for supercapacitors. *J. Power Sources* **2018**, *402*, 281–295. [CrossRef]

19. Gao, X.; Dong, Y.; Li, S.W.; Zhou, J.W.; Wang, L.; Wang, B. MOFs and COFs for Batteries and Supercapacitors. *Electrochem. Energy Rev.* **2020**, *3*, 81–126. [CrossRef]
20. Liu, X.; Shi, C.; Zhai, C.; Cheng, M.; Liu, Q.; Wang, G. Cobalt-based layered metal-organic framework as an ultrahigh capacity supercapacitor electrode material. *ACS Appl. Mater. Interfaces* **2016**, *8*, 4585–4591. [CrossRef] [PubMed]
21. Gao, W.; Chen, D.; Quan, H.; Zou, R.; Wang, W.; Luo, X.; Guo, L. Fabrication of hierarchical porous metal-organic framework electrode for aqueous asymmetric supercapacitor. *ACS Sustain. Chem. Eng.* **2017**, *5*, 4144–4153. [CrossRef]
22. Rong, H.R.; Song, P.; Gao, G.X.; Jiang, Q.Y.; Chen, X.J.; Su, L.X.; Liu, W.L.; Liu, Q. A three-dimensional Mn-based MOF as a high-performance supercapacitor electrode. *Dalton Trans.* **2023**, *52*, 1962–1969. [CrossRef] [PubMed]
23. Gao, S.W.; Sui, Y.W.; Wei, F.X.; Qi, J.Q.; Meng, Q.K.; He, Y.Z. Facile synthesis of cuboid Ni-MOF for high-performance supercapacitors. *J. Mater. Sci. Lett.* **2018**, *53*, 6807–6818. [CrossRef]
24. Zhou, Y.J.; Mao, Z.M.; Wang, W.; Yang, Z.K.; Liu, X. In-Situ Fabrication of Graphene Oxide Hybrid Ni-Based Metal-Organic Framework (Ni-MOFs@GO) with Ultrahigh Capacitance as Electrochemical Pseudocapacitor Materials. *ACS Appl. Mater. Interfaces* **2016**, *8*, 28904–28916. [CrossRef] [PubMed]
25. Rajkumar, S.; Elanthamilan, E.; Evangeline, J.N.I.; Sharmila, L.; Princy, M.J. Enhanced electrochemical behaviour of Co-MOF/PANI composite electrode for supercapacitors. *Inorganica Chim. Acta* **2020**, *502*, 119393.
26. Pan, Y.; Han, Y.H.; Chen, Y.J.; Li, D.; Tian, Z.; Guo, L.; Wang, Y.Z. Benzoic acid-modified 2D Ni-MOF for high-performance supercapacitors. *Electrochim. Acta* **2022**, *403*, 139679. [CrossRef]
27. Chu, X.; Meng, F.; Deng, T.; Lu, Y.; Bondarchuk, O.; Sui, M.; Feng, M.; Li, H.; Zhang, W. Mechanistic insight into bimetallic CoNi-MOF arrays with enhanced performance for supercapacitors. *Nanoscale* **2020**, *12*, 5669–5677. [CrossRef]
28. Liu, X.J.; Zhang, X.L.; Liu, R.M.; Li, C.P.; Xu, C.Y.; Ding, H.H.; Xing, T.; Dai, Z.R.; Zhu, X.D. A Ni-doped Mn-MOF decorated on Ni-foam as an electrode for high-performance supercapacitors. *J. Nanoparticle Res.* **2022**, *24*, 23. [CrossRef]
29. Wei, X.; Li, N.; Liu, N. Ultrathin NiFeZn-MOF nanosheets containing few metal oxide nanoparticles grown on nickel foam for efficient oxygen evolution reaction of electrocatalytic water splitting. *Electrochim. Acta* **2019**, *318*, 957–965. [CrossRef]
30. Tang, P.P.; Lin, X.; Yin, H.; Zhang, D.X.; Wen, H.; Wang, J.J.; Wang, P. Hierarchically nanostructured nickel-cobalt alloy supported on nickel foam as a highly efficient electrocatalyst for hydrazine oxidation. *ACS Sustain. Chem. Eng.* **2020**, *8*, 16583–16590. [CrossRef]
31. Zhang, X.; Qu, N.; Fan, Q.Y.; Yang, H. Surface functionalization of graphene oxide with DBU as electrode materials for supercapacitors. *Mater. Res. Express* **2019**, *6*, 085606. [CrossRef]
32. Zhang, X.; Fan, Q.Y.; Liu, S.; Qu, N.; Yang, H.; Wang, M.; Yang, J. A facile fabrication of 1D/2D nanohybrids composed of NiCo-hydroxide nanowires and reduced graphene oxide for high-performance asymmetric supercapacitors. *Inorg. Chem. Front.* **2020**, *7*, 204–211. [CrossRef]
33. Yue, L.; Chen, L.; Wang, X.; Lu, D.; Zhou, W.; Shen, D.; Yang, Q.; Xiao, S.; Li, Y. Ni/Co-MOF@aminated MXene hierarchical electrodes for high-stability supercapacitors. *Chem. Eng. J.* **2023**, *451*, 138687. [CrossRef]
34. Zhang, X.; Qu, N.; Yang, S.X.; Fan, Q.Y.; Lei, D.; Liu, A.M.; Chen, X. Shape-controlled synthesis of Ni-based metal-organic frameworks with albizia flower-like spheres@nanosheets structure for high performance supercapacitors. *J. Colloid Interface Sci.* **2020**, *575*, 347–355. [CrossRef]
35. Zhao, S.; Zeng, L.; Cheng, G.; Yu, L.; Zeng, H. Ni/Co-based metal-organic frameworks as electrode material for high performance supercapacitors. *Chin. Chem. Lett.* **2019**, *30*, 605–609. [CrossRef]
36. Li, S.; Zhang, Y.; Liu, N.; Yu, C.; Lee, S.J.; Zhou, S.; Fu, R.; Yang, J.; Guo, W.; Huang, H.; et al. Operando revealing dynamic reconstruction of NiCo carbonate hydroxide for high-rate energy storage. *Joule* **2020**, *4*, 673–687. [CrossRef]
37. Zhang, X.; Lu, W.; Tian, Y.H.; Yang, S.X.; Zhang, Q.; Lei, D.; Zhao, Y.Y. Nanosheet-assembled NiCo-LDH hollow spheres as high-performance electrodes for supercapacitors. *J. Colloid. Interface Sci.* **2022**, *606*, 1120–1127. [CrossRef]
38. Gao, S.; Sui, Y.; Wei, F.; Qi, J.; Meng, Q.; Ren, Y.; He, Y. Dandelion-like nickel/cobalt metal-organic framework based electrode materials for high performance supercapacitors. *J. Colloid Interface Sci.* **2018**, *531*, 83–90. [CrossRef] [PubMed]
39. Yan, Y.; Luo, Y.; Ma, J.; Li, B.; Xue, H.; Pang, H. Facile synthesis of vanadium metal-organic frameworks for high-performance supercapacitors. *Small* **2018**, *14*, 1801815. [CrossRef] [PubMed]
40. Jiang, B.; Ban, X.H.; Wang, Q.; Cheng, K.; Zhu, K.; Ye, K.; Wang, G.L.; Cao, D.X.; Yan, J. Anionic P-substitution toward ternary Ni-S-P nanoparticles immobilized graphene with ultrahigh rate and long cycle life for hybrid supercapacitors. *J. Mater. Chem. A* **2019**, *7*, 24374–24388. [CrossRef]
41. Wang, J.W.; Ma, Y.X.; Kang, X.Y.; Yang, H.J.; Liu, B.L.; Li, S.S.; Zhang, X.D.; Ran, F. A novel moss-like 3D Ni-MOF for high performance supercapacitor electrode material. *J. Solid State Chem.* **2022**, *309*, 122994. [CrossRef]
42. Du, P.C.; Dong, Y.N.; Liu, C.; Wei, W.L.; Liu, D.; Liu, P. Fabrication of hierarchical porous nickel based metal-organic framework (Ni-MOF) constructed with nanosheets as novel pseudo-capacitive material for asymmetric supercapacitor. *J. Colloid Interface Sci.* **2018**, *518*, 57–68. [CrossRef] [PubMed]

43. Xiao, Y.; Wei, W.; Zhang, M.J.; Jiao, S.; Shi, Y.C.; Ding, S.J. Facile Surface Properties Engineering of High-Quality Graphene: Toward Advanced Ni-MOF Heterostructures for High-Performance Supercapacitor Electrode. *ACS Appl. Energy Mater.* **2019**, *2*, 2169. [CrossRef]
44. Zheng, W.; Sun, S.; Xu, Y.; Yu, R.; Li, H. Sulfidation of Hierarchical NiAl-LDH/Ni-MOF Composite for High-Performance Supercapacitor. *Chemelectrochem* **2019**, *6*, 3375–3382. [CrossRef]
45. Nanda, O.P.; Ravipati, M.; Durai, L.; Badhulika, S. Ni-Metal organic framework nanosheets and Ni_3C/biomass porous carbon composite based long cycle life asymmetric supercapacitor. *Mater. Res. Bull.* **2023**, *168*, 112488. [CrossRef]
46. Liang, R.; Du, Y.; Lin, J.; Chen, J.; Xiao, P. Facile-Synthesized Ni-Metal-Organic Framework/Nano Carbon Electrode Material for High-Performance Supercapacitors. *Energy Fuels* **2022**, *36*, 7115–7120. [CrossRef]
47. Hang, X.X.; Yang, R.; Xue, Y.D.; Zheng, S.S.; Shan, Y.Y.; Du, M.; Zhao, J.W.; Pang, H. The introduction of cobalt element into nickel-organic framework for enhanced supercapacitive performance. *Chin. Chem. Lett.* **2023**, *34*, 107787. [CrossRef]

molecules MDPI

Article

Surface Microstructure Study on Corona Discharge-Treated Polyethylene Using Positron Annihilation Spectroscopy

Jingjing Li [1], Zhiwei Shen [1], Liuyang Tie [1], Tianyuan Long [1], Qiyue Zhong [1], Xi Chen [1], Chongshan Yin [2], Liguo Liufu [1], Xianhao Huang [1], Bangyun Xiong [1], Xibo Li [3,*], Chongxiong Duan [1] and Chunqing He [4,*]

1 Guangdong Key Laboratory for Hydrogen Energy Technologies, School of Materials Science and Hydrogen Energy, Foshan University, Foshan 528000, China; jjli@fosu.edu.cn (J.L.); shenzw8106@163.com (Z.S.); tieliuyang311@163.com (L.T.); 13631450402@163.com (T.L.); kop0526@outlook.com (Q.Z.); cx18023404833@163.com (X.C.); 18520322773@163.com (L.L.); 13674041967@163.com (X.H.); xiongbangyun@163.com (B.X.); cechxduan@fosu.edu.cn (C.D.)

2 Hunan Provincial Key Laboratory of Flexible Electronic Materials Genome Engineering, School of Physics and Electronic Science, Changsha University of Science and Technology, Changsha 410114, China; c.sh.yin@foxmail.com

3 Department of Physics, Jinan University, Guangzhou 510632, China

4 Key Laboratory of Nuclear Solid-State Physics Hubei Province, School of Physics and Technology, Wuhan University, Wuhan 430072, China

* Correspondence: lxb@csrc.ac.cn (X.L.); hecq@whu.edu.cn (C.H.)

Abstract: The microstructure and chemical properties of the corona discharge process could provide an effective method for predicting the performance of high-voltage cable insulation materials. In this work, the depth profile of the microstructure and chemical characteristics of corona discharge-treated PE were extensively investigated using Doppler broadening of position annihilation spectroscopy accompanied with positron annihilation lifetime spectroscopy, attenuated total reflectance Fourier transform infrared spectra, Raman spectra and contact angle measurement. By increasing corona discharge duration, the oxygen-containing polar groups, including hydroxyl, carbonyl and ester groups, strongly contribute to the deterioration of hydrophobicity and the enhancement of hydrophilicity. And the mean free volume size, with a broadening distribution, decreases slightly. The line shape *S* parameter decreases because of the decrease in free volume elements and the appearance of oxygen-containing groups. Also, the thickness of the degradation layer, determined from the *S* parameter with positron injection depth, increases and diffuses into the PE matrix. A linear *S*-*W* plot within the degradation layer of different corona treatment duration samples indicates the defect type does not change. The *S* parameter decreases and the *W* parameter increases with an increasing corona duration. Using a slow positron beam, the nondestructive probe can be used to profile the microstructure and chemical environment across the corona discharge damage depth, which is beneficial for investigating the surface and interfacial insulation materials.

Keywords: corona discharge; positron annihilation; free volume; *S* parameter; hydrophilicity

1. Introduction

Polyethylene (PE), with chemically inert properties and a low dielectric constant, has been widely used for electrical insulation materials in power transmission systems such as low-voltage, high-voltage and ultrahigh-voltage cable. Electrical aging and dielectric local heating are the main reasons for the premature failure of insulators operating at high voltage. Long-term corona discharge treatment leads to the deterioration or even failure of insulating materials. The main theory of corona discharge contains space charge, electron collision theory, the ozone erosion effect and dielectric heating [1–3]. The physical and chemical properties of a material surface vary significantly after corona discharge treatment, involving changes in free energy and surface wettability [4], surface charge generation and recombination [5], surface degradation characteristics [6] and the dynamic response of space charge inside the dielectric [7]. A study focusing on the microstructure and chemical properties during the corona discharge process could provide an effective method for predicting the performance of insulating materials. Many other probing techniques, including X-ray, neutron scattering electron microscopy, small-angle neutron scattering (SANS), dynamical mechanical analysis (DMA), X-ray diffraction (XRD) and differential scanning calorimetry (DSC), are used to characterize the structure of polymer materials at different lengths and infer changes in the polymer structure in an indirect way [8]. However, positron annihilation spectroscopy adopts a direct way. It can profile the corona discharge damage depth in corona discharge materials, since the penetration depth of positrons can be controlled via controlling the injected positron energy. Positron annihilation technology is a powerful non-destructive probe tool used to investigate the microstructure and chemical characteristics of different materials containing vacancies, vacancy clusters, free volumes, porous structures, interfaces, surfaces, as well as some other films [9–13]. An energetic positron injected into condensed matter will annihilate with an electron or become positronium (Ps) after slowing down to a thermal state through inelastic collisions with surround atomic cores. Ps, a bound state of a positron and an electron, with a Bohr diameter of 0.106 nm, contains triple-state ortho-Ps (*o*-Ps) with parallel spins and singlet-state para-Ps (*p*-Ps) with antiparallel spins. In vacuum, *o*-Ps annihilates into 3γ rays with an intrinsic lifetime of 142 ns, and *p*-Ps annihilates into 2γ rays with a very short intrinsic lifetime of 125 ps. *o*-Ps is highly sensitive to low-electron-density areas such as vacancies, free volumes and holes, and it is easily trapped therein. It picks up an external electron of antiparallel spin on the wall of the hole and annihilates into 2γ rays, with a lifetime shortened to a few nanoseconds, a process known as pick-off annihilation. Generally, for polymer materials, a positron annihilation lifetime spectrum (PALS) is resolved into three or four components, and it is characterized by lifetimes (τ_1: ~0.125, τ_2: 0.3~0.5 and a long lifetime τ_3 of 1~10 ns) and corresponding intensities (I_1, I_2, I_3). The long-lifetime component is closely related to the hole's size and relative quantity. A semiempirical Tao–Eldrup equation [14,15] based on a simple quantum mechanical model clearly describes the relation between the *o*-Ps lifetime (τ_3) and the mean radius (R) of a hole smaller than 1 nm:

$$\tau_3 = \frac{1}{2}\left(1 - \frac{R}{R+\Delta R} + \frac{1}{2\pi}\sin\left(\frac{2\pi R}{R+\Delta R}\right)\right)^{-1}, \tag{1}$$

where R is the average radius of the free volume and ΔR = 1.66 Å (an empirical parameter).

Doppler broadening spectroscopy (DBS) is based on the detection of 511 keV γ energy spectroscopy when positrons are annihilated with either valance or core electrons, which reflects the momentum distribution of the electrons participating. Annihilating with valance electrons can be described by the line shape parameter S, which is defined as the ratio of the annihilation events of the central region to the total area of the 511 keV annihilation peak. The other parameter, named W, which is defined as the ratio of the annihilation events of the wing regions to the total peak, represents the characteristics of positron annihilation with surrounding core electrons. Generally speaking, S and W can be used to detect the variations in sub-nanometer defect sizes and quantities, as well as the chemical characteristics of materials. In most of the polymeric materials, the S parameter is sensitive to the variation of free volumes or holes and their concentration. And it represents the relative contribution of the low-momentum part of the positron–electron annihilation radiation [16]. For materials with more vacancy-type defects and free volume holes, the S parameter will exhibit a higher value [17,18]. Both the increase in the self-annihilation of *p*-Ps and the information probability of Ps contribute to the narrowing of the 511 keV annihilation peak, and an increase in the S parameter will occur [19].

A depth profile study of a polymer sample (surface, films and interface) [17,20,21] was successfully carried out using the slow positron annihilation technique via controlling the injected positron energy after moderating, converging and accelerating. And the mean implantation depth of positrons depending on the injected positron energy in materials is described as follows [22]:

$$Z_m = 40 \times \frac{E^{1.6}}{\rho}, \quad (2)$$

where Z_m is the mean injected depth (nm), E is the injected positron energy (keV) and ρ is the density of polyethylene in this work (ρ of PE is 0.94 g/cm^3).

The main aim was to investigate the depth profile microstructure and chemical properties of the corona discharge process at an early stage. In this study, the PE samples were corona discharge-treated for varying durations using a "needle-plate" system at 5 kV and 50 Hz. And the microstructure, chemical properties and hydrophilicity were extensively studied using ATR-FTIR, Raman spectroscopy, contact angle, PALS and DBS.

2. Results and Discussion

2.1. Chemical Properties of Corona Discharge-Treated PE

To extensively study the chemical properties of corona discharge-treated PE, ATR-FTIR spectrometry was used. As shown in Figure 1, there are four absorption bands of methylene (-CH_2-) for untreated PE. Asymmetric and symmetric stretching vibration as well as the bending and rocking vibration of the C-H bond are observed at 2915 cm^{-1}, 2847 cm^{-1}, 1459 cm^{-1} and 719 cm^{-1}, respectively. These findings are consistent with those reported in the literature [23]. After corona discharge treatment, the absorption band intensity decreases significantly with increasing corona discharge duration. Further, the absorption band at 3700~3000 cm^{-1} is related to the hydroxyl (O-H) stretching vibration. It suggests that the corona discharge treatment contributes to the formation of hydroxyl radicals. The stretching vibration of carbonyl (C=O) appears at 1720 cm^{-1} and 1592 cm^{-1} [24]. And the vibration of ester group (-O-C=O-) appears at 1247 cm^{-1} and 1039 cm^{-1}. With increasing corona discharge treatment duration from 20 h to 50 h, a much weaker carbonyl group is noticed. Another interesting phenomenon is the vibration band of the ester group showing noticeable differences between the 20 h- and 50 h-treated sample. An absorption band can also be observed in the region of 1300–1421 cm^{-1}, indicative of the formation of a C-N bond. Corona discharge damage on the hydrocarbon backbone of PE may produce carbon radicals, which react with nitrogen in air. The results suggest the chemical modification generated by corona discharge treatment, as evidenced by the introduction of oxygen-containing groups or even a C-N bond within the carbon-hydrogen backbone.

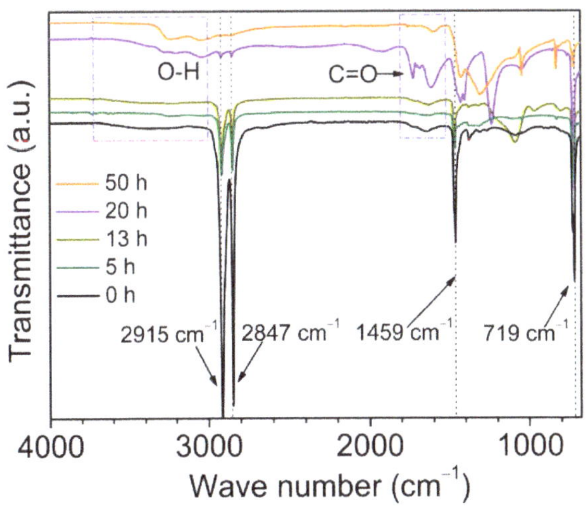

Figure 1. ATR-FTIR spectra of PE before and after different corona discharged duration.

Considering the methylene contributes to poor wettability, these new generated oxygen-containing polar groups suggest that the hydrocarbons chain structure of PE is destroyed by corona discharge treatment and the surface is chemically modified by corona discharge for the formation of carbon radicals, unstable hydroperoxides and the oxygen-containing polar groups [25]. And the wettability will be improved.

Raman spectra of untreated and corona discharge-treated PE with varying duration are shown in Figure 2. Raman shifts appearing at 1060 cm^{-1}, 1126 cm^{-1}, 1168 cm^{-1}, 1292 cm^{-1}, 1336 cm^{-1}, 1414 cm^{-1}, 1436 cm^{-1} and 1456 cm^{-1} are the characteristic peaks of methylene. The Raman shifts lying at 1060 cm^{-1} and 1126 cm^{-1} are attributed to asymmetric (ν_{as}) and symmetric (ν_s) stretching vibration of C-C bonds, respectively. And the Raman shifts located at 1168 cm^{-1} and 1292 cm^{-1} are attributed to the rocking vibration (ρ) and distortion vibration (τ) of methylene, respectively. The Raman shifts that appear at 1366 cm^{-1} and 1414 cm^{-1} are attributed to the wagging vibration (ω) of methylene. The Raman shifts that appear at 1436 cm^{-1} and 1456 cm^{-1} are attributed to the bending vibration (δ) of methylene. With increasing corona discharge duration from 5 h to 50 h at 5 kV, the characteristic Raman band position is unchanged, but the band intensity decreases obviously. The Raman spectra variations in intensity are in accordance with the ATR-FTIR description, which indicates the successful chemical modification of corona discharge treatment.

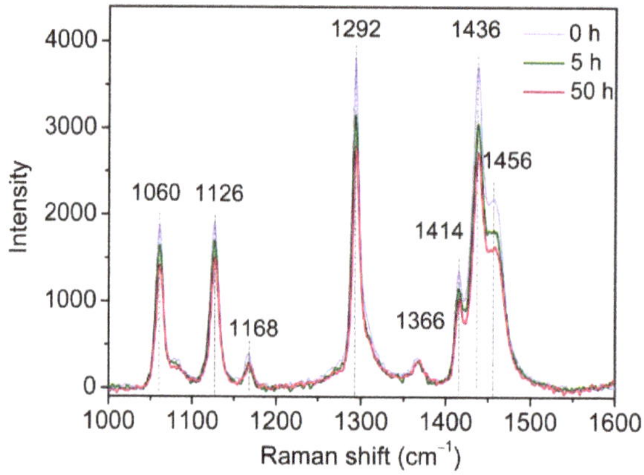

Figure 2. Raman spectra of untreated and corona discharge-treated PE for different duration.

2.2. *Surface Hydrophilicity of Corona Discharge-Treated PE*

Figure 3 depicts the water contact angles of untreated and corona discharge-treated PE samples. For the untreated PE and 5-hour-treated sample, the water contact angles are 104.7° and 98.5°, respectively, which exhibits hydrophobicity. With increasing treatment duration from 5 h to 50 h, the water contact angle decreases to 48.8°. After corona discharge treatment, the decrease in the water contact angle indicates a remarkable increase in hydrophilicity, which is explained by the introduction of oxygen-containing polar groups. In order to meet the requirements of dyeability improvement, Park et al. increased the corona discharge power to introduce a polar group and to induce the graft polymerization of acrylic acid onto the corona-treated low-density PE film surface [25]. The results substantiate the enhanced hydrophilicity caused by corona discharge treatment for the variation in chemical character of olefin chain.

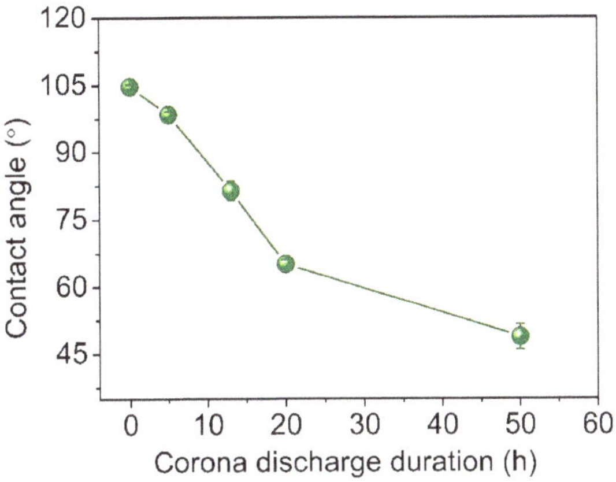

Figure 3. Water contact angle of untreated PE and corona discharge-treated PE.

2.3. *Free Volumes Analysis of Corona Discharge-Treated PE*

Figure 4a shows the PALS analysis for untreated and corona discharge-treated PE. After the source component was subtracted and divided into three lifetime components by an LT program, the third lifetime component was attributed to the *o*-Ps. The τ_3 and I_3 are closely related to free volumes' mean size and the relative quantity, respectively. Figure 4b shows the variation in τ_3 versus corona duration. It is clearly observed that the τ_3 of untreated PE is ~2.529 ns, and the corresponding mean free volume radius calculated from the Tao–Eldrup equation is 0.329 nm. After corona discharge treatment, τ_3 shows a slightly decrease, and it is maintained at ~2.510 ns. For conventional PALS, the detection depth is approximately in the tens of microns. The results exhibit average defect information of the corona aging surface and the virgin matrix inside, which could not reflect the information of corona aging region on the surface accurately. As shown in Figure 4c, for untreated PE, the probability density distribution of τ_3 exhibits a narrow distribution. However, for 5-hour- and 50-hour-treated samples, the probability density distributions of τ_3 widen and decrease clearly. It can be explained that corona discharge treatment destroys the olefin chain structure and changes the stacking structure. This effect contributes to the broadening of distribution and the decrease in free volume size. Figure 4d shows no significant differences in the *o*-Ps intensity (I_3) as a consequence of corona discharge treatment duration. For untreated PE, I_3 is 30.554%. After corona treatment, I_3 keeps at ~30.5%. Due to the energetic positrons, the implantation depth spans the corona aging surface and the virgin matrix inside. Both the surface and the matrix contribute to the observed positron annihilation information. Therefore, it is necessary to investigate the corona discharge-treated surface and the virgin matrix separately. Another effective detection method should be used to study the depth profile of the microstructure within the corona discharge aging area.

2.4. *Microstructure Depth Profile of Corona Discharge-Treated PE*

The positron stopping profile for a variable energy positron beam (E) equaling 2.0, 3.0, 5.0, and 8.0 keV is calculated and illustrated in Figure 5. The inset shows the mean implantation depth (Z_m) as a function of *E*. As explained elsewhere [26], considering the positron diffusion length in PE within a few nanometers, the stopping profile can be used as the deposition depth profile of positrons. After corona discharge treatment, the depth profile of microstructure can be extensively studied by a slow positron beam with variable energy. The line shape *S* parameter is closely related to the character of overall free volumes or holes. Therefore, the microstructure evolution across the corona discharge damage depth can be extensively studied via the variation in *S* parameter.

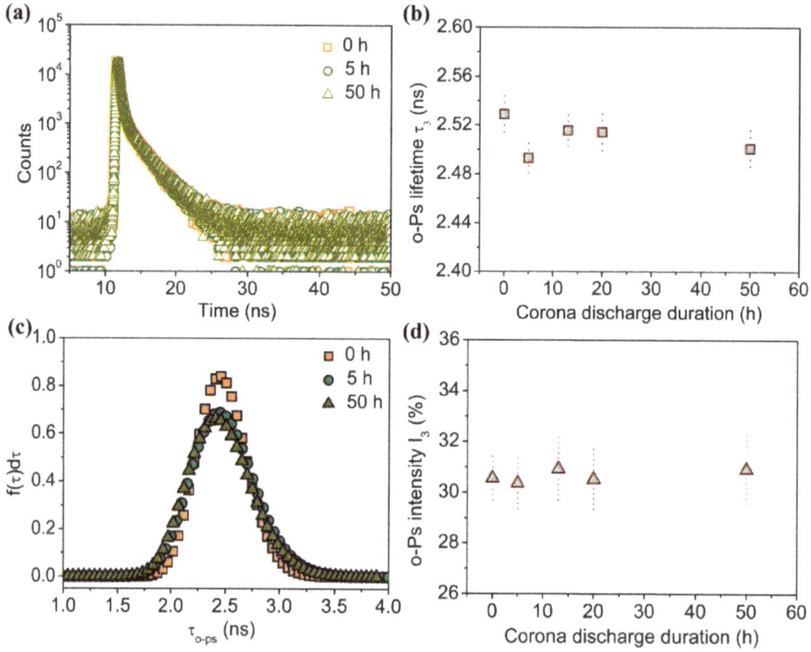

Figure 4. Positron annihilation lifetime spectroscopy analysis of untreated PE and corona discharge-treated PE: (**a**) the positron annihilation lifetime spectra; (**b**) o-Ps lifetime τ_3 against corona discharge duration; (**c**) distribution of o-Ps lifetime τ_3; (**d**) o-Ps intensity I_3 against corona discharge duration.

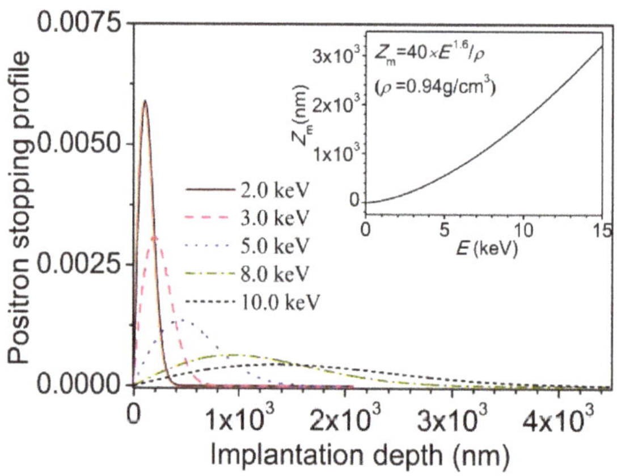

Figure 5. The positron stopping profile in polyethylene with different positron incident energies. The inset shows the relation between positron implantation energy (E) and the mean implantation depth (Z_m) (ρ = 0.94 g cm^{-3}).

Figure 6 shows the depth profile of S parameters for different corona discharge-treated PE samples. As a typical S-E curve, the S parameter exhibits a small value and increases dramatically with increasing positron energy from 0 keV to 2 keV in region I. It is ascribed to the positron annihilation near the surface because of the backing scatting and backing diffusion of the low energy positron or Ps injected close to the surface, which is an intrinsic feature of the slow positron technique [27]. The thickness of the near surface is around 129 nm. And then the S parameter increases gradually and keeps at a constant of 0.526 until the positron energy increases to 6.0 keV at a depth of 748 nm (region II). This is because most positrons are implanted in PE bulk and annihilate therein.

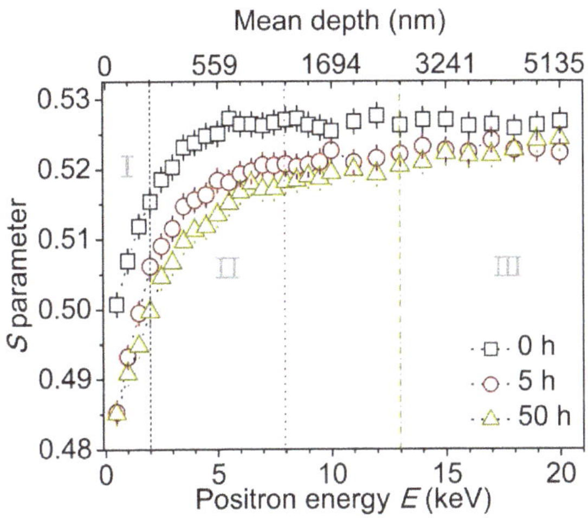

Figure 6. *S* parameters of untreated and corona discharge-treated PE versus the incident positron energy *E* and the mean depth from the surface. The *S-E* curves have been divided into two or three regions (I, II and III) to separate the near surface layer, coronaaging layer and the PE virgin bulk.

After corona discharge treatment, the *S-E* curves possess different characters, and the magnitude of the *S* parameter systematically decreases with increasing corona duration from 5 h to 50 h when positron energy ranges from 0.5 to 20.0 keV. And the *S* parameter reaches a lower plateau after reaching a deeper depth approaching the PE bulk with increasing corona duration. For the 5-hour-treated sample, in region I, similar to the untreated PE and other polymers [28], the *S* parameter exhibits a low value near the surface. In region II, with positron energy increasing up to 8.0 keV, the *S* parameter increases up to a plateau of ~0.520, indicating that much more positrons penetrate into the corona aging layer at a depth of ~1185 nm. In region III, positrons with energy beyond 8.0 keV injected into PE bulk and the *S* parameter keeps at ~0.522, which approaches the value of untreated PE.

For the 50-hour-treated sample, the *S-E* curve presents similar features but also shows a corona-aging region (II) with a thickness of 2577 nm when the positron energy ranges from 2.5 keV to 13.0 keV. Additionally, compared with the other two samples, the lowest *S* parameter indicates that there is a densest structure with smallest free volume size and lowest free volume concentration. The reduction in *S* parameter can be explained as follows: (1) the chain breakage leads to a decrease in free volume size and quantity, so as to a decrease in trapped positronium formation, and (2) the oxygen-containing groups including hydroxyl, carbonyl and ester groups contribute to a decrease in the *S* parameter [27–30]. Considering that the variation in *S* parameter is directly related to the free volume changes in size or concentration, a lower value of *S* parameter means the size or the concentration of free volumes is lower. It suggests that a dense surface structure with a certain thickness different from the PE virgin bulk is formed after corona discharge treatment. With increasing the corona treatment duration from 5 h to 50 h, the *S* parameter decreases, while the thickness of corona aging layer increases from 1185 nm to 2577 nm. During the corona discharge process, the polyethylene chain structure linked by methylene is broken, and the shorter chains possess a more compact structure than that in the untreated PE. Therefore, the resultant lower free volume size and free volume quantity as well as oxygen-containing polar groups contribute to decreases in the *S* parameter. This is consistent with the results of ATR-FTIR and Raman spectroscopy.

To extensively elucidate the structural changes after corona discharge treatment, the reduction in the S parameter, $-\Delta S$, is presented and calculated as follows:

$$-\Delta S = -(S_t - S_0) \tag{3}$$

where S_0 and S_t are the S parameters before and after the specified corona duration, respectively.

Figure 7 shows the $-\Delta S$ of 5-hour and 50-hour-treated samples versus incident mean depth of monoenergetic positrons beam. And the dashed line is well fitted using a two-exponential decay function. It is clearly observed that the $-\Delta S$ is larger near the sample surface within 129 nm, which corresponds to the positron injected energy of 2.0 keV. And it decreases gradually with the depth approaching to the PE bulk. In addition, for the 5-hour-treated sample, it drops to a plateau at the position of ~1.2 μm from the surface. Increasing the corona discharge treatment duration from 5 h to 50 h, the $-\Delta S$ increases, and it keeps consistent with each other beneath the depth of ~2.5 μm from the surface. The profile of $-\Delta S$ reflects the gradient attenuation of corona discharge treatment with depth. And the degradation moves deeper with increasing corona discharge treatment duration.

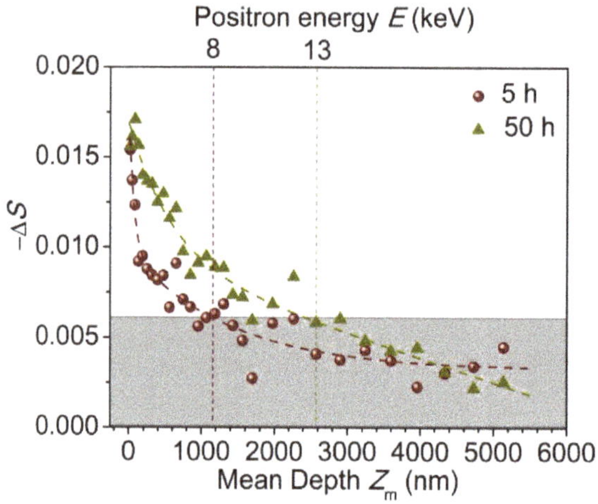

Figure 7. $-\Delta S$ versus the incident mean depth for 5-hour and 50-hour-treated samples.

Generally speaking, the S-W plot indicates the interaction between positions and valence electrons, and it can be used to investigate the chemical environments of the positron annihilation trapping sites. Figure 8a shows the linear plot for the specified (S, W) parameters of a 5-hour-treated sample in different trapping sites along the depth profile. With increasing the incident positron energy, the (S, W) data basically fall on a straight line. It indicates that the defect type where the positrons are annihilated does not change. The (S, W) data of (0.512, 0.0339) and (0.522, 0.0296) are attributable to the annihilation of positrons in the corona-aging region and the PE bulk, respectively. Figure 8b depicts the relation among (S, W) data derived from the annihilation of positrons in the corona-aging region with different corona durations. The S parameter decreases and the W parameter increases gradually with increases in the corona discharge duration. The lowest S value is observed in the 50-hour-treated sample, and the highest value is found in the untreated PE sample. However, these (S, W) data from different corona durations fall on a straight line. This indicates that the tapping site is same where positrons annihilated but has a different chemical environment. With increasing the corona duration from 5 h to 50 h, changes in the microstructure and chemical environment within the PE depth file result in a decrease in the S parameter and an increase in the W value. On one hand, the hydrocarbon chain scission contributes to the formation of small free volumes and low mobility. Due to the intermolecular crosslinking and shorter bond lengths, the S parameter decreases [16]. On the other hand, with increases in the corona duration, the oxygen-containing groups, including hydroxyl, carbonyl and ester groups, are introduced. Compared with the hydrocarbon chain, positron annihilation in the proximity of an oxygen environment contributes to the trapping of positrons. Positron annihilation with high-momentum core electrons belonging to oxygen is enhanced. A lower S parameter is obtained [27]. Therefore, positron is an effective probe to detect the depth profile of microstructures and chemical characteristics in the surface and interfacial study.

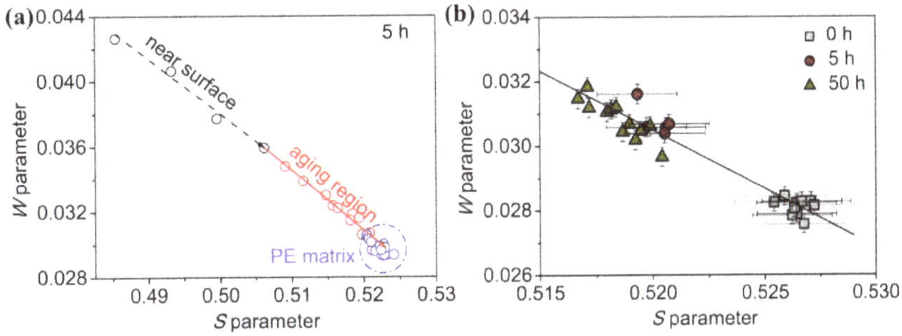

Figure 8. (**a**) *S*-*W* plot of PE corona discharge treated for 5 h. Arrows show the increasing incident positron energy *E* from 0.5 keV to 20 keV. (**b**) *S*-*W* plot of PE matrix and corona-aging region.

3. Materials and Methods

Preparations: PE plate samples were prepared using a hot-pressing method with dimensions of 30 mm × 30 mm × 10 mm. And the density was 0.94 g·cm^{-3}. As illustrated in Figure 9, the corona discharge setup consisted of a "needle-plate" electrode system, as described in our previous studies [31,32]. Both the needle and plate electrodes were made of 304 stainless steels. The diameter of the needle electrode was 3 mm, with a curvature radius of the needle tip measuring 50±3 um. The cylinder plate electrode had a diameter of 20 mm and a thickness of 5 mm. During the corona discharge treatment, the experiments were conducted in air at atmospheric pressure and room temperature, with humidity below 50%. An alternating voltage of 5 kV at 50 Hz was applied to the needle electrode while the plate electrode was grounded. Samples were positioned at the center of the plate electrode with a distance of 2 mm between the specimen surface and the needle. The durations for corona discharge treatment were 5 h, 13 h, 20 h and 50 h, respectively.

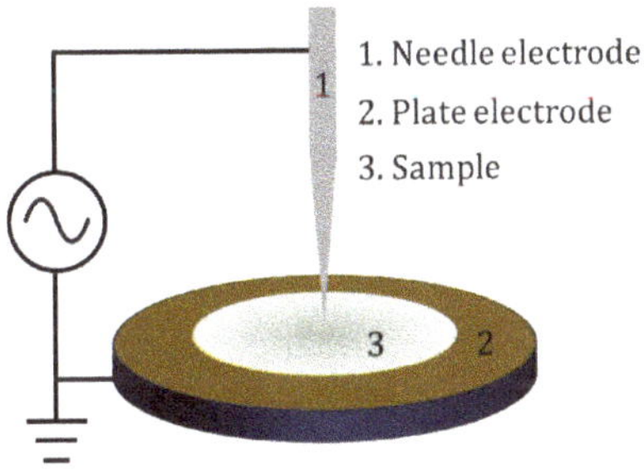

Figure 9. The "needle-plate" electrodes system.

Characterizations: The chemical characteristics of untreated and corona discharge-treated PE samples were investigated using ATR-FTIR and Raman spectroscopy. The ATR-FTIR spectra were acquired using an FTIR spectrophotometer (Nicolet 170 SXIR; Waltham, MA, USA). Raman spectra were acquired using a Laser Confocal Raman Microspectroscopy (LabRAM HR800, Waltham, MA, USA) with a laser wavelength of 488 nm over the range of 4000 cm^{-1}–500 cm^{-1}. The spot diameter was 1~2 μm, and the CCD exposure time was 10 s. Water contact angle (WCA) measurements were utilized to characterize the hydrophilicity variations. The water contact angle of each sample was measured by a sessile drop method using a goniometer (SL200 KB Kino, Boston, MA, USA) with an accuracy of $\pm 3\theta$. And the volume of each droplet of deionized water was 3 μL. Five locations were examined on each sample. All the PALS measurements in this work were conducted under ambient conditions and were carried out utilizing a conventional fast–fast coincidence system with a time resolution of 280 ps (full width at half-maximum). During the measurements, two identical samples with a thickness of ~1 mm were sandwiched together with a 20 μCi positron source ($^{22}NaCl$) sealed with 7 μm thick Kapton films, and each lifetime spectrum was accumulated with a total of one million counts. All of the PALS spectra were evaluated using the LT routine [33] after source correction using Ni signal crystal as a reference. All the spectra were decomposed into three lifetime components without fixing any lifetime or intensity, and the variances of the fits were in the range of 0.95~1.10. The depth profile of microstructure was investigated by DBS with a high-purity germanium detector with a monoenergetic positron beam ranging from 0.5 keV to 20 keV at the Institute of High Energy Physics Chinese Academy of Science, Beijing, China. A spectrum with 5×10^5 counts was collected for each incident positron energy. The line shape parameter S for 511 keV γ-annihilation peak is defined as the ratio of the annihilation events of the central region (511 keV $\pm$ 0.75 keV) to the total area of the peak (499.5–522.5 keV). The line shape parameter W is defined as the ratio of the annihilation events of the wing regions (505.10–508.40) keV and (513.6–516.9) keV to the total area of the peak.

4. Conclusions

In conclusion, the depth profile of microstructures and chemical properties of the corona discharge-treated polyethylene were investigated using positron annihilation spectroscopy accompanied with ATR-FTIR, Raman spectroscopy and water contact angle. By increasing corona discharge duration, the formation of the oxygen-containing polar groups including hydroxyl, carbonyl, and ester groups, together with the breakage of the hydrocarbon chain, strongly contribute to the deterioration of hydrophobicity and the enhancement of hydrophilicity. In addition, the mean free volume size, with a broadening of the probability density distribution, decreases with increases in the corona duration. Also, the thickness of the degradation layer, determined from the S parameter with positron injection depth, increases and diffuses into the PE matrix. Increasing the corona discharge duration from 5 h to 50 h, the thickness of corona degradation layer ranges from ~1.2 μm to ~2.5 μm, which is proved by the S-E curves and the fitting results from the $-\Delta S$-Z_m curves. At the same time, a linear S-W plot within the degradation layer of different corona discharge-treated samples indicates the defect type does not change. The S parameter decreases and the W parameter increases with an increasing corona duration because of the hydrocarbon chain's scission and the oxygen-containing groups. The chemical variation during the corona discharge process can be clearly identify by positron annihilation. Using a slow positron beam, the nondestructive probe can be used to profile the microstructure and chemical environment across the corona discharge damage depth, which is beneficial to investigate the surface and interfacial insulation materials a function of depth from the surface.

Author Contributions: Methodology and validation, J.L.; formal analysis, Z.S., L.T., T.L., Q.Z., X.C., L.L. and X.H.; investigation, J.L. and Z.S.; resources and data curation, C.H.; writing—original draft preparation, J.L. and Z.S.; writing—review and editing, C.Y., B.X., X.L., C.D. and C.H.; funding acquisition, J.L., C.Y., C.D. and C.H. All authors have read and agreed to the published version of the manuscript.

Funding: This research was funded by the National Natural Science Foundation of China, grant number 12105048, 12105027, 12075172, and 12375288. National Key R&D Program of China, grant number 2019YFA0210003. Guangdong Basic and Applied Basic Research Foundation, grant number

2023A1515011881. Guangdong Undergraduate Training Program for Innovation and Entrepreneurship, grant number S202311847048.

Institutional Review Board Statement: Not applicable.

Informed Consent Statement: Not applicable.

Data Availability Statement: Data are contained within the article.

Conflicts of Interest: The authors declare no conflicts of interest.

References

1. Yin, W. Failure mechanism of winding insulations in inverter-fed motors. *IEEE Electr. Insul. Mag.* **1997**, *13*, 18–23.
2. Metzler, D.; Hake, J. The effect of mechanical deformation and wire size on the ozone erosion resistance of inverter-duty magnet wires. In Proceedings of the Electrical Insulation Conference and Electrical Manufacturing and Coil Winding Conference, Rosemont, IL, USA, 25–25 September 1997; pp. 389–395. [CrossRef]
3. Bellomo, J.P.; Lebey, T.; Oraison, J.; Peltier, F. Electrical aging of stator insulation of low voltage rotating machines supplied by inverters. In Proceedings of the Conference Record of the 1996 IEEE International Symposium on Electrical Insulation, Montreal, QC, Canada, 16–19 June 1996; pp. 210–213. [CrossRef]
4. Pascual, M.; Balart, R.; Sánchez, L.; Fenollar, O.; Calvo, O. Study of the aging process of corona discharge plasma effects on low density polyethylene film surface. *J. Mater. Sci.* **2008**, *43*, 4901–4909. [CrossRef]
5. Molinié, P. A review of mechanisms and models accounting for surface potential decay. *IEEE Trans. Plasma Sci.* **2012**, *40*, 167–176. [CrossRef]
6. Kireche, N.; Rondot, S.; Bitam-Megherbi, F.; Jbara, O.; Gilliot, M.; Patat, J.M. Experimental study of polymethyl methacrylate damage under corona discharge. *Eur. Phys. J. Appl. Phys.* **2018**, *82*, 31301. [CrossRef]
7. Li, Y.; Yasuda, M.; Takada, T. Pulsed electroacoustic method for measurement of charge accumulation in solid dielectrics. *IEEE Trans. Dielectr. Electr. Insul.* **1994**, *1*, 188–195.
8. Sharma, S.K.; Pujari, P.K. Role of free volume characteristics of polymer matrix in bulk physical properties of polymer nanocomposites: A review of positron annihilation lifetime studies. *Prog. Polym. Sci.* **2017**, *75*, 31–47. [CrossRef]
9. Volfová, L.; Andrei Irimiciuc, S.; Chertopalov, S.; Hruška, P.; Čížek, J.; Vondráček, M.; Novotný, M.; Butterling, M.; Liedke, M.O.; Wagner, A.; et al. Tailoring pulsed laser deposition fabricated copper oxide film by controlling plasma parameters. *Appl. Surf. Sci.* **2023**, *608*, 155128. [CrossRef]
10. Liu, Z.; Zhou, Y.; Ping, Y.; Qian, L.; Li, J.; Liu, L.; Liu, Y.; Ito, K.; Wei, C.; He, C. Enhancement in photoelectric properties of ITO films by regulating defects and dopants with supercritical fluid treatment. *Appl. Surf. Sci.* **2021**, *565*, 150551. [CrossRef]
11. Tsuji, K.; Nakaya, M.; Uedono, A.; Hotta, A. Enhancement of the gas barrier property of polypropylene by introducing plasma-treated silane coating with SiOx-modified top-surface. *Surf. Coat. Tech.* **2015**, *284*, 377–383. [CrossRef]
12. Yang, L.; Qian, L.B.; Cao, D.W.; Li, X.; Guo, J.C.; Zhang, H.L.; Zhang, D.W.; He, C.Q. Surface collapse of fine powders ground from HKUST-1 big crystals investigated by positron annihilation lifetime spectroscopy. *Appl. Surf. Sci.* **2024**, *646*, 158937. [CrossRef]
13. Chen, Y.; Sun, K.; Zhang, T.; Zhou, J.; Liu, Y.; Zeng, M.; Ren, X.; Feng, R.; Yang, Z.; Zhang, P.; et al. TiO_2-modified montmorillonite-supported porous carbon-immobilized Pd species nanocomposite as an efficient catalyst for sonogashira reactions. *Molecules* **2023**, *28*, 2399. [CrossRef] [PubMed]
14. Eldrup, M.; Lightbody, D.; Sherwood, J.N. The temperature dependence of positron lifetimes in solid pivalic acid. *Chem. Phys.* **1981**, *63*, 51–58. [CrossRef]
15. Tao, S.J. Positronium Annihilation in Molecular Substances. *J. Chem. Phys.* **1972**, *56*, 5499–5510. [CrossRef]
16. Mallon, P.E.; Greyling, C.J.; Vosloo, W.; Jean, Y.C. Positron annihilation spectroscopy study of high-voltage polydimethylsiloxane (PDMS) insulators. *Radiat. Phys. Chem.* **2003**, *68*, 453–456. [CrossRef]
17. Wang, Z.; Luo, Y.; Zheng, F.; Zhang, N.; Yin, C.S.; Li, J.J.; He, C.Q.; Peng, X.Y.; Huang, Z.; Fang, P.F. Study on surface structure of plasma-treated polydimethylsiloxane (PDMS) elastomer by slow positron beam. *Surf. Interface Anal.* **2018**, *50*, 819–826. [CrossRef]
18. Uedono, A.; Chen, Z.Q.; Suzuki, R.; Ohdaira, T.; Mikado, T.; Fukui, S.; Shiota, A.; Kimura, S. Nanoporous structure of methylsilsesquioxane films using monoenergetic positron beams. *J. Appl. Phys.* **2001**, *90*, 2498–2503. [CrossRef]
19. Yu, R.S.; Ito, K.; Hirata, K.; Zheng, W.; Kobayashi, Y. Effects of coexistent pores and paramagnetic defects on positron annihilation in silicon oxide thin films. *J. Appl. Phys.* **2003**, *93*, 3340–3344. [CrossRef]
20. He, C.; Hamada, E.; Suzuki, T.; Kobayashi, H.; Kondo, K.; Shantarovich, V.P.; Ito, Y. Characterization of polymer sub-surface using slow positron beam. *J. Radioanal. Nucl. Ch.* **2003**, *255*, 431–435. [CrossRef]
21. Peng, X.Y.; Wang, Z.; Lin, H.S.; Fang, P.F.; Huang, Z. Surface structure of aged composite insulator studied by slow positron beam. *J. Wuhan Univ. Technol.-Mater. Sci. Ed.* **2019**, *34*, 1008–1012. [CrossRef]
22. Schultz, P.J.; Lynn, K.G. Interaction of positron beams with surfaces, thin films, and interfaces. *Rev. Mod. Phys.* **1988**, *60*, 701. [CrossRef]
23. Seyhan, A.; Gunaydin, B.N.; Polat, Y.; Kilic, A.; Demir, A.; Avci, H. Improvement of polyethylene fiber wettability and mechanical properties through an environmentally sustainable spinning process. *Int. J. Adhes. Adhes.* **2022**, *119*, 103250. [CrossRef]

24. Mirabedini, S.M.; Rahimi, H.; Hamedifar, S.; Mohsen Mohseni, S. Microwave irradiation of polypropylene surface: A study on wettability and adhesion. *Int. J. Adhes. Adhes.* **2004**, *24*, 163–170. [CrossRef]
25. Park, S.J.; Jin, J.S. Effect of corona discharge treatment on the dyeability of low-density polyethylene film. *J. Colloid. Inter. Sci.* **2001**, *236*, 155–160. [CrossRef] [PubMed]
26. Li, J.J.; Xiong, B.Y.; Yin, C.S.; Zhang, X.W.; Zhou, Y.W.; Wang, Z.; Fang, P.F.; He, C.Q. Free volume characteristics on water permeation and salt rejection of polyamide reverse osmosis membranes investigated by a pulsed slow positron beam. *J. Mater. Sci.* **2018**, *53*, 16132–16145. [CrossRef]
27. Suzuki, T.; He, C.; Shantarovich, V.; Kondo, K.; Hamada, E.; Matso, M.; Ma, L.; Ito, Y. The influence of radiation and light on Ps formation in PMMA and PE studied by coincidence Doppler-broadening spectroscopy. *Radiat. Phys. Chem.* **2003**, *66*, 161–166. [CrossRef]
28. Ghasemifard, M.; Ghamari, M. Characterization of polar groups in polyvinyl chloride-ethyl vinyl acetate blends using coincidence Doppler broadening spectroscopy. *J. Appl. Polym. Sci.* **2023**, *141*, e54985. [CrossRef]
29. Xiong, B.; Mao, W.; Tang, X.; He, C. Positron annihilation characteristics in mesostructural silica films with various porosities. *J. Appl. Phys.* **2014**, *115*, 094303. [CrossRef]
30. Xiong, B.Y.; Mao, W.F.; Yue, J.; Xu, X.S.; He, C.Q. Mesoporosity in silica films studied by a slow positron beam and ellipsometry. *Phys. Lett. A* **2014**, *378*, 249–253. [CrossRef]
31. Wang, Z.; Li, J.J.; Zhou, Y.M.; Zhang, N.; He, C.Q.; Peng, X.Y.; Huang, Z.; Cao, X.Z.; Wang, B.Y.; Fang, P.F. Investigation of the surface microstructure evolution of silicone rubber during corona discharge via slow positron beam and electrochemical impedance spectroscopy. *Plasma Process Polym.* **2019**, *16*, 1900057. [CrossRef]
32. Yang, Y.; Wang, Z.; Peng, X.Y.; Huang, Z.; Fang, P.F. Influence of crosslinking extent on free volumes of silicone rubber and water diffusion after corona discharge. *Materials* **2022**, *15*, 6833. [CrossRef]
33. Kansy, J. Microcomputer program for analysis of positron annihilation lifetime spectra. *Nucl. Instrum. Methods Phys. Res. Sect. A* **1996**, *374*, 235–244. [CrossRef]

molecules

MDPI

Review

Advances in Active Materials of Enzymatic Electrochemical Sensors for Detecting Organophosphorus Pesticides

Sijie Ma [1], Zihang Chen [1], Fuxiong Yang [1], Ting Yao [1], Suo Wang [2,*], Yi Yu [3], Liangbin Xiong [4], Xiaodong Hong [1,*] and Guangjin Wang [1,*]

[1] School of Materials and Energy, Foshan University, Foshan 528000, China; 15899654506@163.com (S.M.); 15729508936@163.com (Z.C.); 15207853481@163.com (F.Y.); 18381662518@163.com (T.Y.)

[2] The Fifth Electronic Research Institute of the Ministry of Industry and Information Technology, Guangzhou 510610, China

[3] Hubei Key Laboratory of Advanced Technology for Automotive Components, Wuhan University of Technology, Wuhan 430070, China; michael.yy2024@gmail.com

[4] School of Optoelectronic Engineering, Guangdong Polytechnic Normal University, Guangzhou 510665, China; xiongliangbin@gpnu.edu.cn

* Correspondence: wangs@ceprei.org (S.W.); hongxiaodong@lntu.edu.cn (X.H.); wgj501@163.com (G.W.)

Abstract

Organophosphorus pesticides (OPs) have been widely employed to increase food production and alleviate the increasingly serious food crisis. However, excessive use of these pesticides has seriously affected human health and even caused death due to significant pesticide residues in food. Therefore, enzymatic electrochemical sensors have been developed to monitor OP residues in food. The electrochemical detection performance of these sensors is determined by the physicochemical properties of electrochemical active materials in their active layers. The definition and classification of OPs are first introduced in this review, then the components of enzymatic electrochemical sensors, including electrodes, electrochemical active layer and bioactive enzyme layer, are analyzed in detail. Furthermore, this review emphatically discusses the recent development of enzymatic electrochemical sensors based on various electrochemical active materials: carbon-based, polymer-based, metal-based, metallic compound-based, metal organic framework-based and covalent organic framework-based materials. Finally, probable research directions for developing enzymatic electrochemical sensors with high sensitivity, excellent stability and good reproducibility are outlined to accelerate rapid, effective and low-cost on-site detection OPs in food. This review is expected to provide inspiration for the design and preparation of the high-performance enzymatic electrochemical sensors.

Keywords: organophosphorus pesticides; acetylcholinesterase; enzymatic electrochemical sensors; active materials

1. Introduction

Over the past twenty years, the population of the world has increased by 1.6 billion, which has caused a serious global food crisis [1]. For relieving this problem, pesticides have been employed to eliminate weeds and pests and thus increase food production. According to the data from the China Pesticide Information Network [2], the worldwide total pesticide usage reached about 4 million tons in 2023, which was 2% lower than last year and 14% more than a decade ago. During the same period, the total pesticide usage of China was about 230 thousand tons. Among these used pesticides, the usage amount of organophosphorus pesticides (OPs) has accounted for more than 30% of total pesticide usage due to their low cost, degradability and effectiveness against various weeds and pests [3]. Unfortunately, only a small fraction of OPs is used to eliminate weeds and pests, and majority enters soil and drinking water, thus resulting in food pollution. It is well known that ingesting food containing residual OPs can cause poisoning or even death in humans [4]. Consequently, developing accurate, rapid and sensitive detection technologies to monitor OPs residual in food is of great importance to ensuring human safety.

At present, the most common detection technologies for OP residues are chromatographic methods [5,6], including gas chromatography, gas chromatography–mass spectrometry, liquid chromatography, liquid chromatography–mass spectrometry and high-performance liquid chromatography coupled with ultraviolet detector. These approaches exhibit ultrahigh accuracy, excellent sensitivity and good selectivity for detecting OPs, but they require expensive professional instruments and well-trained technicians. Furthermore, their operation processes are considerably complicated, and the pretreatment of their detection samples takes a long time. Most importantly, these methods are unsuitable for on-site detection of OP residues and only work in laboratories. Therefore, it is urgent to develop accurate, rapid, stable, reliable and simple operational detection technologies for achieving highly sensitive and selective on-site detection of OP residues in food.

Due to their advantages in miniaturization, rapid response and suitability for on-site detection, sensor technologies such as electrochemiluminescence [7], colorimetric [8], fluorescent [9] and electrochemical [10] sensors have been widely employed to detect OP residues in food. Among them, electrochemical sensing technologies including enzymatic/nonenzymatic electrochemical sensors exhibit low cost, simple operation, fast detection, high reliability, good repeatability and selectivity, thus attracting much attention in OP residue detection [11]. Compared with non-enzymatic electrochemical sensors, enzymatic electrochemical sensors are able to simultaneously detect multiple OPs. For this reason, many efforts have been devoted to developing high-performance enzymatic electrochemical sensors. As we all know, the detection performance of enzymatic electrochemical sensors is directly determined by the physicochemical properties of active materials in their electrochemical active layers [12]. Accordingly, the development of active materials is significant for the construction of high-performance enzymatic electrochemical sensors. After continued and unremitting efforts by researchers, studies on active materials for enzymatic electrochemical sensors have made significant process. Therefore, it is significant to summarize these advances to provide inspiration for the design and preparation of the high-performance enzymatic electrochemical sensors.

Recently, Zeng et al. [13] summarized the research progress in the development of acetylcholinesterase-based sensors for detection of OPs in food and environment samples, focusing on the immobilization strategies of acetylcholinesterase (AChE) and the development progress of AChE-based optical and capillary electrophoresis sensing detection technologies. However, the systematic summaries on the recent advances in active materials for high-performance enzymatic electrochemical sensors are still lacking. Hence, we will initially discuss the definition and classification of OPs, then detail the components of enzymatic electrochemical sensors, including electrodes, electrochemical active layer and bioactive enzyme layer. The recent development of enzymatic electrochemical sensors based on various electrochemical active materials such as carbon-based, polymer-based, metal-based, metallic compound-based, metal organic framework (MOF)-based and covalent organic framework (COF)-based materials will be emphatically summarized. Finally, probable research directions for developing enzymatic electrochemical sensors with high sensitivity, excellent stability and good reproducibility are outlined to accelerate rapid, effective and low-cost on-site detection of OPs in food.

https://doi.org/10.3390/molecules31040717

2. The Definition and Classification of OPs

OPs are one of the important pesticides for controlling weeds and pests [3]. Their active agents are organic phosphorus compounds, where a phosphorus atom binds to oxygen, nitrogen, sulfur, halogen atoms and organic groups such as alkoxy, alkyl, aryl and amino groups via covalent bonds. The general molecular formula of OPs is shown in Figure 1. R1 and R2 represent organic groups (alkoxy, alkyl, aryl and amino, etc.), while X denotes leaving groups. These leaving groups are prone to hydrolysis by hydrolases or substitute through phosphorylation of AChE. Until now, OPs have been divided into different types based on the different classified criteria. In terms of chemical structure and application [14], 71 common OPs are used for eradicating weeds, fungi and pests. Simultaneously, according to toxicological mechanism [15], OPs can be classified into 13 subtypes, including phosphates, phosphinates, phosphonofluoridates, phosphonates, phosphorothioates, phosphoramidothioates, phosphorodithioates, phosphoramidates, s-substituted phos-phorothioates, phosphonothioates, phosphorofluoridates, s-substituted phosphonothioates and s-substituted phosphoramidothioates. In addition, based on toxicity level, OPs can be divided into low, middle and high toxicity [16].

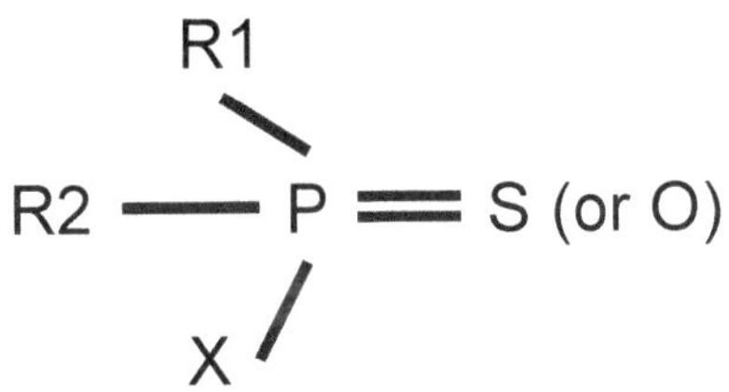

Figure 1. The general molecular formula of OPs.

Based on data from the National Center for Biotechnology Information of National Library of Medicine [17], Table 1 summarizes the chemical name (based on International Union of Pure and Applied Chemistry, IUPAC), molecular formula and physicochemical properties of OPs which have been reported in Section 4 of this paper. As shown in Table 1, most OPs are hydrolysis in alkaline condition, but trichlorfon is rapid isomerization in alkaline condition. The majority of OPs are stable in neutral condition, and a small fraction are stable in weakly acidic condition. As such, the application and detection of OPs are recommended under neutral or weakly acidic conditions. On the one hand, almost all OPs are hardly volatile, but dichlorvos is a highly volatile and toxic at ambient temperature and pressure, requiring careful handling during spraying. On the other hand, most OPs are photosensitive and thermosensitive, necessitating storage in dark and room-temperature conditions. It is worth noting that phoxim with high photosensitivity is suitable for underground pest control. On the contrary, acephate, glyphosate, pirimiphos methyl, trichlorfon and profenofos display high photostability and thermostability, making them problematic due to long-term persistence in the food chain. For solubility, all OPs are soluble in organic solvent, and some are slightly dissolved in water, nonpolar solutions, aliphatic hydrocarbons or mineral oils. A few OPs such as acephate, glyphosate, trichlorfon, methamidophos, methomyl, and monocrotophos are extremely soluble in water. Therefore, based on toxicity, stability and solubility, acephate, glyphosate and trichlorfon may be the ideal OP models for evaluating the detection performance of enzymatic electrochemical sensors.

https://doi.org/10.3390/molecules31040717

Table 1. The chemical name, molecular formula and physicochemical properties of commonly used OPs.

OPs	Chemical Name	Molecular Formula	Property
Acephate	N-[methoxy-(methylsulfanyl)-phosphoryl]acetamide	$C_4H_{10}NO_3PS$	White crystalline powder, low toxicity and systematic insecticide, slow hydrolysis in alkaline condition and extremely stable in acidic/neutral condition, photostable and thermotolerant, low volatility, extremely soluble in water, soluble in organic solvent, slight soluble in non-polar solvent
Glyphosate	2-(phosphonomethyl-amino)acetic acid	$C_3H_8NO_5P$	White crystalline powder, low toxicity and systematic herbicide, slow hydrolysis in alkaline condition and stable in acidic/neutral condition, photostable, low volatility, extremely soluble in water, insoluble in organic and non-polar solvent
Malathion	diethyl 2-dimethoxy-phosphinothioylsulfanyl-butanedioate	$C_{10}H_{19}O_6PS_2$	Colorless to pale yellow oily liquid, low toxicity, rapid hydrolysis in alkaline condition and stable in acidulous/neutral condition, photosensitive, low volatility, soluble in organic solvent, slight soluble in water and mineral oils
Phoxim	N-diethoxy-phosphinothioyloxy-benzenecarboximidoyl cyanide	$C_{12}H_{15}N_2O_3PS$	Light yellow oily liquid, low toxicity, rapid hydrolysis in alkaline condition and stable in acidulous/neutral condition, high photosensitive, low volatility, soluble in organic solvent, slight soluble in aliphatic hydrocarbons and insoluble in water
Pirimiphos methyl	4-dimethoxy-phosphinothioyloxy-N,N-diethyl-6-methylpyrimidin-2-amine	$C_{11}H_{20}N_3O_3PS$	Colorless crystalline solid, low to middle toxicity, slow hydrolysis in alkaline condition and stable in acidic/neutral condition, photostable and thermotolerant, low volatility, soluble in aliphatic hydrocarbons and organic solvent, insoluble in water
Trichlorfon	2,2,2-trichloro-1-dimethoxyphosphorylethanol	$C_4H_8Cl_3O_4P$	White crystalline powder, low to middle toxicity, rapid isomerization in alkaline condition and stable in acidic condition, photostable and thermotolerant, low volatility, extremely soluble in water, soluble in organic solvent, slight soluble in nonpolar solvent
Chlorpyrifos	diethoxy-sulfanylidene-[(3,5,6-trichloro-2-pyridinyl)oxy]-lambda5-phosphane	$C_9H_{11}C_{13}NO_3PS$	White crystalline solid, middle toxicity and broad-spectrum insecticide, hydrolysis in alkaline condition, photosensitive, low volatility, soluble in organic solvent, insoluble in water
Dimethoate	2-dimethoxy-phosphinothioylsulfanyl-N-methylacetamide	$C_5H_{12}NO_3PS_2$	White crystalline powder, middle toxicity and systematic insecticide, rapid hydrolysis in alkaline condition and stable in acidic/neutral condition, photosensitive, low volatility, soluble in organic solvent, slight soluble in aliphatic hydrocarbons and water

https://doi.org/10.3390/molecules31040717

Table 1. *Cont.*

OPs	Chemical Name	Molecular Formula	Property
Fenitrothion	dimethoxy-(3-methyl-4-nitrophenoxy)-sulfanylidene-lambda5-phosphane e	$C_9H_{12}NO_5PS$	Light yellow crystalline solid, middle toxicity, hydrolysis in alkaline condition, photosensitive, low volatility, soluble in organic solvent, slight soluble in aliphatic hydrocarbons, insoluble in water
Profenofos	4-bromo-2-chloro-1-[ethoxy(propylsulfanyl)-phosphoryl]oxybenzene	$C_{11}H_{15}BrClO_3PS$	Light yellow liquid, middle toxicity and broad-spectrum insecticide, rapid hydrolysis in alkaline condition and stable in acidulous/neutral condition, photostable and thermotolerant, low volatility, soluble in organic solvent, insoluble in water
Diazinon	diethoxy-(6-methyl-2-propan-2-ylpyrimidin-4-yl)oxy-sulfanylidene-lambda5-phosphane	$C_{12}H_{21}N_2O_3PS$	Colorless to pale yellow oily liquid, middle toxicity and broad-spectrum insecticide, rapid hydrolysis in alkaline condition and stable in acidulous/neutral condition, photosensitivity, low volatility, soluble in organic solvent, insoluble in water
Dichlorvos	2,2-dichloroethenyl dimethyl phosphate	$C_4H_7C_{12}O_4P$	Colorless to pale yellow transparent liquid, high toxicity and broad-spectrum insecticide and fumigant, rapid hydrolysis in alkaline condition and stable in acidulous/neutral condition, photosensitive and thermosensitive, high volatility, extremely soluble in organic solvent, soluble in water
Ethion	diethoxyphosphinothioyl-sulfanylmethylsulfanyl-diethoxy-sulfanylidene-lambda5-phosphane	$C_9H_{22}O_4P_2S_4$	Colorless to pale yellow oily liquid, high toxicity and broad-spectrum insecticide and acaricide, rapid hydrolysis in alkaline condition and stable in acidulous/neutral condition, photosensitive and thermosensitive, low volatility, soluble in organic solvent, insoluble in water
Isocarbophos	propan-2-yl 2-[amino(methoxy)-phosphinothioyl]oxybenzoate	$C_{11}H_{16}NO_4PS$	Colorless and transparent oily liquid, high toxicity and systematic insecticide, rapid hydrolysis in alkaline condition and stable in acidulous condition, photosensitive, low volatility, soluble in organic solvent, insoluble in water
Methamido-phos	[amino(methylsulfanyl)-phosphoryl]oxymethane	$C_2H_8NO_2PS$	White needle-like crystal, high toxicity and broad-spectrum insecticide, rapid hydrolysis in alkaline condition and stable in acidulous condition, photosensitive and thermosensitive, low volatility, extremely soluble in water, soluble in organic solvent, slight soluble in non-polar solvent
Methomyl	methyl N-(methylcarbamoyloxy)-ethanimidothioate	$C_5H_{10}N_2O_2S$	White crystalline powder, high toxicity and carbamate insecticide, rapid hydrolysis in alkaline condition, photosensitive and thermosensitive, low volatility, extremely soluble in water, soluble in organic solvent, slight soluble in non-polar solvent

https://doi.org/10.3390/molecules31040717

Table 1. *Cont.*

OPs	Chemical Name	Molecular Formula	Property
Methyl Parathion	dimethoxy-(4-nitrophenoxy)-sulfanylidene-lambda5-phosphane	$C_8H_{10}NO_5PS$	Colorless crystal, high toxicity, rapid hydrolysis in alkaline condition, photosensitive, low volatility, extremely soluble in organic solvent, insoluble in water
Monocroto-phos	dimethyl [(E)-4-(methylamino)-4-oxobut-2-en-2-yl] phosphate	$C_7H_{14}NO_5P$	Colorless crystalline solid, high toxicity and systematic insecticide, rapid hydrolysis in alkaline condition and stable in acidulous condition, photosensitive and thermosensitive, low volatility, extremely soluble in water, soluble in organic solvent
Omethoate	2-dimethoxy-phosphorylsulfanyl-N-methylacetamide	$C_5H_{12}NO_4PS$	Colorless and transparent oily liquid, high toxicity and systematic insecticide, rapid hydrolysis in alkaline condition and stable in acidulous/neutral condition, photosensitive, low volatility, soluble in water and organic solvent, slight soluble in non-polar solvent
Parathion	diethoxy-(4-nitrophenoxy)-sulfanylidene-lambda5-phosphane	$C_{10}H_{14}NO_5PS$	Colorless needle-like crystal, high toxicity, rapid hydrolysis in alkaline condition, photosensitivity, low volatility, soluble in organic solvent, insoluble in water
Paraoxon	diethyl (4-nitrophenyl) phosphate	$C_{10}H_{14}NO_6P$	Colorless to pale yellow crystalline solid, high toxicity, rapid hydrolysis in alkaline condition and stable in acidulous/neutral condition, photosensitive, volatility, soluble in organic solvent, polar and non-polar solvent, insoluble in water

3. The Construction of Enzymatic Electrochemical Sensors

As shown in Figure 2, typical enzymatic electrochemical sensors consist of three core components: (1) electrodes include a reference, auxiliary and working electrode in traditional three-electrode system; (2) electrochemical active layer contains non-active site blocking agents, adhesion agents and electrochemical active materials; and (3) bioactive enzyme layer refers to AChE, butyrylcholinesterase and organophosphorus hydrolase layer. Each component will be discussed in the following section.

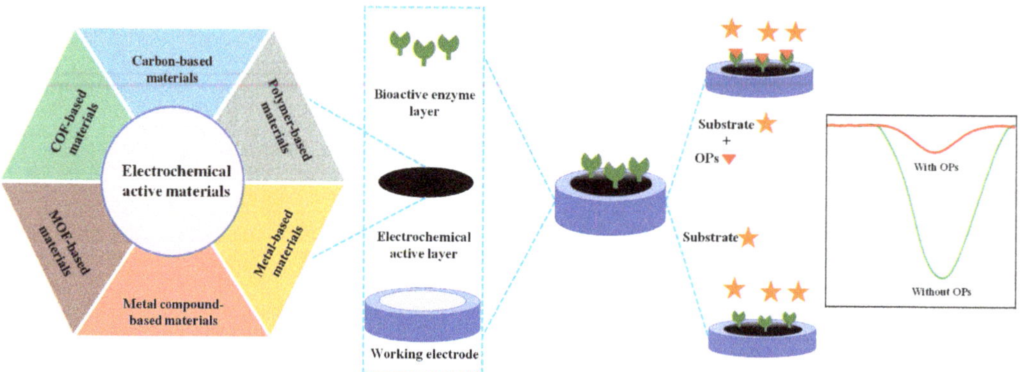

Figure 2. The construction of enzymatic electrochemical sensors and their detection mechanism.

https://doi.org/10.3390/molecules31040717

3.1. Electrodes

Enzymatic electrochemical sensors based on traditional three-electrode systems contain reference, auxiliary and working electrodes [18]. Therein, the reference electrode is able to provide the stable, reversible and reproducible reference potential for calibrating the applied potential of working electrode. The ideal reference electrode requires stable potential, low polarization and good compatible with tested system. The auxiliary electrode forms a close current circuit with the working electrode for conducting the reverse current of the working electrode and ensuring the stable and continuous process of the target electrochemical reaction on the working electrode. The ideal auxiliary electrode requires good conductivity, high stability and chemical inertness. The working electrode is capable of transforming the charge change in the electrochemical reaction of the target OPs into detectable current, potential and resistance signal. The electrochemical properties of the electrochemical active material modified working electrode directly determine the detection performances of three-electrode enzymatic electrochemical sensors. The requirements for an ideal working electrode are the same as an auxiliary electrode, that is, high stability, excellent conductivity and chemical inertness. According to the summarized literature in Section 4 of this paper, the common working electrodes for the constructure of enzymatic electrochemical sensors contain glass carbon electrode (GCE), scree-printed electrode (SPE), boron-doped diamond electrode (BDDE), indium tin oxide electrode (ITOE), fluorine-doped tin oxide electrode (FTOE) and gold electrode (GE).

3.2. Electrochemical Active Layer

The electrochemical active layer on the working electrode is the most crucial functional layer in enzymatic electrochemical sensors which connects to the working electrode and the bioactive enzyme layer and gives assistance to the generation, transmission and amplification of detection signals. As such, the detection performances of electrochemical sensors for target OPs are directly determined by the electrical performances of the electrochemical active layer. The main functions of the electrochemical active layer are providing catalytic active sites for the electrochemical reaction of target OPs, accelerating the charge transfer between the working electrode and the bioactive enzyme layer and boosting the fixation of bioactive enzymes. Accordingly, the rational optimization of the electrochemical active layer is significant for providing enzymatic electrochemical sensors with distinguished electrochemical performances for the detection of target OPs.

It is well known that the electrochemical active layer in enzymatic electrochemical sensors is made up of non-active site blocking agents, adhesion agents and electrochemical active materials. According to the summarized literature in Section 4 of this paper, the common non-active site blocking agent is bovine serum albumin (BSA), which has abundant amino and carboxyl groups [19]. Additionally, the most common adhesion agents refer to chitosan (Chi), glutaraldehyde and nafion. Therein, Chi is the deacetylation product of chitin, which has abundant amino and carboxyl groups in its molecular chain [20]. These functional groups contribute to the incorporation with bioactive enzymes by weak physicochemical interactions such as hydrogen bond or electrostatic effect, achieving the mild fixation of bioactive enzymes and avoiding the inactivation of bioactive enzymes caused by very slight physical adsorption and strong chemical crosslinking. What is more important is that the hydrophilic microenvironment from Chi is conducive to maintaining the spatial conformation of bioactive enzymes and enhancing the catalytic activity and storage stability of bioactive enzymes. Additionally, there is an aldehyde group in each end of glutaraldehyde molecular chain. Due to the strong chemical interaction between the aldehyde group of glutaraldehyde and the amino group of bioactive enzymes, the bioactive enzymes are strongly anchored into the three-dimensional network of the electrochemical

https://doi.org/10.3390/molecules31040717

active layer by glutaraldehyde [21]. Their strong chemical interaction is likely to cause the change in the spatial conformation of bioactive enzymes, which in turn gives rise to the inaction of bioactive enzymes. Unlike with glutaraldehyde, the nafion composing the polytetrafluoroethylene framework and sulfonic acid group immobilizes bioactive enzymes on the electrochemical active layer through slight physical adsorption [22]. This very slight physical adsorption may cause the loss of bioactive enzymes on the electrochemical active layer, thereby resulting in the reduction in detection performance of enzymatic electrochemical sensors.

The active materials in the electrochemical active layer are the core functional materials of enzymatic electrochemical sensors. The active materials, which are coated on the working electrode surface, directly participate the electrochemical reaction of target OPs and promote the amplification of detectable response signals. To prepare high-performance enzymatic electrochemical sensors, it is necessary to exploit the electrochemical active materials with excellent conductivity, outstanding electrochemical catalytic activity, high specific surface area, controllable porous structure, considerable electro-/chemical stability, strong anti-interference ability and good biocompatibility. Consequently, various electrochemical active materials such as carbon-, polymer-, metal-, metallic compound-, metal organic framework- and covalent organic framework-based materials have attracted much attention in the development of high-performance enzymatic electrochemical sensors. After the continuous efforts of researchers, studies on these aforementioned electrochemical active materials for the development of high-performance enzymatic electrochemical sensors have made many breakthroughs which will be emphatically discussed in Section 4 of this paper.

3.3. *Bioactive Enzyme Layer*

The bioactive enzyme layer referring to AChE, butyrylcholinesterase and the organophosphorus hydrolase layer is the signal recognition element in enzymatic electrochemical sensors. It is well known that AChE is one of the most important serine hydrolases extracted from various organisms or tissues such as marine organisms, insects, plants and human serum [23]. The primary structure of AChE consists of over 500 amino acid residues, and its catalytic domain is the serine (Ser203)–histidine (His447)–glutamic acid (Glu334) triplet. In the secondary structure of AChE, α/β hydrolase folding is the core architecture of the catalytic domain; in the tertiary structure, the catalytic domain folds into a spherical conformation with a diameter of about 45 Å. The quaternary structure of AChE is assembled from catalytic and anchoring subunits. The hydroxyl groups of Ser203 can specifically bind with the phosphoric groups of OPs through robust covalent bonds, forming stable phosphorylated enzyme complexes and causing the irreversible inactivation of AChE [24]. Analogous to AChE, butyrylcholinesterase also belongs to the cholinesterase family but is mainly derived from animal serum. Their structural difference lies in the number of amino acid residues and the composition of the active triplet [25], yet they interact with OPs in the same manner. Therefore, AChE and butyrylcholinesterase are employed as signal recognition elements to build inhibited enzymatic electrochemical sensors for OP detection. For inhibited enzymatic electrochemical sensors, both bioactive enzymes could catalyze the hydrolysis of acetylsalicycholine to produce thiocholine, which is oxidized on the electrochemical active layer to generate detectable response current. In the presence of OPs in the electrolyte solution, the response current intensity decreased due to the catalytic activity of enzyme inhibition by OPs. The electrochemical detection performance of these sensors can be evaluated based on the linear relationship between the response current intensity of sensors and the OP concentration.

Unlike the inhibitory effect of AChE and butyrylcholinesterase, organophosphorus hydrolase extracted from pseudomonas, xanthobacter or *Escherichia coli* (colon bacillus) can

https://doi.org/10.3390/molecules31040717

directly catalyze the cleavage of P-O/S bonds in OPs, producing p-nitrophenol. This product is oxidized on the electrochemical active layer of catalytic enzymatic electrochemical sensors to generate a detectable response current [26]. The electrochemical detection performance of these sensors can be evaluated on the basis of the linear relationship between the response current of sensors and the OP concentration. Furthermore, organophosphorus hydrolase can be reused and is suitable for continuous online monitoring because of its stable catalytic activity. However, compared to AChE and butyrylcholinesterase, organophosphorus hydrolase has several notable limitations, including high cost, poor stability, short half-life, low reproducibility, complex purification, insufficient broad-spectrum detection of OPs and significant loss of catalytic activity after immobilization [27–29]. For this reason, the following section will summarize and review the research progress of electrochemical active materials in inhibited enzymatic electrochemical sensors.

4. Electrochemical Active Materials

Due to their excellent electrical conductivity, superior catalytic activity, high specific surface area, remarkable electro-/chemical stability, strong anti-interference capability and outstanding biocompatibility, carbon-based, polymer-based, metal-based, metal compound-based, MOF-based and COF-based materials have attracted extensive attention as active materials for constructing the electrochemical active layer in the development of high-performance enzymatic electrochemical sensors. Therefore, in this section, we comprehensively summarize and review the detection performance of enzymatic electrochemical sensors based on these active materials.

4.1. Carbon-Based Materials

Owing to their excellent electrical conductivity, favorable chemical and electrochemical stability, high specific surface area, robust mechanical properties, versatile surface functionalization and low density [30], carbon-based materials such as graphene (Gra), carbon nanotubes (CNTs), carbon black (CB) and porous carbon (PC) have been widely investigated for the preparation of enzymatic electrochemical sensors. For instance, Xie et al. [31] coated parathion and BSA on the surface of GCE modified with Gra and Chi composites (Gra-Chi) ink. They exploited the phosphorylation competition of AChE between the parathion immobilized on the modified electrode and target organophosphorus pesticides (OPs) in the electrolyte solution to achieve OP detection, as illustrated in Figure 3a. Under optimized electrochemical conditions, differential pulse voltammetry (DPV) analysis revealed that the peak response current of the modified electrode for chlorpyrifos, dichlorvos, methamidophos, dimethoate, phoxim, parathion, omethoate, pirimiphos methyl, diazinon, fenthion, trichlorfon and methomyl increased with increasing analyte concentration. All target OPs exhibited a same linear concentration range of 1–1500 $ng \cdot mL^{-1}$ and limit of detection (LOD) spanning from 0.012 to 0.23 $ng \cdot mL^{-1}$.

To eliminate the adverse effects of crosslinking agents on the detection properties of enzymatic electrochemical sensors and improve their electron transfer efficiency, Thakkar et al. [29] first modified GCE with carboxylic group-decorated multiwalled carbon nanotubes (MWCNTs). Subsequently, AChE was fixed on the modified electrode by the amidation reaction between carboxylic groups and amine groups of AChE. Under optimal reaction conditions, cyclic voltammetry (CV) analysis demonstrated that the peak response current of the modified electrode decreased with the increase in the concentration of paraoxon, with a linear concentration range of 10–50 nM and an LOD of 0.1 nM. The modified electrodes exhibited high replication, stability and reactivation for detecting paraoxon. In contrast, Sundramoorthy et al. [32] used glutaraldehyde as a crosslinker to immobilize AChE on single-walled carbon nanotubes' (SWCNTs) modified GCE surface,

https://doi.org/10.3390/molecules31040717

preparing enzymatic electrochemical sensors for detection of methyl parathion, as shown in Figure 3b. Under optimized detection conditions, square wave voltammetry (SWV) analysis showed that the response current peak decreased with the increase in the concentration of methyl parathion. The linear concentration range was 1×10^{-10}–5×10^{-6} M and the LOD was 3.75×10^{-11} M. The sensors displayed high stability, excellent selectivity and good reproducibility. Furthermore, Han et al. [33] prepared porous Gra oxide (PGO) via a traditional hydrothermal approach and fixed AChE on PGO-modified GCE to build enzymatic electrochemical sensors for detecting paraoxon. Under optimal electrochemical conditions, DPV analysis showed that the inhibition rate of the modified electrode increased with the increase in the concentration of paraoxon. The linear concentration range was 10–45 ng mL^{-1} and the LOD was 1.58 ng mL^{-1}. The recovery for paraoxon was 96.9–100.4% in vegetables samples.

In addition, Guo et al. [34] reported the use of antimony tin oxide, Chi and mesoporous carbon composites (ATO-Chi-MC) to decorate SPE, followed by AChE immobilization for the detection of chlorpyrifos and methamidophos, as shown in Figure 3c. Under optimal reaction conditions, DPV analysis confirmed that the sensors exhibited superior electrochemical detection performances, good storage stability and excellent repeatability. In the chlorpyrifos concentration range of 0.01–105 μg L^{-1}, the response current peak of sensors decreased with the increase in its concentration with a corresponding LOD of 0.01 μg L^{-1}. In the same linear concentration range, the LOD of sensors for detecting methamidophos was 1 μg L^{-1}. More importantly, enzymatic electrochemical sensors can also be used for simultaneously detecting carbamate pesticides and OPs. For example, Zergioti et al. [35] immobilized AChE on the surface of SPE modified with CB to prepare enzymatic electrochemical sensors for the simultaneous detection of chlorpyrifos and carbofuran. Under optimal conditions, the sensors displayed a linear relationship between their response current peak and the concentration of both pesticides. The linear concentration range was 0.7×10^{-9}–1.4×10^{-8} mol L^{-1} for chlorpyrifos (LOD = 6×10^{-10} mol L^{-1}) and 1.1×10^{-9}–2.3×10^{-8} mol L^{-1} for carbofuran (LOD = 4×10^{-10} mol L^{-1}).

Moreover, Wei et al. [36] modified BDDE with nitrogen-doped PC (N-PC, pore diameter of 150 nm), which was prepared by a hard template method using 1-butyl-3-methylimidazolium dicyanamide ionic liquid as precursors and silica spheres as a template. AChE was then coated on the modified electrode to construct enzymatic electrochemical sensors for detection of OPs such as dichlorvos and fenitrothion. Under optimal electrochemical conditions, chronoamperometry analysis showed that the inhibition rate of the modified electrode increased with the increase in the concentration of OPs, with the same linear concentration range in 10^{-10}–10^{-5} g L^{-1} for both OPs. The LOD was 1.50 pg·L^{-1} for dichlorvos and 4.42 pg·L^{-1} for fenitrothion. The modified electrodes exhibited excellent repeatability reproducibility, high stability, good selectivity and remarkable recoveries for both OPs in real samples. The superior electrochemical performance of the modified electrodes was attributed to the high conductivity of N-PC (~79 Ω) which is smaller than that of bare BDDE (~365 Ω) and PC-modified BDDE (~199 Ω).

Although carbon-based enzymatic electrochemical sensors have demonstrated outstanding performance for detecting various OPs such as ultralow sensitivity, good stability and considerable recovery in real samples, their practical applications are hindered by their intrinsic drawbacks, including insufficient dispersibility, weak interaction with bioactive enzymes, limited selectivity and low reproducibility. Therefore, strategies such as surface modification, heteroatom doping and composite formation with crosslinking agents can be employed to improve the detection performance of carbon-based enzymatic electrochemical sensors.

https://doi.org/10.3390/molecules31040717

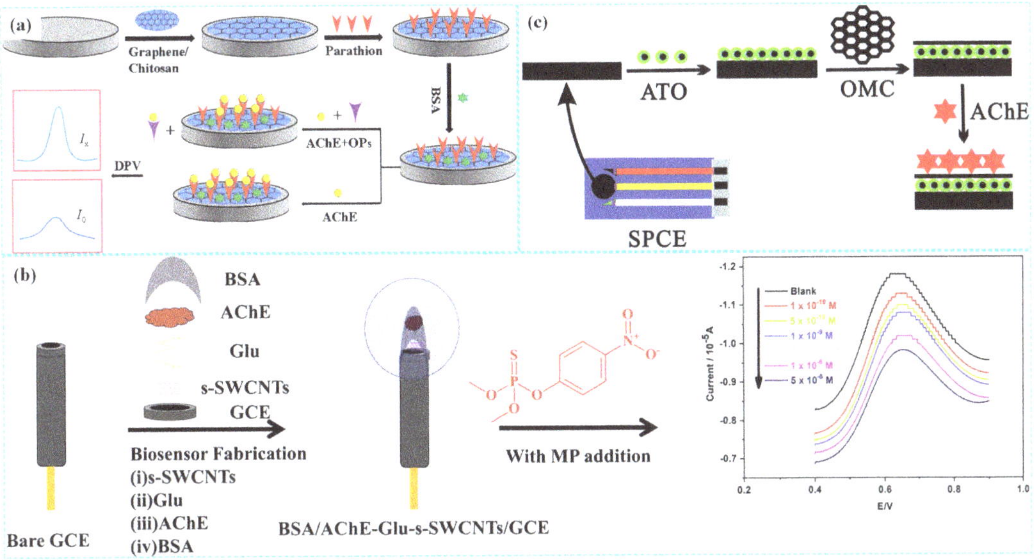

Figure 3. The preparation of Gra–Chi–based ((**a**), reprinted with permission from Ref. [31], Copyright 2021, Elsevier), SWCNT–based ((**b**), reprinted with permission from Ref. [32], Copyright 2019, Elsevier) and ATO–Chi–MC–based ((**c**), reprinted with permission from Ref. [34], Copyright 2019, Royal Society of Chemistry) enzymatic electrochemical sensors.

4.2. Polymer-Based Materials

Benefiting from their excellent conductivity derived from π conjugated structure, outstanding biocompatibility, favorable porosity, considerable mechanical stability, remarkable chemical stability and high specific surface area [37], polymer-based materials such as polyaniline (PANI), polyimide (PI), polypyrrole (PPY), poly 3,4-ethylene dioxy thiophene (PEDOT), polystyrene sulfonate (PSS) and 4, 7-di (furan-2-yl) benzo [1,2,5] thiadiazole (FBThF), are widely employed as bioactive enzyme immobilization carriers to the fabrication of enzymatic electrochemical sensors for detecting OPs.

As shown in Figure 4a, Kumar et al. [38] modified ITOE with Ag-CuO nanoparticles and PANI composites (Ag-CuO-PANI) with an electrodeposited approach, followed by glutaraldehyde-mediated AChE immobilization to prepare enzymatic electrochemical sensors. Under optimal reaction conditions, DPV analysis revealed the peak response current decreased with the increase in with the concentration of paraoxon, with a linear concentration range of 5–100 pM, a sensitivity of 0.5536 μA $(pM)^{-1}$ cm^{-2} and an LOD of 11.35 pM. The recoveries were 92–111% for paraoxon in banana, tomato and soil samples. The excellent electrochemical performance was related to the uniform distribution of catalytic active constituents within the PANI matrix. Based on the unique conduction characteristics of doping/dedoping PANI [39], Wang et al. [40] developed enzymatic resistive sensors by coating composition films consisting of Chi, AChE, PANI nanofibers and CNTs (Chi-PANI NFs-CNTs) on GE, as illustrated in Figure 4b. These sensors were combined with smartphones for on-site detection of paraoxon. Under optimized conditions, a linear relationship was observed between the response resistance of portable sensors and the concentration of paraoxon-methyl, in a concentration range of 1 ppt–100 ppb with an LOD of 0.304 ppt. The portable sensor achieved satisfactory recovery rates, which were comparable to those of traditional mass spectrometry methods for real sample analysis.

Furthermore, Jia et al. [41] used the composites of PI and reduced Gra oxide (rGO) as a flexible working electrode and fabricated enzymatic electrochemical sensors by the

https://doi.org/10.3390/molecules31040717

photoreduction of Au nanoparticles (Au NPs) and the fixation of monolayer MoS_2 and AChE (AChE-Au NPs-MoS_2-rGO-PI), as shown in Figure 4c. Under optimized reaction conditions, DPV analysis demonstrated that the linear relationship between the peak response current of sensors and the concertation of paraoxon existed in the concentration range of 0.005–0.150 $\mu g\ mL^{-1}$ with a sensitivity of 4.44 $\mu A\ (\mu g\ mL^{-1})^{-1}$ and an LOD of 0.0014 $\mu g\ mL^{-1}$.

Owing to the promotion of Au NPs for the nucleation of PPY at applied potential [42], Santos et al. [43] integrated Au NPs into indigo carmine doped PPY (Au NPs-IC-PPY) by electrochemical synthesis methods using FTOE as working electrodes. AChE was then immobilized to construct enzymatic electrochemical sensors for detecting methyl parathion, as shown in Figure 4d. Under optimized detection conditions, CV analysis showed that the difference between anodic and cathodic response peak currents of sensors was linearly dependent on the concentration of methyl parathion, in the concentration range of 1.3×10^{-7}–1.0×10^{-3} mol L^{-1}, with an LOD of 24 fmol L^{-1}. The outstanding electrochemical performance of sensors was ascribed to the synergistic effect between the high biocompatibility of indigo carmine doped PPY chains and the good conductivity of Au NPs. Subsequently, Santos et al. [44] introduced dodecyl sulfate and indigo carmine into the PPY matrix (DS-IC-PPY) to further improve its conductivity, stability and electroactivity. Using the same preparation method, enzymatic electrochemical sensors for detecting carbaryl were constructed. Under optimal reaction conditions, the sensors showed high sensitivity with a linear carbaryl concentration of 0.05–0.25 ng mL^{-1}, an LOD of 0.033 ng $cm^2\ mL^{-1}$ and a limit of quantification (QOD) of 0.11 ng $cm^2\ mL^{-1}$.

In addition, Dechtrirat et al. [45] incorporated Gra sheets into PEDOT and PSS mixed matrices, followed by Au NP decoration (Au NPs-Gra-PEDOT-PSS). The prepared composites were inkjet-printed on SPE, and AChE was immobilized to construct enzymatic electrochemical sensors for detecting chlorpyrifos (Figure 4e). Under optimal reaction conditions, DPV analysis indicated the peak response current of sensors decreased with the concentration of chlorpyrifos. The linear concentration range was 0.1–10 nM, and the LOD was 0.07 nM. The excellent electrochemical performance was correlated with the high conductivity of each component in the composites. Moreover, Pan et al. [46] prepared FBThF films on GCE using electro-polymerization, followed by modification with the mixtures of Ag nanoparticles (Ag NPs) and amino group-decorated rGO (Ag NPs-rGO-FBThF) and AChE fixation to fabricate enzymatic electrochemical sensors, as shown in Figure 4f. Under optimal detection conditions, CV analysis showed that the inhibition rate of sensors was directly proportional to the concentration of OPs. The linear concentration range was 0.099–9.9 $\mu g\ L^{-1}$ for malathion and 0.0206–2.06 $\mu g\ L^{-1}$ for trichlorfon, respectively. Their corresponding LOD was about 0.032 $\mu g\ L^{-1}$ and 0.001 $\mu g\ L^{-1}$. The sensors displayed remarkable reproducibility, good stability and high recovery for both OPs in real samples.

Enzymatic electrochemical sensors based on polymer-based materials have exhibited excellent electrochemical performances for detecting OPs. However, their detection performance is limited by inherent drawbacks, such as structural degradation under sensor operating conditions (leading to rapid conductivity loss), non-specific interactions with OPs (reducing sensor selectivity), reliance on high-purity monomers and specific dopants for synthesis and uneven physicochemical properties caused by complex preparation processes. Therefore, strategies including composite formation with other inorganic and carbon-based materials, surface functional modification, preparation process optimization and structural design refinement can be applied to improve the detection performance of polymer-based enzymatic electrochemical sensors.

https://doi.org/10.3390/molecules31040717

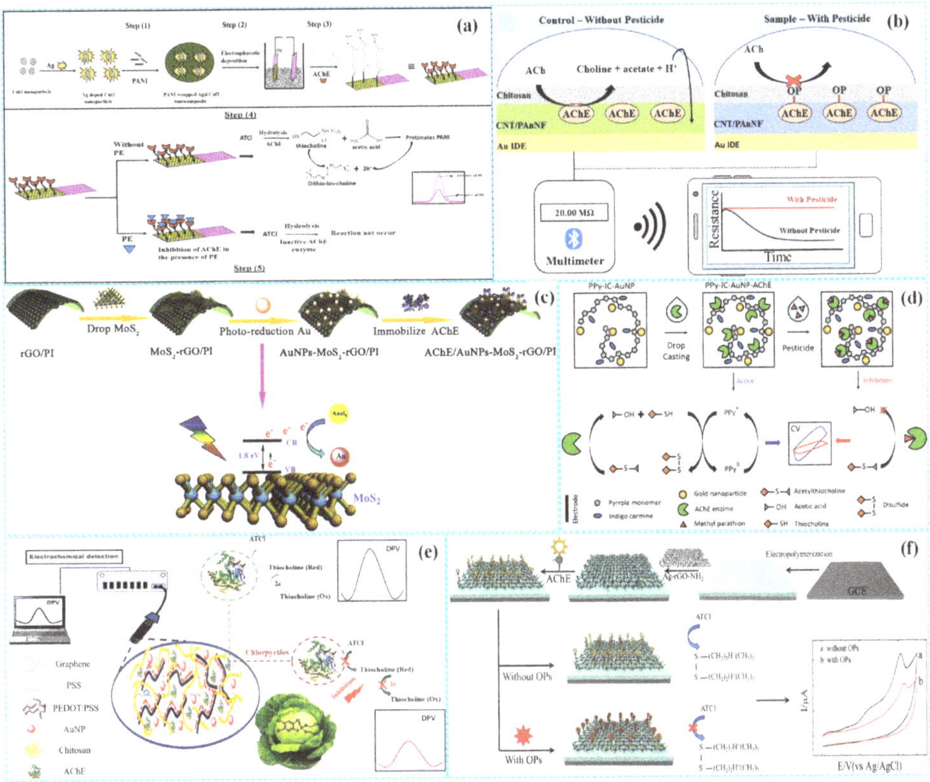

Figure 4. The preparation of Ag−CuO−PANI−based ((**a**), reprinted with permission from Ref. [38], Copyright 2023, Elsevier), Chi−PANI NFs−CNT−based ((**b**), reprinted with permission from Ref. [40], Copyright 2023, Elsevier), Au NPs−MoS_2−rGO−PI−based ((**c**), reprinted with permission from Ref. [41], Copyright 2020, Elsevier), Au NPs−PPY−based ((**d**), reprinted with permission from Ref. [43], Copyright 2025, Elsevier), Au NPs−Gra−PSS−PDOT−based ((**e**), reprinted with permission from Ref. [45], Copyright 2023, Elsevier) and Ag NPs−rGO−FBThF−based ((**f**), reprinted with permission from Ref. [46], Copyright 2019, Elsevier) enzymatic electrochemical sensors.

4.3. Metal-Based Materials

It is well established that metal-based materials (e.g., monometals such as Au, Ag and Cu; bimetals such as AuPt, AuPd and PdRh) have been widely used to construct enzymatic electrochemical sensors due to their excellent electron transfer capability, good catalytic activity and stability, outstanding biocompatibility, high sensitivity and repeatability [47]. Based on the specific interaction between Au and the amino group of AChE, Dhull et al. [48] fixed AChE on SWCNTs using nafion as a crosslinker, followed by coating the composites on Au wire (AW) electrode-decorated by Au NPs and MWCNTs to constructing enzymatic electrochemical sensors (AChE-SWCNTs-Au NPs-MWCNTs-AW) for detecting multiple OPs. A linear relationship between the response current of sensors and the concentration of OPs was observed. The concentration range was 1 nM–46 μM for methyl parathion, 1 nM–52 μM for monocrotophos, 1 nM–52 μM for chlorpyrifos and 20 nM–130 μM for endosulfan. Their corresponding LOD were 1.9 nM, 2.3 nM, 2.2 nM and 2.5 nM, respectively. Using MWCNTs as conductive additives, Citterio et al. [49] modified SPE with Chi and Au NPs composites and immobilized AChE to fabricate enzymatic electrochemical sensors (AChE-Au NPs-MWNTs-SPE) for detecting paraoxon. Under optimized conditions,

https://doi.org/10.3390/molecules31040717

amperometry analysis revealed two linear relationships between the response current and the concentration of paraoxon. One concentration range was 0.01–10 μg L^{-1}, and the other was 10–100 μg L^{-1}. The LOD of the sensors was 0.03 μg L^{-1}. The high detection performance of sensors was attributed to the synergistic effect between the high conductivity of Au NPs and the high specific surface area of MWCNTs. Additionally, Hou et al. [50] prepared enzymatic electrochemical sensors based on the composites of plant esterase, Chi, Gra and Au NPs (Au NPs-PE-Chi-Gra) for detecting methyl parathion and malathion, as shown in Figure 5a. Under optimal reaction conditions, DPV analysis showed the linear concentration range was 0–200 ppb for methyl parathion (LOD = 0.19 nM) and 0–500 ppb for malathion (LOD = 1.51 nM), respectively. To further improve the specific surface area of Au, Guo et al. [51] designed Au nanocages to load the composites of GO and Chi for SPE modification (Au NCs-GO-Chi), followed by the fixation of AChE to construct enzymatic electrochemical sensors, as shown in Figure 5b. Under optimized conditions, the sensors exhibited high electrochemical performance for detecting chlorpyrifos with a linear concentration range of 0.01–500 μg L^{-1} and an LOD of 3 ng L^{-1}. The outstanding detection performance of sensors was related to the synergistic effect between the high specific surface area, good dispersibility and remarkable adhesion of GO and the high specific surface area and excellent electron conductivity of Au nanocages. Furthermore, Cui et al. [52] incorporated the Au NPs and mesoporous SiO_2 core-shell composites into the matrix of Chi and TiO_2 hydrogel (Au NPs-SiO_2-TiO_2) to modify GCE, followed by the fixation of AChE to construct enzymatic electrochemical sensors, as shown in Figure 5c. Under optimized electrochemical conditions, the sensors displayed high sensitivity for detecting dichlorvos and fenthion. The linear concentration range was 0.018–13.6 μM for both OPs, but the LOD was 5.3 nM for dichlorvos and 1.3 nM for fenthion, respectively.

The hydrolysis of indoxyl acetate catalyzed by AChE produces hydroxyindole, which can promote the electrochemical reduction of Ag^+ to metallic Ag, generating a detectable response current. Based on this principle, Li et al. [53] designed enzymatic electrochemical sensors for detecting chlorpyrifos by correlating AChE catalytic activity with chlorpyrifos concentration in the electrolyte solution, using the response current from Ag electrodeposition on GE surfaces as a detection signal. Under optimal detection conditions, linear sweep voltammetry (LSV) analysis showed that the peak response current of sensors was positively related with the concentration of chlorpyrifos, in the concentration range of 10 pM–10 nM with an LOD of 4.0 pM. The recovery for chlorpyrifos was 92–104% in lake water. In addition, using the mixtures of Ag nanofibers (Ag NFs) and Gra as conductive additives, Zhang et al. [54] immobilized AChE on GCE modified with TiO_2 hydrogel and Chi composites to construct enzymatic electrochemical sensors (AChE-TiO_2-Chi-Ag NFs-Gra-Chi-GCE), as shown in Figure 5d. Under optimal reaction conditions, DPV analysis showed a linear relationship between the response current peak and the concentration of dichlorvos, in the concentration range of 0.036–22.63 mM with an LOD of 7.4 nM. Furthermore, based on the interaction between AChE and amino group-modified CNTs, Gao et al. [55] used Ag NP-decorated nitrogen-fluorine co-doped MoS_2 (Ag NPs-N-F-MoS_2) as carriers to immobilize the composites of AChE and CNTs, preparing enzymatic electrochemical sensors. Under optimal detection conditions, DPV analysis confirmed that the sensors exhibited high sensitivity, good stability and excellent reproducibility and selectivity. In the concentration range of $10–10^{-6}$ mg mL^{-1} for monocrotophos, the peak response current of sensors decreased linearly with the increase in its concentration. However, two linear relationships were observed for chlorpyrifos. One concentration range was 5×10^{-8}–10^{-7} mg·mL^{-1}, and the other was 10^{-7}–10^{-4} mg mL^{-1}. The LOD was 0.05 pg·mL^{-1} for monocrotophos and 1 pg·mL^{-1} for chlorpyrifos, respectively. Additionally, Gao et al. [56] used three-dimensional Ag tree-like nanostructure (3D Ag TN) as AChE carriers to prepare enzymatic

https://doi.org/10.3390/molecules31040717

electrochemical sensors for detecting omethoate, as shown in Figure 5e. Under optimal reaction conditions, DPV analysis showed that the peak response current of sensors was inversely proportional to the concentration of omethoate. The linear concentration range was 10^{-13}–10^{-7} M. The LOD of sensors was 1.26×10^{-14} M. The high detection performance of sensors was attributed to the high conductivity, ultrahigh hydrophilic and biocompatibility of tree-like Ag structures.

As shown in Figure 5f, Kulchat et al. [57] coated the composites of Cu nanowires (Cu NWs), rGO and AChE on the SPE surface to construct enzymatic electrochemical sensors (AChE-Cu NWs-rGO-SPE). Under optimized electrochemical conditions, CV analysis showed the response anodic peak current of sensors decreased with the increase in the concentration of chlorpyrifos. The linear concentration range was 10 μg L^{-1}–200 μg L^{-1}. The LOD and LQD were 3.1 μg L^{-1} and 12.5 μg L^{-1}, respectively. The sensors displayed strong anti-interference capacity against metal ions and satisfactory recovery rates for chlorpyrifos in drinking water and orange juice samples. The excellent electrochemical performance of sensors was ascribed to the high conductivity of Cu NWs and rGO, which facilitated the amplification of the detection signal.

In addition, bimetallic materials have been used to prepare high-performance enzymatic electrochemical sensors for detecting OPs, owing to their synergistic effects. For instance, Huang et al. [58] immobilized the mixtures of AChE and deacetylated Chi on GCE modified with self-interconnected AuPd bimetallic nanowires (AuPd NWs), fabricating enzymatic electrochemical sensors for detecting malathion. Under optimal detection conditions, DPV analysis showed the sensors displayed high sensitivity, good reproducibility, remarkable selectivity and good stability. The peak response current of sensors decreased with the increase in the concentration of malathion, in the concentration range of 0.1 PM to 100 nM. The LOD was 0.037 pM. The high performance of sensors was related to the unique structure of bimetallic nanowires, which facilitated the formation of multiple electron transfer paths and increased the electrochemical active area. Ping et al. [59] coated AuPd bimetallic nanoparticles in MXene (AuPd NPs-MXene)-modified SPE and used glutaraldehyde as crosslinker for AChE immobilization to construct enzymatic electrochemical sensors for the detection of paraoxon, as shown in Figure 5g. Under optimal conditions, amperometry analysis showed that the inhibition rate of the sensors was positively correlation with the concentration of paraoxon. The linear concentration range was 0.1–1000 μg L^{-1}. The LOD was 1.75 ng L^{-1}. The recovery for paraoxon was 87.93–111.02% in pear and cucumber samples. Benefiting from the ultralow density, abundant porous structure and ultrahigh specific surface area of hydrogels, Wu et al. [60] prepared the composites of polydopamine-coated AuPt bimetallic hydrogels (PD-AuPt) with one-step wet chemical methods to modify GCE, followed by the fixation of AChE to build enzymatic electrochemical sensors, as shown in Figure 5h. Under optimized electrochemical conditions, DPV analysis showed a linear relationship between the response current of sensors and the concentration of paraoxon. The linear concentration range was 0.5–1000 ng L^{-1}, and the LOD was 0.185 ng L^{-1}. Furthermore, Gao et al. [61] utilized one-dimensional PdRh bimetallic nanotubes (1D PdRh NTs, average diameter of 12 nm and pore size of 8–50 nm) to modify GCE, followed by Chi-mediated AChE immobilization to build enzymatic electrochemical sensors for detection of carbaryl, as shown in Figure 5i. Chronoamperometry analysis showed that the sensitivity of the sensors was 0.279 μA nM. Differential pulse voltammetry analysis displayed a linear relationship between the peak response current of sensors and the concentration of carbaryl from 9.44×10^{-8}–0.944 mg L^{-1}, and the calculated LOD was 9.44×10^{-8} mg L^{-1}. The sensors exhibited high selectivity for inorganic salt ions. The recovery was 94.01–102.80% for carbaryl in tap and lake water samples. The excellent electrochemical performance of sensors was attributed to the large specific surface area,

https://doi.org/10.3390/molecules31040717

abundant catalytic activity sites and rapid electron transfer properties of one-dimensional bimetallic nanotubes.

In conclusion, owing to the specific interactions between metal-based materials and bioactive enzymes, metal-based materials have been widely used as enzyme carriers to fabricate enzymatic electrochemical sensors, which exhibit high sensitivity, good selectivity, excellent stability and considerable reproducibility for detecting OPs. However, most reported metal-based materials are precious metals. Therefore, the development of high-performance non-precious metal-based materials is crucial for constructing low-cost enzymatic electrochemical sensors for OP detection. Additionally, rational nanostructure design of metal-based materials is necessary to improve their mass transfer performance and enzyme immobilization capacity, thereby enhancing the electrochemical detection of metal-based enzymatic electrochemical sensors.

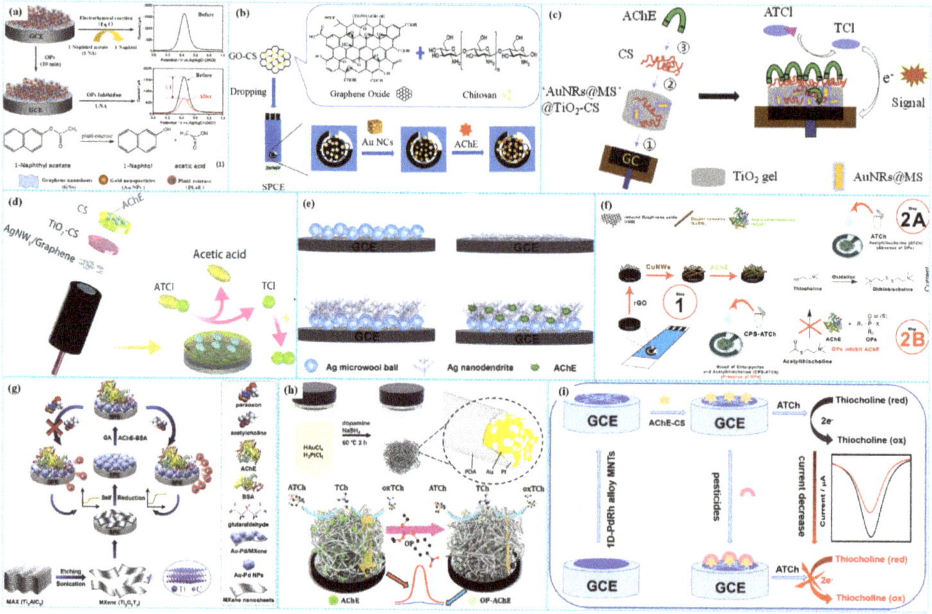

Figure 5. The preparation of Au NPs−PE−Chi−Gra−based ((**a**), reprinted with permission from Ref. [50], Copyright 2015, American Chemical Society), Au NPs−GO−Chi−based ((**b**), reprinted with permission from Ref. [51], Copyright 2019, Royal Society of Chemistry), Au NPs−SiO_2−TiO_2−based ((**c**), reprinted with permission from Ref. [52], Copyright 2019, Elsevier), TiO_2−Chi−Ag NFs−Gra−Chi−based (**d**), 3D Ag TN−based ((**e**), reprinted with permission from Ref. [56], Copyright 2020, Elsevier), Cu NWs−rGO−based ((**f**), reprinted with permission from Ref. [57], Copyright 2024, Elsevier), AuPd NPs−MXene−based ((**g**), reprinted with permission from Ref. [59], Copyright 2020, Elsevier), PD−AuPt−based ((**h**), reprinted with permission from Ref. [60], Copyright 2019, John Wiley and Sons) and 1D PdRh NT−based ((**i**), reprinted with permission from Ref. [61], Copyright 2022, Elsevier) enzymatic electrochemical sensors.

4.4. Metal Compound-Based Materials

It has been reported that the abundant redox active sites of metal compound-based materials can accelerate electron transfer between bioactive enzymes and working electrodes, thus reducing response time and improving sensor sensitivity [62]. In addition, surface functional groups of metal compound-based materials facilitate bioactive enzyme immobilization through covalent bonding [63]. For these reasons, metal compound-based materials have been applied to build enzymatic electrochemical sensors for detecting OPs. For instance, Gao et al. [64] modified GCE with Au NP-decorated three-dimensional flower-like α-Fe_2O_3 and CNT composites (Au-NPs-3D Fe_2O_3-CNTs), followed by the fixation of AChE to construct enzymatic electrochemical sensors. Under optimized detection conditions, DPV analysis showed that the sensors displayed excellent sensitivity, high selectivity, considerable stability and remarkable reproducibility for detecting omethoate. A linear relationship between the peak response current of sensors and omethoate concentration was observed in the concentration rang of 10^{-13}–10^{-7} M with an LOD of 1.9×10^{-15} M. The recoveries were 96.04–100.99% for omethoate in tomato samples. The sensors also displayed ultrahigh sensitivity for detection of other OPs, with an LOD of 1.02×10^{-15} M for malathion, 1.34×10^{-14} M for fenitrothion and 10^{-13} M for chlorpyrifos, respectively. The outstanding electrochemical performance was attributed to the synergistic effect between the large specific surface area and good biocompatibility of α-Fe_2O_3 and the high conductivity of Au NPs and CNTs. In addition, Niu et al. [65] immobilized AChE on GCE modified with magnetic Fe_3O_4 and mesoporous carbon sphere core-shell composites (Fe_3O_4-MCS) to build enzymatic electrochemical sensors. Under optimized reaction conditions, two linear relationships were observed between the peak response current and the concentration of malathion. One concentration range was 0.01–50 ppb, the other was 50–600 ppb. The calculated LOD was 0.0182 ppb.

It is well known that Ti atoms of TiO_2 can bind to the carboxyl/amino groups of AChE [66] and react with the phosphate groups of OPs [67]. For this reason, Guo et al. [68] utilized graphitized CNF-supported TiO_2 nanoparticles (TiO_2-g-CNFs) to modify GCE, followed by AChE fixation to construct enzymatic electrochemical sensors (Figure 6a). Under optimal detection conditions, the response current of sensors was positively correlated with the concentration of paraoxon. The linear paraoxon concentration range was 10^{-13}–10^{-8} M, and the LOD was 3.3 fM. The sensors exhibited good selectivity, remarkable reproducibility, high stability and satisfactory recovery rates. The outstanding electrochemical performance was related to the high conductivity derived from the graphitized structure of carbon nanofibers. In addition, Yang et al. [69] incorporated TiO_2 nanoparticles into Chi matrices (TiO_2 NPs-Chi) to decorate GCE, followed by coating the mixtures of Chi and AChE on modified electrodes, building patterned enzymatic electrochemical sensors. Under optimized conditions, DPV analysis showed that the peak response current decreased with the increase in the concentration of dichlorvos in the concentration range of 1.13 nM–22.6 μM with an LOD of 0.23 nM. Furthermore, based on electrodeposited composites of Chi and TiO_2 hydrogels (TiO_2 H-Chi), Cui et al. [70] immobilized AChE on modified GCE to prepare enzymatic electrochemical sensors, as shown in Figure 6b. Under optimal reaction conditions, the sensors exhibited high sensitivity for detecting dichlorvos. There were two linear relationships between the peak response current and the concentration of dichlorvos. One concentration range was 0.036–0.453 μM, and the other was 0.453–22.6 μM. The calculated LOD of sensors was 29 nM.

Based on the synergistic effect between p- and n-type semiconductors, Niu et al. [71] fixed AChE on GCE decorated with p-n heterojunctions of n-type ZnO nanoparticles and p-type rGO (ZnO-rGO) to build enzymatic electrochemical sensors for the detection of methyl parathion, as shown in Figure 6c. Under optimized detection conditions, DPV analysis indicated that the peak response current decreased with the increase in the concentration of methyl parathion. The linear concentration range was 10^{-5}–1 μg mL^{-1} with a sensitivity of 0.004638 μA (ng mL^{-1})$^{-1}$ and an LOD of 0.11871 ng mL^{-1}. The recovery for methyl parathion was 90.92–108.4% in apple and cucumber samples. The high detection performance was correlated with the formation of p-n heterojunction, which improved electrical conductivity and reduced the redox reaction potential. In addition, Liu et al. [72] immobilized AChE on GCE modified with core-shell hollow ZnO-CoO and N-PC composites (ZnO-CoO-N-PC).

https://doi.org/10.3390/molecules31040717

Under optimal electrochemical conditions, the sensors displayed high sensitivity for detecting chlorpyrifos and parathion-methyl. The concentration range was 7.6×10^{-15}–7.6×10^{-6} M for parathion-methyl (LOD = 7.6×10^{-15} M) and 2.74×10^{-13}–2.74×10^{-6} M for chlorpyrifos (LOD = 2.74×10^{-13} M), respectively. Huo et al. [73] prepared enzymatic electrochemical sensors by immobilizing AChE on GCE modified with the composites of three-dimensional Gra and flower-like CuO nanoparticles (3D Gra-CuO NPs), as shown in Figure 6d. Under optimized electrochemical conditions, SWV analysis showed that the peak response current decreased with increasing the concentration of malathion. A linear relationship was observed in the concentration of 1 ppt–15.555 ppb with an LOD of 0.31 ppt. The sensors displayed good stability, excellent reproducibility and considerable selectivity.

Thiocholines, the hydrolysis products of acetylthiocholine, can accelerate the decomposition of MnO_2. Based on this mechanism, Tang et al. [74] used thionine as a multiple electron mediator to fabricate split-type homogeneous enzymatic electrochemical sensors for detecting dichlorvos. In these sensors, OP-induced inhibition of bioactive enzyme catalytic activity reduces thionine release, leading to a decrease in the response current change. Under optimal detection conditions, differential pulse voltammetry analysis showed that the peak response current decreased with the increase in the concentration of dichlorvos. The concentration range was 10^{-6}–10^{-10} M, and the LOD was 3×10^{-10} M. In addition, it is well known that the electrochemical hydrolysis of H_2O_2 generates a response current, and the oxidation of o-phenylenediamine produces a color change. Since the specific binding between NiCoFeS nano-enzymes and thiocholines can inhibit their catalytic activity for H_2O_2 electrochemical hydrolysis, Zou et al. [75] designed dual-signal (colorimetric and electrochemical) homogeneous enzymatic sensors for detection of trichlorfon based on the controllable synthesis of NiCoFeS nano-enzymes, as shown in Figure 6e. Under optimal conditions, differential pulse voltammetry analysis revealed a linear relationship between the response current change and the concentration of trichlorfon, in the concentration range of 12.5 fg mL^{-1}–1.25 ng mL^{-1} with an LOD of 9.74fg mL^{-1}. Colorimetric analysis showed a linear correlation between the absorbance and the concentration of trichlorfon, in the concentration range of 0.25 pg mL^{-1}–125 pg mL^{-1} with an LOD of 0.12 pg mL^{-1} under the same detection conditions.

In addition, combining the multiple metallic redox sites, abundant catalytic active centers and high specific surface area of $CuInS_2$ with the high conductivity of GO, Hasin et al. [76] used glutaraldehyde to fix the composites of $CuInS_2$ and GO ($CuInS_2$-GO) on SPE, followed by coating the mixtures of AChE and BAS on the modified electrode to build enzymatic electrochemical sensors for detecting chlorpyrifos (Figure 6f). Under optimized conditions, LSV analysis showed that two linear relationships between the response current and the concentration of chlorpyrifos were observed. One linear concentration range was 0.5–230 ng mL^{-1}, and the other was 240–470 ng mL^{-1}. The LOD was 0.023 ng mL^{-1}. The sensors displayed good reproducibility, excellent selectivity and high recovery for chlorpyrifos in vegetable samples. Furthermore, bimetallic oxides have also been used to construct enzymatic electrochemical sensors for OP detection. For example, Gao et al. [77] prepared three-dimensional hierarchical flower-like Co-based bimetallic oxide nanosheets (3D CoXO) using a hydrothermal approach, followed by AChE immobilization on modified GCE to fabricate enzymatic electrochemical sensors. Under optimal conditions, DPV analysis showed that the sensors displayed high sensitivity, good selectivity and excellent stability for detecting diazinon and omethoate. The linear concentration range was 3.88×10^{-6}–3.88 μM for diazinon (LOD = 0.36 pM) and 4.69×10^{-7}–0.469 μM for omethoate (LOD = 0.033 pM), respectively.

In addition, Prussian blue (PB) has been used as an active material for constructing enzymatic electrochemical sensors. For instance, Teng et al. [78] used nafion as a crosslinker

https://doi.org/10.3390/molecules31040717

to immobilize AChE on the surface of SPE modified by PB, constructing dual-channel enzymatic electrochemical sensors. Under optimized detection conditions, the dual-channel sensors displayed high sensitivity for isocarbophos, chlorpyrifos and trichlorfon with an LOD of 10^{-7} g mL^{-1} for all three analytes. Moreover, Cinti et al. [79] modified SPE with PB and CB bioactive mixtures (PB-CB) and immobilized butyrylcholinesterase to construct glove-based portable enzymatic electrochemical sensors for on-site detection dichlorvos on apples and oranges, as shown in Figure 6g. Under optimized parameters, the glove-based sensors achieved an LOD as low as 8 nM. Their sensitivity was significantly higher than other portable sensors, although it remained lower than that of traditional laboratory-based analytical techniques.

In conclusion, various metal compound-based materials have been employed to construct enzymatic electrochemical sensors. Although these sensors display high sensitivity, good stability and considerable reproducibility, their limited intrinsic conduction reduces response speed and sensitivity. Additionally, the uneven distribution of surface functional groups can cause the leakage of immobilized bioactive enzymes. Therefore, to develop high-performance enzymatic electrochemical sensors, it is critical to fabricate hybrid composites to improve the conductivity of metal compound-based materials and to perform surface functional modification to enhance bioactive enzyme immobilization capacity.

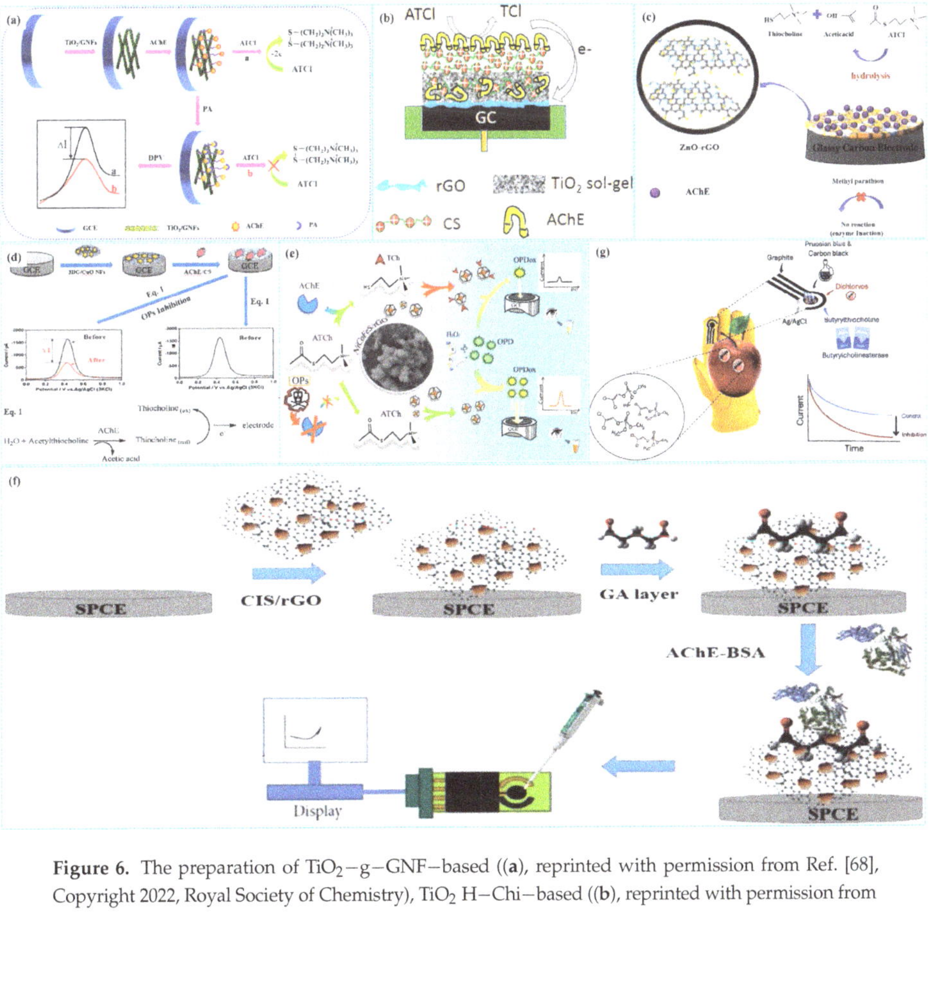

Figure 6. The preparation of TiO_2−g−GNF−based ((**a**), reprinted with permission from Ref. [68], Copyright 2022, Royal Society of Chemistry), TiO_2 H−Chi−based ((**b**), reprinted with permission from

https://doi.org/10.3390/molecules31040717

Ref. [70], Copyright 2018, Elsevier), ZnO−rGO−based ((**c**), reprinted with permission from Ref. [71], Copyright 2024, Elsevier), 3D Gra−CuO NP−based ((**d**), reprinted with permission from Ref. [73], Copyright 2019, Elsevier), NiCoFeS−based ((**e**), reprinted with permission from Ref. [75], Copyright 2025, Elsevier), $CuInS_2$−GO−based ((**f**), reprinted with permission from Ref. [76], Copyright 2021, Elsevier) and PB−BC−based ((**g**), reprinted with permission from Ref. [79], Copyright 2025, Elsevier) enzymatic electrochemical sensors.

4.5. MOF-Based Materials

As we all know, MOFs are constructed by the coordination bonding of metal ions (or clusters) and organic ligands [80]. In October 2025, the Royal Swedish Academy of Sciences awarded the Nobel Prize in Chemistry to Susumu Kitagawa, Richard Robson and Omar Yaghi for their pioneering contributions in the synthesis and functionalization of MOFs, also referred to as porous coordination polymers [81]. In 1989, Robson et al. [82] proposed the concept of porous coordination polymers based on the crystal structure of diamond. In 1995, Yaghi et al. [83] first synthesized MOF-5, a material with a stable three-dimensional open framework structure, using Zn_4O clusters as secondary building units and terephthalic acid as organic ligands. Due to its unique structural characteristics, MOF-5 has been exploited as the active material of enzymatic electrochemical sensors. For example, Basu et al. [84] adopted an in situ electrochemical approach to deposit anisotropic Au nanorods on the surface of ITOE modified with MOF-5. Cysteamine was used as a crosslinker to immobilize AChE on the modified electrode surface, thereby constructing enzymatic electrochemical sensors for detecting OPs, as shown in Figure 7a. Under optimized reaction conditions, CV analysis revealed a linear relationship between the response anodic current peak of sensors and the concentration of chlorpyrifos, which existed in the concentration range of 30 ng L^{-1}–600 ng L^{-1}. The sensitivity of sensors was 2.04 μA ng^{-1} L cm^{-2} with an LOD of 3 ng L^{-1}. Moreover, the sensors exhibited high sensitivity for detection of five OPs and their mixtures. In addition, the detection performance of these sensors for parathion, methyl parathion, malathion, ethion and chlorpyrifos in abelmoschus esculentus, solanum melongena, capsicum annuum and momordica charantia linn samples was comparable to those of a gas chromatography–electron capture detector.

Furthermore, the Zn^{2+}–imidazole coordination moieties in ZIF-8 MOFs endow the material with good biocompatibility, making it suitable for supporting bioactive enzymes [85]. Utilizing Chi as link binders and Gra as conductive additives, Wang et al. [86] coated AChE-Chi-Gra-ZIF-8 composites on GCE to assemble enzymatic electrochemical sensors for detection of isocarbophos, as shown in Figure 7b. Under optimized experimental conditions, differential pulse voltammetry analysis demonstrated that the inhibition ratio of the enzymatic sensors increased with the increase in the concentration of isocarbophos, with a linear concentration range of 1.73–345.7 nM and an LOD of 0.62 nM. Additionally, Zhang et al. [87] employed carboxymethylation materials as link binders to immobilize AChE on amino-functionalized Zn-based MOFs (Zn-MOFs), constructing enzymatic electrochemical sensors for glyphosate detection. Under optimized reaction conditions, DPV analysis indicated that the peak response current decreased with the increase in the concentration of glyphosate, displaying a linear range of 10^{-15}–10^{-9} M, LOD of 1.24×10^{-13} M and LOQ of 4.13×10^{-13} M, respectively. The integration of these sensors with near-field communication chips enabled on-site detection of glyphosate. The outstanding detection performance of the enzymatic sensors is related to the charge redistribution within Zn-based MOFs, which facilitates charge separation between these sensors and glyphosate [88]. This phenomenon is induced by the positively charged amine groups and the abundant hydrogen bond acceptor/donor in the ligand of 5-(5-Amino-3-carboxypyridin-2-yl) isophthalic acid ligands.

https://doi.org/10.3390/molecules31040717

In addition, the strong interaction between the Zr ion centers of Zr-based MOFs and the P=O or P=S groups of OPs enable these MOFs to serve as recognition elements of non-enzymatic electrochemical sensors for detecting OPs [89]. Simultaneously, the large specific surface area of Zr-based MOFs makes them ideal carriers for bioactive enzymes in enzymatic electrochemical sensors. To this end, Bagheri et al. [90] anchored AChE onto GCE modified with Ce-UiO-66-MWCNT composites to prepare enzymatic electrochemical sensors for detecting paraoxon, as depicted in Figure 7c. Under optimal detection conditions, DPV analysis showed a linear relationship between the inhibition rate of the enzymatic sensors and the paraoxon concentration, in the concentration range of 0.01–150 nM with an LOD of 0.004 nM. The enzymatic sensors had acceptable stability, good selectivity, remarkable reproducibility and satisfactory recovery rates for paraoxon in cabbage and spinach samples. Their outstanding electrochemical performance for detection of paraoxon was ascribed to the synergistic effect of multiple factors: the multiple valence states and good oxygen affinity of Ce ions [91], the high conductivity of MWCNTs and the high chemical and thermal stability of UiO-66.

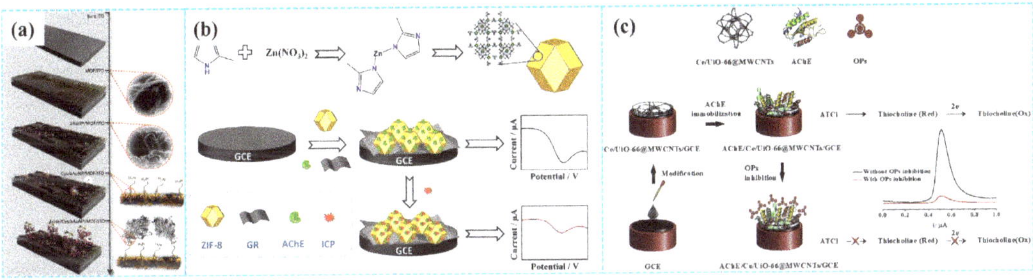

Figure 7. The preparation of MOF−5−based ((**a**), reprinted with permission from Ref. [84], Copyright 2020, Royal Society of Chemistry), Chi−Gra−ZIF−8−based (**b**) and Ce−UiO−66−MWCNT−based ((**c**), reprinted with permission from Ref. [90], Copyright 2019, Elsevier) enzymatic electrochemical sensors.

In brief, owing to their ultrahigh specific surface area and abundant catalytic active sites, MOFs as active materials for enzymatic electrochemical sensors exhibit high sensitivity and good stability. To further enhance the detection performance of MOF-based enzymatic electrochemical sensors, it is imperative to introduce conductive materials such as PC, Gra and conductive polymers to form hybrid composites, which can accelerate the electron transfer between bioactive enzymes and the working electrode. Additionally, optimizing the pore structure of MOF-based materials can promote the diffusion of substrates and expand the linear detection range of electrochemical sensors.

4.6. COF-Based Materials

Due to their large specific surface area, high porosity, well-defined crystalline structure, considerable chemical and electrochemical stability and remarkable biocompatibility, COFs exhibit great potential for immobilizing bioactive enzymes [92]. However, enzymatic electrochemical sensors using TCh and $[Fe(CN)_6]^{3-}$ as signal generators (that is "turn off" mechanism) suffer from inadequate selectivity and repeatability [93]. To address these problems, Ma et al. [94] developed "turn on" low potential enzymatic electrochemical sensors by immobilizing AChE on COF NF-modified GCEs, with $[Ru(bpy)_3]^{2+}$ as a signal generator (Figure 8a). Compared to sensors based on TCh and $[Fe(CN)_6]^{3-}$, these sensors showed nearly identical catalytic activity and stability for dichlorvos detection while demonstrating significantly enhanced selectivity and repeatability. Alternatively, Wang et al. [95] synthe-

https://doi.org/10.3390/molecules31040717

sized core-shell COFs-MWCNT composites using 1,3,5-Tris(4-aminophenyl) benzene and 2,5-divinylterephthalaldehyde as organic building blocks. These composites were used for modifying GCE, followed by AChE immobilization for malathion detection. Electrochemical analysis revealed a linear relationship between the oxidation current of acetylthiocholine chloride and its concentration under optimized conditions. The linear concentration range was 1 nM–10 μM, and the LOD was 0.5 nM. The recovery was 96.0–101.6% for tap water and 98.0–105.0% for spinach samples, respectively.

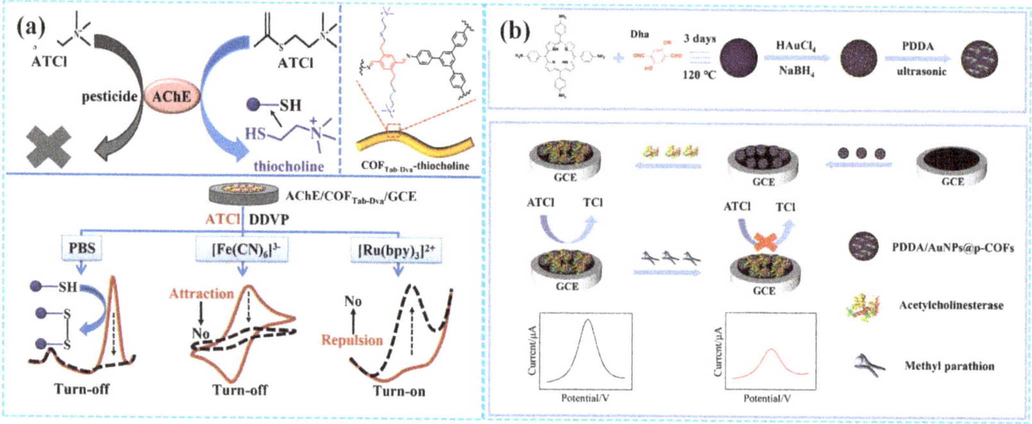

Figure 8. The detection mechanism of the "turn off" and "turn on" strategy ((**a**), reprinted with permission from Ref. [94], Copyright 2022, Elsevier); the preparation of the PDDA-Au NPs-COF-based enzymatic electrochemical sensor ((**b**), reprinted with permission from Ref. [96], Copyright 2024, Elsevier).

Taking advantage of the synergistic effect of the conductivity of Au NPs, the conjugated π-electron macrocycles of porphyrin-based COFs and the electrostatic interaction between the positively charged poly diallyldimethylammonium chloride (PDDA) and the negatively charged AChE, Zhang et al. [96] immobilized AChE on PDDA-Au NPs-COF composite-modified GCE to fabricate enzymatic electrochemical sensors for methyl parathion detection (Figure 8b). DPV analysis revealed that these sensors exhibited remarkable sensing performances for methyl parathion detection, including outstanding electrochemical activity, good stability, excellent repeatability and high selectivity. A linear relationship between the inhibition efficiency of enzymatic electrochemical sensors and the concentration of methyl parathion was observed in the concentration range of 1.9×10^{-9}–3.8×10^{-5} M with an LOD of 2.3×10^{-10} M. The recovery for methyl parathion was 95.7–107.9% in tomato and strawberry samples. The introduction of Au NPs and PDDA reduced the electron transfer resistance of modified electrodes, thus improving the detection ability of enzymatic electrochemical sensors. Separately, Song et al. [97] prepared enzymatic electrochemical sensors based on the composites of biomass PC and COFs (BPC-COFs) for trichlorfon detection. Under optimized conditions, DPV analysis showed the response current peak of the sensors decreased with the increase in the concentration of trichlorfon, in the concentration range of 0.2–19 ng mL^{-1}. The calculated LOD was 0.067 ng mL^{-1}.

In conclusion, COF-based enzymatic electrochemical sensors have displayed high sensitivity for OP detection. However, their performance is constrained by four key limitations: (1) poor intrinsic conductivity of COF-based materials hinders electron transfer between the bioactive enzyme and the working electrode, resulting in the weak response

https://doi.org/10.3390/molecules31040717

signals; (2) bioactive enzyme-induced channel blockage of COF-based materials causes the inhomogeneous distribution of bioactive enzymes and hinders the diffusion of substrates; (3) the complex preparation process of COF-based materials requires toxic organic solvent; and (4) the residual organic solvent and monomers in COF-based materials may inactivate immobilized bioactive enzymes. Therefore, addressing these challenges is significant for developing high-performance COF-based enzymatic electrochemical sensors.

Overall, based on the aforementioned literature, the GCE is the most commonly preferred working electrode for fabricating lab-scale enzymatic electrochemical sensors, attributed to its high chemical inertness, low cost and facile modifiability. However, GCE is difficult to integrate with alternative supporting substrates such as paper- or glove-based substrates, which limits its applicability for achieving on-site detection of OPs. From this point of view, as compared with other working electrodes, SPE fabricated via screen-printing technology represents the ideal candidate for building practical enzymatic electrochemical sensors and enabling on-site OP detection. The primary challenge hindering the practical application of SPE-based enzymatic electrochemical sensors lies in reducing contact resistance and enhancing electrical conductivity.

In addition, Table 2 comprehensively summarizes the electrochemical active materials, analytical techniques and electrochemical detection performances of reported enzymatic electrochemical sensors for OP analysis. As shown in Table 2, a variety of analytical techniques have been employed to evaluate the detection performance of enzymatic electrochemical sensors, including DPV, CV, LSV, SWV, electrochemical impedance spectroscopy, amperometry and chronoamperometry. Among them, DPV is the most widely used technique for OP detection. This method combines linear scanning voltammetry with pulse technology, endowing it with numerous advantages: (1) a lowered LOD, making it suitable for trace-level analysis; (2) a well-resolved redox potential window, enabling simultaneous detection of multiple analytes; (3) excellent anti-interference capacity against inert electrolytes and trace impurities in test solutions, accompanied by a stable baseline; (4) a favorable balance between detection sensitivity and efficiency, rendering it applicable for batch sample analysis; and (5) the ability to detect reversible/semi-reversible electroactive species. Despite these merits, DPV also has several inherent drawbacks: low sensitivity toward irreversible electroactive species and inapplicability for rapid dynamic detection. It is therefore necessary to select an appropriate analytical technique based on the practical application scenarios of enzymatic electrochemical sensors.

Table 2. The electrochemical active materials, analytical techniques and electrochemical detection performance of enzymatic electrochemical sensors.

Electrochemical Active Materials	Analytical Techniques	Electrochemical Detection Performance	Ref.
Gra-Chi	DPV	**Linear concentration range** for 11 OPs: 1–1500 ng·mL^{-1}; **LOD**: 0.012–0.23 ng·mL^{-1}; **Repeatability**: relative standard deviation (RSD) range of 1.35–6.69% (3 times); **Stability**: standard deviation (SD) of 6.7% (10 days, 37.4 °C); **Recovery rate** in apple samples: 87.24–110.2%.	[31]
MWCNTs	CV	**Linear concentration range** for paraoxon: 10–50 nM; **LOD**: 0.1 nM; **Anti-interference ability**: ascorbic acid, uric acid, nitrobenzene, nitrophenol, inorganic ions, fructose and sucrose; **Repeatability**: SD of 0.071 (3 times); **Stability**: 54.52% of initial activity (32 days, 4 °C); **Reactivation**: 90.68% of initial activity (15 min, 1.0 mM pralidoxime iodide solution); **Recovery rate** in potato and tap water samples: 97.28–103.56%.	[29]

https://doi.org/10.3390/molecules31040717

Table 2. *Cont.*

Electrochemical Active Materials	Analytical Techniques	Electrochemical Detection Performance	Ref.
SWCNTs	SWV	**Linear concentration range** for methyl parathion: 1×10^{-10}–5×10^{-6} M; **LOD**: 3.75×10^{-11} M; **Anti-interference ability**: glucose, uric acid, oxalic acid, ascorbic acid and orthophosphoric acid; **Repeatability**: RSD of 0.89% (3 times); **Stability**: 93.14% of initial activity (33 days, 4 °C); **Reproducibility**: RSD of 4.68% (3 electrodes); **Recovery rate** in strawberry and apple samples: 98.31–102.22%.	[32]
PGO	DPV	**Linear concentration range** for paraoxon: 10–45 ng mL^{-1}; **LOD**: 1.58 ng mL^{-1}; **Anti-interference ability**: citric acid, oxalic acid, glucose, p-toluenesulfonic acid, toluene, p-nitrophenol and carbofuran; **Repeatability**: RSD of 3.6% (6 times); **Stability**: 97.2% of initial activity (30 days, 4 °C); **Reproducibility**: RSD of 1.1% (6 electrodes); **Reactivation**: 96.74% of initial activity (20 min, 0.1 M PBS solution); **Recovery rate** in lettuce and cabbage samples: 96.9–100.4%.	[33]
ATO-Chi-MC	DPV	**Linear concentration range** for chlorpyrifos and methamidophos: 0.01–10^5 μg L^{-1}; **LOD**: 0.01 μg L^{-1} (chlorpyrifos) and 1 μg L^{-1} (methamidophos); **Stability**: 80% of initial activity (30 days, 4 °C); **Reproducibility**: RSD of 6.3% (5 electrodes); **Recovery rate** in oilseed rape, lettuce and Chinese cabbage samples: 91–107%.	[34]
CB	Amperometry	**Linear concentration range** for OPs: 0.7×10^{-9}–1.4×10^{-8} mol L^{-1} (chlorpyrifos) and 1.1×10^{-9}–2.3×10^{-8} mol L^{-1} (carbofuran); **LOD**: 6×10^{-10} mol L^{-1} (chlorpyrifos) and 4×10^{-10} mol L^{-1} (carbofuran); Stability: 86.6% of initial activity (14 days, 4 °C); **Reproducibility**: RSD range of 2.3–8.6%; **Recovery rate** in olive oil samples: 50–80%.	[35]
N-PC	Chrono-amperometry	**Linear concentration range** for dichlorvos and fenitrothion: 10^{-10}–10^{-5} g L^{-1}; **LOD**: 1.50 pg L^{-1} (dichlorvos) and 4.42 pg L^{-1} (fenitrothion); **Anti-interference ability**: inorganic ions; **Repeatability**: RSD of 1.07% (6 times); **Stability**: 92.3% of initial activity (30 days, 4 °C); **Reproducibility**: RSD of 5.38% (6 electrodes); **Reactivation**: 93.0% of initial activity (8 min, 5.00 mmol·L^{-1} pralidoxime iodide solution); **Recovery rate** in real samples: 96.1–100% (dichlorvos, lettuce leaves) and 90.8–107% (fenitrothion, shanghaiqing leaves).	[36]
Ag-CuO-PANI	DPV	**Linear concentration range** for paraoxon: 5–100 pM; **LOD**: 11.35 pM; **Sensitivity**: 0.5536 μA (pM)$^{-1}$ cm^{-2}; **Anti-interference ability**: inorganic ions, ascorbic acid, dopamine and uric acid; **Stability**: 71.3% of initial activity (20 days, 4 °C); **Reproducibility**: RSD of 1.74% (5 electrodes); **Reactivation**: 93.0% of initial activity (8 min, 5.00 mmol·L^{-1} pralidoxime iodide solution); **Recovery rate** in banana, tomato and soil samples: 92–111%.	[38]
Chi-PANI NFs-CNTs	Electrochemical impedance spectroscopy	**Linear concentration range** for paraoxon-methyl: 1 ppt–100 ppb; **LOD**: 0.304 ppt; **Anti-interference ability**: inorganic ions, amino acids, citric acid and glucose; **Repeatability**: RSD of less than 5% (6 times); **Reproducibility**: RSD of 1.74% (5 electrodes); **Recovery rate** in river water, well water, orange juice, grape juice, milk and beef samples: 90–112%.	[40]
Au NPs-MoS_2-rGO-PI	DPV	**Linear concentration range** for paraoxon: 0.005–0.150 μg mL^{-1}; **LOD**: 0.0014 μg mL^{-1}; **Sensitivity**: 4.44 μA (μg mL^{-1})$^{-1}$; **Stability**: 96.0% of initial activity (7 days, 4 °C); **Reproducibility**: RSD of 4.68% (5 electrodes); **Recovery rate** in vegetable leave samples: 99–104%.	[41]
Au NPs-IC-PPY	CV	**Linear concentration range** for methyl parathion: 1.3×10^{-7}–1.0×10^{-3} mol L^{-1}; **LOD**: 24 fmol L^{-1}; **Sensitivity**: 14 μA cm^{-2} g^{-1} mL; **Stability**: 84% of initial activity (7 days, −15 °C).	[43]
DS-IC-PPY	Chrono-amperometry	**Linear concentration range** for carbaryl: 0.05–0.25 ng mL^{-1}; **LOD**: 0.033 ng mL^{-1} cm^{-2}; **Sensitivity**: -59.5×10^3 A cm^{-2} mL g^{-1}; LOQ: 0.11 ng cm^2 mL^{-1}; **Repeatability**: RSD of 1.8% (4 times); **Stability**: 40% of initial activity (30 days, 4 °C); **Reproducibility**: RSD of 3.7% (3 electrodes); **Recovery rate** in tap water samples: 114–120%	[44]

https://doi.org/10.3390/molecules31040717

Table 2. *Cont.*

Electrochemical Active Materials	Analytical Techniques	Electrochemical Detection Performance	Ref.
Au NPs-Gra-PEDOT-PSS	DPV	**Linear concentration range** for chlorpyrifos: 0.1–10 nM; **LOD**: 0.07 nM; **Anti-interference ability**: inorganic ions, ascorbic acid, citric acid, uric acid and glucose; **Stability**: 91.94% of initial activity (56 days, 4 °C); **Reproducibility**: RSD of 2.45% (9 electrodes); **Recovery rate** in cabbage samples: ~106%.	[45]
Ag NPs-rGO-FBThF	CV	**Linear concentration range** for OPs: 0.099–9.9 μg L^{-1} (malathion) and 0.0206–2.06 μg L^{-1} (trichlorfon); **LOD**: 0.032 μg L^{-1} (malathion) and 0.001 μg L^{-1} (trichlorfon); **Stability**: 82.7% of initial activity (30 days, 4 °C); **Reproducibility**: RSD of 2.30% (malathion) and 3.44% (trichlorfon) (5 electrodes); **Recovery rate** in tap water samples: 95.2–108%.	[46]
SWCNTs-Au NPs-MWCNTs	Amperometry	**Linear concentration range** for OPs: 1 nM–46 μM (methyl parathion), 1 nM–52 μM (monocrotophos), 1 nM–52 μM (chlorpyrifos) and 20 nM–130 μM (endosulfan); **LOD**: 1.9 nM (methyl parathion), 2.3 nM (monocrotophos), 2.2 nM (chlorpyrifos) and 2.5 nM (endosulfan); **Anti-interference ability**: inorganic ions, glucose, fructose, sucrose and ascorbic acid; **Stability**: 70% of initial activity (60 days).	[48]
Au NPs-MWCNTs	Amperometry	**Linear concentration range** for paraoxon: 0.01–10 μg L^{-1} and 10–100 μg L^{-1}; **LOD**: 0.03 μg L^{-1}; **Stability**: 83% of initial activity (49 days, 4 °C); **Reproducibility**: RSD of 3.02% (10 electrodes); **Recovery rate** in spinach samples: −104%.	[49]
Au NPs-PE-Chi-Gra	DPV	**Linear concentration range** for OPs: 0–200 ppb (methyl parathion) and 0–500 ppb (malathion); **LOD**: 0.19 nM (methyl parathion) and 1.51 nM (malathion); **Anti-interference ability**: inorganic ions, glucose and citric acid; **Stability**: 73% of initial activity (15 days, 4 °C); **Recovery rate** in carrot and apple samples: 92–107.40% (methyl parathion) and 93–109.50% (malathion).	[50]
Au NCs-GO-Chi	DPV	**Linear concentration range** for chlorpyrifos: 0.01–500 μg L^{-1}; **LOD**: 3 ng L^{-1}; **Anti-interference ability**: inorganic ions; **Stability**: 81% of initial activity (30 days, 4 °C); **Reproducibility**: RSD of 2.8% (10 electrodes); **Recovery rate** in spinach, oilseed rape and lettuce samples: 96.12–101.68%.	[51]
Au NPs-SiO_2-TiO_2	DPV	**Linear concentration range** for dichlorvos and fenthion: 0.018–13.6 μM; **LOD**: 5.3 nM (dichlorvos) and 1.3 nM (fenthion); **Stability**: 98% of initial activity (30 days, −4 °C); **Reproducibility**: RSD of 3.2% (4 electrodes); **Recovery rate** in cabbage juice samples: −106%.	[52]
Electrodeposited Ag	LSV	**Linear concentration range** for chlorpyrifos: 10 pM–10 nM; **LOD**: 1 4.0 pM; **Reproducibility**: RSD of 7.8% (10 electrodes); **Recovery rate** in lake water samples: 92–104%.	[53]
TiO_2-Chi-Ag NFs-Gra-Chi	DPV	**Linear concentration range** for dichlorvos: 0.036–22.63 mM; **LOD**: 7.4 nM; **Anti-interference ability**: inorganic ions and glucose; **Reproducibility**: RSD of 2.128% (3 electrodes); **Recovery rate** in lettuce juice samples: 92.7–110.9%.	[54]
Ag NPs-N-F-MoS_2	DPV	**Linear concentration range** for OPs: 10–10^{-6} mg mL^{-1} (monocrotophos), 5×10^{-8}–10^{-7} mg·mL^{-1} and 10^{-7}–10^{-4} mg mL^{-1} (chlorpyrifos); **LOD**: 0.05 pg·mL^{-1} (monocrotophos) and 1 pg·mL^{-1} (chlorpyrifos); **Anti-interference ability**: inorganic ions, ascorbic acid, glucose, citric acid, carbaryl and 2-nitrophenol; **Repeatability**: RSD of 2.4% (6 times); **Stability**: 93.7% of initial activity (30 days, 4 °C); **Recovery rate** in fruit samples: 94.6–104.2%.	[55]
3D Ag TN	DPV	**Linear concentration range** for omethoate: 10^{-13}–10^{-7} M; **LOD**: 1.26×10^{-14} M; **Anti-interference ability**: inorganic ions and glucose; **Repeatability**: RSD of 2.4% (7 times); **Stability**: 91.3% of initial activity (15 days, 4 °C); **Reproducibility**: RSD of 3.9% (7 electrodes); **Recovery rate** in tap water samples: 97.2–102%	[56]
Cu NWs-rGO	CV	**Linear concentration range** for chlorpyrifos: 10 μg L^{-1}–200 μg L^{-1}; **LOD**: 3.1 μg L^{-1}; LQD: 12.5 μg L^{-1}; **Anti-interference ability**: inorganic ions; **Recovery rate** in water and orange juice samples: 96.67–105.65%	[57]

https://doi.org/10.3390/molecules31040717

Table 2. *Cont.*

Electrochemical Active Materials	Analytical Techniques	Electrochemical Detection Performance	Ref.
AuPd NWs	DPV	**Linear concentration range** for malathion: 0.1 PM–100 nM; **LOD**: 0.037 pM; **Anti-interference ability**: inorganic ions, carbamide, glucose and citric acid; **Repeatability**: RSD of 1.78% (5 times); **Stability**: 91% of initial activity (30 days, 4 °C); **Reproducibility**: RSD of 3.86% (6 electrodes); **Recovery rate** in tap water samples: 97.8–103%.	[58]
AuPd NPs-MXene	Amperometry	**Linear concentration range** for paraoxon: 0.1–1000 μg L^{-1}; **LOD**: 1.75 ng L^{-1}; **Anti-interference ability**: inorganic ions, ethyl-paraoxon, ediphenphos, fenitrothion, phenol, ascorbic acid and glucose; **Stability**: 95% of initial activity (7 days, 4 °C); **Recovery rate** in pear and cucumber samples: 87.93–111.02%.	[59]
PD-AuPt	DPV	**Linear concentration range** for paraoxon: 0.5–1000 ng L^{-1}; **LOD**: 0.185 ng L^{-1}; **Anti-interference ability**: inorganic ions, urea and glucose; **Recovery rate** in tap water and lake water samples: 94.63–103.46%.	[60]
1D PdRh NTs	DPV	**Linear concentration range** for carbaryl: 9.44×10^{-8}–0.944 mg L^{-1}; **LOD**: 9.44×10^{-8} mg L^{-1}; **Sensitivity**: 0.279 μA nM; **Anti-interference ability**: inorganic ions, glucose, ascorbic acid, citric acid and urea; **Repeatability**: RSD of 1.65% (8 times); **Stability**: 91.1% of initial activity (30 days, 4 °C); **Reproducibility**: RSD of 2.29% (8 electrodes); **Recovery rate** in tap and lake water samples: 94.01−102.80%.	[61]
Au NPs-3D Fe_2O_3-CNTs	DPV	**Linear concentration range** for omethoate, malathion, fenitrothion and chlorpyrifos: 10^{-13}–10^{-7} M; **LOD**: 1.9×10^{-15} M (omethoate), 1.02×10^{-15} (malathion), 1.34×10^{-14} M (fenitrothion) and 10^{-13} M (chlorpyrifos); **Anti-interference ability**: inorganic ions, glucose, urea and ascorbic acid; **Repeatability**: RSD of 1.69% (5 times); **Stability**: 91.3% of initial activity (30 days, 4 °C); **Reproducibility**: RSD of 2.25% (5 electrodes); **Recovery rate** in real samples: 96.04–100.99%.	[64]
Fe_3O_4-MCS	DPV	**Linear concentration range** for malathion: 0.01–50 ppb and 50–600 ppb; **LOD**: 0.0182 ppb; **Anti-interference ability**: inorganic ions; **Repeatability**: RSD of 6.9% (6 times); **Stability**: 79% of initial activity (30 days, 4 °C); **Reproducibility**: RSD of 6.4% (6 electrodes); **Recovery rate** in pear samples: 97.80–104.10%.	[65]
TiO_2-g-CNFs	DPV	**Linear concentration range** for paraoxon: 10^{-13}–10^{-8} M; **LOD**: 3.3 fM; **Anti-interference ability**: inorganic ions and glucose; **Stability**: 96.5% of initial activity (30 days); **Reproducibility**: RSD of 3.7% (6 electrodes); **Recovery rate** in lake water samples: 96–101%.	[68]
TiO_2 NPs-Chi	DPV	**Linear concentration range** for dichlorvos: 1.13 nM–22.6 μM; **LOD**: 0.23 nM; **Anti-interference ability**: inorganic ions and glucose; **Recovery rate** in cabbage juice samples: 97.4–110%.	[69]
TiO_2 H-Chi	DPV	**Linear concentration range** for dichlorvos: 0.036–0.453 μM and 0.453–22.6 μM; **LOD**: 29 nM; **Repeatability**: RSD of 1.49% (4 times); **Stability**: 100% of initial activity (30 days); **Recovery rate** in cabbage juice samples: −103%.	[70]
ZnO-rGO	DPV	**Linear concentration range** for methyl parathion: 10^{-5}–1 μg mL^{-1}; **LOD**: 0.11871 ng mL^{-1}; **Sensitivity**: 0.004638 μA (ng $mL^{-1})^{-1}$; **Recovery rate** in apple and cucumber samples: 90.92–108.4%.	[71]
ZnO-CoO-N-PC	DPV	**Linear concentration range** for OPs: 7.6×10^{-15}–7.6×10^{-6} M (parathion-methyl) and 2.74×10^{-13}–2.74×10^{-6} M (chlorpyrifos); **LOD**: 7.6×10^{-15} M (parathion-methyl) and 2.74×10^{-13} M (chlorpyrifos); **Anti-interference ability**: inorganic ions, glucose, urea and sodium citrate; **Repeatability**: RSD of 2.22% (6 times); **Stability**: 90% of initial activity (15 days, 4 °C); **Reproducibility**: RSD of 3.07% (6 electrodes).	[72]
3D Gra-CuO NPs	SWV	**Linear concentration range** for malathion: 1 ppt–15.555 ppb; **LOD**: 0.31 ppt; **Anti-interference ability**: inorganic ions, glucose, methyl parathion and deltamethrin; **Stability**: 93.70% of initial activity (20 days, 4 °C); **Reproducibility**: RSD of 4.7% (5 electrodes); **Recovery rate** in water samples: 94.33–106.83%.	[73]

https://doi.org/10.3390/molecules31040717

Table 2. *Cont.*

Electrochemical Active Materials	Analytical Techniques	Electrochemical Detection Performance	Ref.
MnO_2-based homogeneous sensor	DPV	**Linear concentration range** for dichlorvos: 10^{-6}–10^{-10} M; **LOD**: 3×10^{-10} M; **Anti-interference ability**: inorganic ions and ascorbic acid; **Recovery rate** in cucumber and pear juice samples: 96–104%.	[74]
NiCoFeS-based homogeneous sensor	DPV	**Linear concentration range** for trichlorfon: 12.5 fg mL^{-1}–1.25 ng mL^{-1}; **LOD**: 9.74 fg mL^{-1}; **Anti-interference ability**: inorganic ions, ascorbic acid, glucose, tryptophan and leucine; **Repeatability**: RSD of 1.07% (6 times); **Stability**: more than 95% of initial activity (21 days); **Reproducibility**: RSD of less than 5% (5 electrodes); **Recovery rate** in plum, watermelon, lettuce and cucumber samples: 92.8–107.4%.	[75]
$CuInS_2$-GO	LSV	**Linear concentration range** for chlorpyrifos: 0.5–230 ng mL^{-1} and 240–470 ng mL^{-1}; **LOD**: 0.023 ng mL^{-1}; **Anti-interference ability**: inorganic ions, p-nitrophenol and o-nitrophenol; **Repeatability**: RSD of 3.07% (5 times); **Reproducibility**: up to 90.8% (5 electrodes); **Recovery rate** in real samples: 96–106%.	[76]
3D CoXO	DPV	**Linear concentration range** for OPs: 3.88×10^{-6}–3.88 μM (diazinon) and 4.69×10^{-7}–0.469 μM (omethoate); **LOD**: 0.36 pM (diazinon) and 0.033 pM (omethoate); **Anti-interference ability**: inorganic ions, glucose, urea, citric acid, chlorpyrifos, methamidophos and carbaryl; **Repeatability**: RSD of 2.98% (5 times); **Stability**: 90.3% of initial activity (30 days, 4 °C); **Reproducibility**: RSD of 2.53% (5 electrodes); **Recovery rate** in real samples: 95.4–104.1%.	[77]
PB	Amperometry	**Linear concentration range** for isocarbophos, chlorpyrifos and trichlorfon: 1×10^{-7} g mL^{-1}–5×10^{-6} g mL^{-1}; **LOD**: 10^{-7} g mL^{-1}; **Reproducibility**: RSD of 5.4% (3 electrodes); **Recovery rate** in flowering cabbage samples: 84.4–117.4% (chlorpyrifos), 77.1–109.6% (isocarbophos) and 79.1–119.3% (trichlorfon).	[78]
PB-CB	chrono-amperometry	**Linear concentration range** for dichlorvos in apple and orange samples: 0–50 ppb; **LOD**: 8 nM (in solution) and 40 nM (dry) (apple), 12 nM (in solution) and 24 nM (dry) (orange); **Repeatability**: RSD of less than 10% (4 times); **Recovery rate**: 83–95% (in solution) and 97–106% (dry) (apple), 81–104% (in solution) and 91–115% (dry) (orange).	[79]
MOF-5	CV	**Linear concentration range** for chlorpyrifos: 30 ng L^{-1}–600 ng L^{-1}; **LOD**: 3 ng L^{-1}; **Sensitivity**: 2.04 μA ng^{-1} L cm^{-2}; **Anti-interference ability**: inorganic ions, calcium phosphate, aflatoxinB1 and diisopropyl fluorophosphate; **Stability**: 90.0% of initial activity (15 days).	[84]
Chi-Gra-ZIF-8	DPV	**Linear concentration range** for isocarbophos: 1.73–345.7 nM; **LOD**: 0.62 nM; **Anti-interference ability**: inorganic ions, glucose and urea; **Stability**: 98.09% of initial activity (6 days, 4 °C); **Reproducibility**: RSD of 5.05% (5 electrodes); **Recovery rate** in cabbage and tap water samples: 88.1–122.4%.	[86]
Zn-MOFs	DPV	**Linear concentration range** for OPs: 10^{-15}–10^{-9} M (glyphosate), 10^{-15}–10^{-10} M (trichlorfon), 10^{-15}–10^{-9} M (fenitrothion), 10^{-15}–10^{-11} M (iprobenfos), 10^{-15}–10^{-12} M (acephate) and 10^{-15}–10^{-9} M (methyl-parathion); **LOD**: 1.24×10^{-13} M(glyphosate), 8.75×10^{-14} M(trichlorfon), 4.40×10^{-14} M(fenitrothion), 3.36×10^{-13} M(iprobenfos), 7.57×10^{-14} M(acephate) and 4.31×10^{-13} M(methyl-parathion); **LOQ**: 4.13×10^{-13} M (glyphosate), 1.92×10^{-13} M (trichlorfon), 1.47×10^{-13} M (fenitrothion), 1.12×10^{-13} M (iprobenfos), 2.25×10^{-13} M (acephate) and 1.44×10^{-13} M (methyl-parathion); **Anti-interference ability**: inorganic ions, glucose, uric acid, ascorbic acid, cypermethrin, 2,4-d-butyl ester and propoxur; **Recovery rate** in strawberry, cucumber, oilseed rape and tomato samples: 87.1–109.6%.	[87]
Ce-UiO-66-MWCNTs	DPV	**Linear concentration range** for paraoxon: 0.01–150 nM; **LOD**: 0.004 nM; **Anti-interference ability**: inorganic ions, oxalic acid, citric acid, glucose, carbaryl, 4-nitrophenol, carbofuran, malathion and diazinon; **Repeatability**: RSD of 4.3% (5 times); **Stability**: 85% of initial activity (20 days, 4 °C); **Recovery rate** in cabbage and spinach samples: 95–102%.	[90]

https://doi.org/10.3390/molecules31040717

Table 2. *Cont.*

Electrochemical Active Materials	Analytical Techniques	Electrochemical Detection Performance	Ref.
COF NFs	DPV	**Linear concentration range** for dichlorvos: 10^{-10}–10^{-5} g L^{-1}; **LOD**: 1.50 pg·L^{-1}; **Anti-interference ability**: inorganic ions, ascorbic acid and dopamine; **Stability**: almost unchanged (30 days); **Reproducibility**: RSD of 4.2% (6 electrodes); **Recovery rate** in real samples: 85.6–136%.	[94]
COFs-MWCNTs	DPV	**Linear concentration range** for malathion: 1 nM–10 μM; **LOD**: 0.5 nM; **Anti-interference ability**: inorganic ions; **Stability**: 82.9% of initial activity (4 °C); **Reproducibility**: RSD of 2.19% (10 electrodes); **Recovery rate** in real samples: 96–101.6% (tap water) and 98–105% (spinach).	[95]
PDDA-AuNPs-COFs	DPV	**Linear concentration range** for methyl parathion: 1.9×10^{-9}–3.8×10^{-5} M; **LOD**: 2.3×10^{-10} M; **Anti-interference ability**: inorganic ions, citric acid, ascorbic acid, glucose, urea, acetamiprid and indoxacarb; **Repeatability**: RSD of 2.4% (5 times); **Stability**: 92.0% of initial activity (14 days, 4 °C); **Reproducibility**: RSD of 2.8% (5 electrodes); **Recovery rate** in tomato and strawberry samples: 95.7–107.9%.	[96]
BPC-COFs	DPV	**Linear concentration range** for trichlorfon: 0.2–19 ng mL^{-1}; **LOD**: 0.067 ng mL^{-1}; **Anti-interference ability**: inorganic ions, nitrophenol, catechol and hydroquinone; **Stability**: 94% of initial activity (30 days, 4 °C); **Reproducibility**: RSD of 3.9% (5 electrodes); **Reactivation**: up to five times (10 min, 5 mM praldoximin chloride solution); **Recovery rate** in schisandra chinensis samples: 96.1–105%.	[97]

As is well established, the key parameters for evaluating the detection performance of enzymatic electrochemical sensors toward OP include the linear concentration range, LOD, LOQ, sensitivity, selectivity, storage stability, repeatability, reproducibility, reusability and recovery rate in real samples, all of which are also systematically summarized in Table 2. Specifically, the linear concentration range and LOD represent the fundamental metrics for the performance evaluation of enzymatic electrochemical sensors. Through the unremitting efforts of researchers, enzymatic electrochemical sensors have been demonstrated to exhibit a broad linear concentration range down to the nanomolar or micromolar level, accompanied by an ultralow LOD reaching the picomolar scale. These sensors have also shown excellent repeatability and reproducibility, with the RSD consistently below 10%, suggesting their preparation process is well optimized and suitable for batch production. In addition, enzymatic electrochemical sensors have displayed robust anti-interference capability against inorganic ions, low-molecular-weight organic acids and sugars. However, their selectivity toward OPs remains relatively poor, which is attributable to the intrinsic broad-spectrum catalytic activity of the enzymes employed for OP detection. Therefore, the development of selective sensors is of significance for the specific detection of OPs. Furthermore, storage stability represents another crucial performance parameter of enzymatic electrochemical sensors. As shown in Table 2, the vast majority of storage stability assessments have been conducted at 4 °C over a 30-day period. Even under such optimized storage conditions, the storage stability of these sensors was still unsatisfactory, as some sensors retained merely 40–50% of their initial activity after storage. Although reactivation treatments could restore the electrochemical activity of these sensors, the recovery efficiency was only approximately 90%. Accordingly, it is imperative to develop effective strategies to enhance the storage stability of enzymatic electrochemical sensors. Finally, the recovery rates of these sensors in real samples exhibit a wide distribution, so it is necessary to suppress this distribution for obtaining detection performance comparable to that of chromatographic techniques.

https://doi.org/10.3390/molecules31040717

5. Conclusions and Outlooks

This review comprehensively discusses the recent advances in enzymatic electrochemical sensors for the detection of OPs. First, the fundamental information of commonly used OPs is summarized. Based on their toxicity, stability and solubility, acephate, glyphosate and trichlorfon are identified as ideal OP models for evaluating the detection performance of enzymatic electrochemical sensors. The structure of such sensors, including electrodes, electrochemical active layers and bioactive enzyme layers, is systematically reviewed. Particularly, active materials (such as carbon-based, polymer-based, metal-based, metal compound-based, MOF-based and COF-based materials), which serve as the core components of enzymatic electrochemical sensors, are emphatically discussed. It is found that enzymatic electrochemical sensors based on these active materials exhibit ultrahigh sensitivity (with a broad linear concentration range down to the nanomolar or micromolar level and an LOD reaching the picomolar scale), acceptable stability (retaining approximately 85–95% of their initial activity after 30 days of storage at 4 °C), excellent repeatability and reproducibility (with the RSD consistently below 10%) and satisfactory recovery rates for OP detection (ranging roughly from 90% to 115%). In addition, due to their inherently high specific area, well-defined crystalline and porous structures, MOF- and COF-based materials are considered as one of the most important active materials for enzymatic electrochemical sensors. However, research on MOF- and COF-based enzymatic electrochemical sensors is still in its infancy. To further promote the practical application of such sensors, future efforts should focus on developing green and eco-friendly synthesis methods to prepare MOF- and COF-based materials and improving their intrinsic conductivity. Furthermore, since the detection mechanism relies on the inhibition of AChE activity by the phosphate groups of OPs, this type of biosensor demonstrates broad-spectrum applicability to all OPs. The specificity for detecting different OPs can be enhanced by employing alternative biomolecules as signal recognition elements. On the other hand, from the perspective of storage stability, the development of non-enzymatic electrochemical sensors is of great significance for OP detection.

Author Contributions: Writing—original draft, writing—review and editing: S.M. and Z.C.; collecting the literature, data curation: F.Y. and T.Y.; writing—review and editing: Y.Y. and L.X.; data curation, writing—review and editing: S.W.; conceptualization, supervision, writing—review and editing, funding acquisition: X.H. and G.W. All authors have read and agreed to the published version of the manuscript.

Funding: This research was funded by the National Natural Science Foundation of China [grant number 32072304], Guangdong Basic and Applied Basic Research Foundation (grant numbers 2023A1515140180, 2022A1515140159, 2024A1515012826), Key Project in the Field of Biomedicine and Health (Food) by Department of Education of Guangdong Province (grant number 2025ZDZX2067) and Guangdong Province Key Construction Discipline Research Ability Improvement Project (grant number 2024ZDJS027).

Data Availability Statement: Data will be made available on request.

Conflicts of Interest: The authors declare no conflicts of interest.

References

1. Zorbas, C.; Resnick, D.; Jones, E.; Suri, S.; Iruhiriye, E.; Headey, D.; Martin, W.; Vos, R.; Menon, P. From promises to action: Analyzing global commitments to tackle hunger and food insecurity. *Food Policy* **2025**, *136*, 102968. [CrossRef]
2. China Pesticide Information Network. Available online: http://www.chinapesticide.org.cn/zwb/detail/31197 (accessed on 16 September 2025).

https://doi.org/10.3390/molecules31040717

3. Mali, H.; Shah, C.; Raghunandan, B.H.; Prajapati, A.S.; Patel, D.H.; Trivedi, U.; Subramanian, R.B. Organophosphate pesticides an emerging environmental contaminant: Pollution, toxicity, bioremediation progress, and remaining challenges. *J. Environ. Sci.* **2023**, *127*, 234–250. [CrossRef] [PubMed]
4. Jokanović, M.; Oleksak, P.; Kuca, K. Multiple neurological effects associated with exposure to organophosphorus pesticides in man. *Toxicology* **2023**, *484*, 153407. [CrossRef] [PubMed]
5. Li, L.; Zhou, S.; Jin, L.; Zhang, C.; Liu, W. Enantiomeric separation of organophosphorus pesticides by high-performance liquid chromatography, gas chromatography and capillary electrophoresis and their applications to environmental fate and toxicity assays. *J. Chromatogr. B* **2010**, *878*, 1264–1276. [CrossRef] [PubMed]
6. Liu, Y.; Gong, S.; Ye, L.; Li, J.; Liu, C.; Chen, D.; Fang, M.; Letcher, R.J.; Su, G. Organophosphate (OP) diesters and a review of sources, chemical properties, environmental occurrence, adverse effects, and future directions. *Environ. Int.* **2021**, *155*, 106691. [CrossRef]
7. Irkham; Putra, C.P.; Kharismasari, C.Y.; Zakiyyah, S.N.; Rahmawati, I.; Anggraningrum, I.T.; Wahyuni, W.T.; Valenti, G.; Paolucci, F.; Hartati, Y.W. Advancements in electrochemiluminescence-based sensors for ultra-sensitive pesticide residue detection. *Sens. Bio-Sens. Res.* **2024**, *46*, 100708. [CrossRef]
8. Xu, X.; Zhang, W.; Huang, J.; Xu, H. Colorimetric sensors for detection of organophosphorus pesticides in food: From sensing strategies to chemometrics driven discrimination. *Trends Food Sci. Technol.* **2024**, *152*, 104683. [CrossRef]
9. Song, K.; Zhang, L.; Meng, Y.; Ding, J.; Li, X.; Liu, Y.; Ma, H. Recent development in fluorescent probes for monitoring organophosphorus pollutants. *Chin. Chem. Lett.* **2026**, *37*, 111902. [CrossRef]
10. Gupta, A.; Sharma, K.; Pandey, G.; Lawaniya, S.D.; Pandey, H.; Awasthi, K.K.; Awasthi, K.; Awasthi, A. Combatting organophosphate toxicity: Novel strategies for detection, analysis, and sensor development for public health protection. *TrAC Trends Anal. Chem.* **2025**, *193*, 118451. [CrossRef]
11. Dube, A.; Malode, S.J.; Ali Alshehri, M.; Shetti, N.P. Recent advances in the development of electrochemical sensors for detecting pesticides. *J. Ind. Eng. Chem.* **2025**, *144*, 77–99. [CrossRef]
12. Kumaravel, A.; Chandrasekaran, M. Electrochemical determination of chlorpyrifos on a nano-TiO_2/cellulose acetate composite modified glassy carbon electrode. *J. Agric. Food Chem.* **2015**, *63*, 6150–6156. [CrossRef] [PubMed]
13. Zhang, Q.; Yu, J.; Wei, Y.; Yang, R.; Liu, J.; Su, Y.; Gao, D.; Zeng, J. Advances in acetylcholinesterase-based biosensing technologies for organophosphorus pesticide detection: A comprehensive review (2020–2024). *Food Chem.* **2025**, *496*, 146769. [CrossRef] [PubMed]
14. Eto, M. Organophosphorus Insecticides. In *Encyclopedia of Agrochemicals*; John Wiley & Sons: New York, NY, USA, 2003.
15. Cheng, Y.; Lv, Y.; Zhao, X.; Lu, Y.; Jiao, T.; Ma, T.; Fu, Y. Multidimensional toxicity of organophosphate pesticides and mitigation strategies for agricultural sustainability. *J. Agric. Food Chem.* **2025**, *73*, 18047–18062. [CrossRef] [PubMed]
16. Karaboga, S.; Severac, F.; Collins, E.-M.S.; Stab, A.; Davis, A.; Souchet, M.; Hervé, G. Organophosphate toxicity patterns: A new approach for assessing organophosphate neurotoxicity. *J. Hazard. Mater.* **2024**, *470*, 134236. [CrossRef]
17. National Center for Biotechnology Information. Available online: https://pubchem.ncbi.nlm.nih.gov/#query=acephate (accessed on 25 March 2005).
18. Zinovicius, A.; Balciunas, E.; Rozene, J.; Petroniene, J.J.; Bogusevice, A.; Ino, K.; Striska, L.; Mockaitis, T.; Morkvenaite, I. Composites-based electrodes in enzymatic electrochemical glucose biosensors. *J. Electroanal. Chem.* **2025**, *998*, 119505. [CrossRef]
19. Huang, L.; Qu, L.; Jia, S.; Ding, S.; Zhao, J.; Li, F. The interaction of allicin with bovine serum albumin and its influence on the structure of protein. *Process Biochem.* **2022**, *112*, 139–144. [CrossRef]
20. Tsauria, Q.D.; Gareso, P.L.; Tahir, D. Structure–property–application framework for multifunctional chitosan across biomedical, environmental, and technological domains: A review. *Int. J. Biol. Macromol.* **2025**, *331*, 148380. [CrossRef]
21. Strauss, J.D.; Wagenknecht, T. Structure of glutaraldehyde cross-linked ryanodine receptor. *J. Struct. Biol.* **2013**, *181*, 300–306. [CrossRef]
22. Ke, Y.; Yuan, W.; Zhou, F.; Guo, W.; Li, J.; Zhuang, Z.; Su, X.; Lu, B.; Zhao, Y.; Tang, Y.; et al. A critical review on surface-pattern engineering of nafion membrane for fuel cell applications. *Renew. Sustain. Energy Rev.* **2021**, *145*, 110860. [CrossRef]
23. Sussman, J.L.; Harel, M.; Frolow, F.; Oefner, C.; Goldman, A.; Toker, L.; Silman, I. Atomic structure of acetylcholinesterase from torpedo californica: A prototypic acetylcholine-binding protein. *Science* **1991**, *253*, 872–879. [CrossRef]
24. Bagrowska, W.; Karasewicz, A.; Góra, A. Comprehensive analysis of acetylcholinesterase inhibitor and reactivator complexes: Implications for drug design and antidote development. *Drug Discov. Today* **2024**, *29*, 104217. [CrossRef]
25. Nicolet, Y.; Lockridge, O.; Masson, P.; Fontecilla-Camps, J.C.; Nachon, F. Crystal structure of human butyrylcholinesterase and of its complexes with substrate and products. *J. Biol. Chem.* **2003**, *278*, 41141–41147. [CrossRef]
26. Kwak, Y.; Lee, S.-E.; Shin, J.-H. Expression of organophosphorus hydrolase in *Escherichia coli* for use as whole-cell biocatalyst. *J. Mol. Catal. B Enzym.* **2014**, *99*, 169–175. [CrossRef]
27. Kumaran, A.; Vashishth, R.; Singh, S.; U, S.; James, A.; Velayudhaperumal Chellam, P. Biosensors for detection of organophosphate pesticides: Current technologies and future directives. *Microchem. J.* **2022**, *178*, 107420. [CrossRef]

https://doi.org/10.3390/molecules31040717

28. Patel, H.; Rawtani, D.; Agrawal, Y.K. A newly emerging trend of chitosan-based sensing platform for the organophosphate pesticide detection using Acetylcholinesterase—A review. *Trends Food Sci. Technol.* **2019**, *85*, 78–91. [CrossRef]
29. Thakkar, J.B.; Gupta, S.; Prabha, C.R. Acetylcholine esterase enzyme doped multiwalled carbon nanotubes for the detection of organophosphorus pesticide using cyclic voltammetry. *Int. J. Biol. Macromol.* **2019**, *137*, 895–903. [CrossRef] [PubMed]
30. Ayanda, O.S.; Mmuoegbulam, A.O.; Okezie, O.; Durumin Iya, N.I.; Mohammed, S.E.; James, P.H.; Muhammad, A.B.; Unimke, A.A.; Alim, S.A.; Yahaya, S.M.; et al. Recent progress in carbon-based nanomaterials: Critical review. *J. Nanopart. Res.* **2024**, *26*, 106. [CrossRef]
31. Xie, X.; Zhou, B.; Zhang, Y.; Zhao, G.; Zhao, B. A multi-residue electrochemical biosensor based on graphene/chitosan/parathion for sensitive organophosphorus pesticides detection. *Chem. Phys. Lett.* **2021**, *767*, 138355. [CrossRef]
32. Kumar, T.H.V.; Sundramoorthy, A.K. Electrochemical biosensor for methyl parathion based on single-walled carbon nanotube/glutaraldehyde crosslinked acetylcholinesterase-wrapped bovine serum albumin nanocomposites. *Anal. Chim. Acta* **2019**, *1074*, 131–141. [CrossRef]
33. Li, Y.P.; Zhao, R.X.; Han, G.Y.; Xiao, Y.M. Novel acetylcholinesterase biosensor for detection of paraoxon based on holey graphene oxide modified glass carbon electrode. *Electroanalysis* **2018**, *30*, 2258–2264. [CrossRef]
34. Hou, W.; Zhang, Q.; Dong, H.; Li, F.; Zhang, Y.; Guo, Y.; Sun, X. Acetylcholinesterase biosensor modified with ATO/OMC for detecting organophosphorus pesticides. *New J. Chem.* **2019**, *43*, 946–952. [CrossRef]
35. Soulis, D.; Trigazi, M.; Tsekenis, G.; Chandrinou, C.; Klinakis, A.; Zergioti, I. Facile and low-cost SPE modification towards ultra-sensitive organophosphorus and carbamate pesticide detection in olive oil. *Molecules* **2020**, *25*, 4988. [CrossRef]
36. Wei, M.; Feng, S. Amperometric determination of organophosphate pesticides using a acetylcholinesterase based biosensor made from nitrogen-doped porous carbon deposited on a boron-doped diamond electrode. *Microchim. Acta* **2017**, *184*, 3461–3468. [CrossRef]
37. Rahman, M.A.; Kumar, P.; Park, D.-S.; Shim, Y.-B. Electrochemical sensors based on organic conjugated polymers. *Sensors* **2008**, *8*, 118–141. [CrossRef] [PubMed]
38. Paneru, S.; Kumar, D. Ag-doped-CuO nanoparticles supported polyaniline (PANI) based novel electrochemical sensor for sensitive detection of paraoxon-ethyl in three real samples. *Sens. Actuators B Chem.* **2023**, *379*, 133270. [CrossRef]
39. Virji, S.; Huang, J.; Kaner, R.B.; Weiller, B.H. Polyaniline nanofiber gas sensors: Examination of response mechanisms. *Nano Lett.* **2004**, *4*, 491–496. [CrossRef]
40. Maanaki, H.; Xu, T.; Chen, G.; Du, X.; Wang, J. Development of integrated smartphone/resistive biosensor for on-site rapid environmental monitoring of organophosphate pesticides in food and water. *Biosens. Bioelectron. X* **2023**, *15*, 100402. [CrossRef]
41. Jia, L.; Zhou, Y.; Wu, K.; Feng, Q.; Wang, C.; He, P. Acetylcholinesterase modified AuNPs-MoS_2-rGO/PI flexible film biosensor: Towards efficient fabrication and application in paraoxon detection. *Bioelectrochemistry* **2020**, *131*, 107392. [CrossRef]
42. Holade, Y.; Hickey, D.P.; Minteer, S.D. Halide-regulated growth of electrocatalytic metal nanoparticles directly onto a carbon paper electrode. *J. Mater. Chem. A* **2016**, *4*, 17154–17162. [CrossRef]
43. Loguercio, L.F.; Griep, J.; Demingos, P.G.; Morawski, R.; Brolo, A.G.; Andrade, G.F.S.; Santos, J.F.L. Enhanced enzymatic electrochemical detection of an organophosphate Pesticide: Achieving Wide linearity and femtomolar detection via gold nanoparticles growth within polypyrrole films. *Talanta* **2025**, *281*, 126714. [CrossRef]
44. Loguercio, L.F.; Thesing, A.; Demingos, P.; de Albuquerque, C.D.L.; Rodrigues, R.S.B.; Brolo, A.G.; Santos, J.F.L. Efficient acetylcholinesterase immobilization for improved electrochemical performance in polypyrrole nanocomposite-based biosensors for carbaryl pesticide. *Sens. Actuators B Chem.* **2021**, *339*, 129875. [CrossRef]
45. Theansun, W.; Sriprachuabwong, C.; Chuenchom, L.; Prajongtat, P.; Techasakul, S.; Tuantranont, A.; Dechtrirat, D. Acetylcholinesterase modified inkjet-printed graphene/gold nanoparticle/poly(3,4-ethylenedioxythiophene):poly(styrenesulfonate) hybrid electrode for ultrasensitive chlorpyrifos detection. *Bioelectrochemistry* **2023**, *149*, 108305. [CrossRef] [PubMed]
46. Zhang, P.; Sun, T.; Rong, S.; Zeng, D.; Yu, H.; Zhang, Z.; Chang, D.; Pan, H. A sensitive amperometric AChE-biosensor for organophosphate pesticides detection based on conjugated polymer and Ag-rGO-NH_2 nanocomposite. *Bioelectrochemistry* **2019**, *127*, 163–170. [CrossRef] [PubMed]
47. Arivazhagan, M.; Mohan, B.; Jakmunee, J. Nanostructured metallic enzymes mimic for electrochemical biosensing of glucose. *Green Anal. Chem.* **2024**, *10*, 100127. [CrossRef]
48. Dhull, V. A Nafion/AChE-cSWCNT/MWCNT/Au-based amperometric biosensor for the determination of organophosphorous compounds. *Environ. Technol.* **2020**, *41*, 566–576. [CrossRef]
49. Hua, Q.T.; Ruecha, N.; Hiruta, Y.; Citterio, D. Disposable electrochemical biosensor based on surface-modified screen-printed electrodes for organophosphorus pesticide analysis. *Anal. Methods* **2019**, *11*, 3439–3445. [CrossRef]
50. Bao, J.; Hou, C.; Chen, M.; Li, J.; Huo, D.; Yang, M.; Luo, X.; Lei, Y. Plant esterase–chitosan/gold nanoparticles–graphene nanosheet composite-based biosensor for the ultrasensitive detection of organophosphate pesticides. *J. Agric. Food Chem.* **2015**, *63*, 10319–10326. [CrossRef]

https://doi.org/10.3390/molecules31040717

51. Yao, Y.; Wang, G.; Chu, G.; An, X.; Guo, Y.; Sun, X. The development of a novel biosensor based on gold nanocages/graphene oxide–chitosan modified acetylcholinesterase for organophosphorus pesticide detection. *New J. Chem.* **2019**, *43*, 13816–13826. [CrossRef]
52. Cui, H.-F.; Zhang, T.-T.; Lv, Q.-Y.; Song, X.; Zhai, X.-J.; Wang, G.-G. An acetylcholinesterase biosensor based on doping Au nanorod@SiO_2 nanoparticles into TiO_2-chitosan hydrogel for detection of organophosphate pesticides. *Biosens. Bioelectron.* **2019**, *141*, 111452. [CrossRef]
53. Liu, Z.; Xia, X.; Zhou, G.; Ge, L.; Li, F. Acetylcholinesterase-catalyzed silver deposition for ultrasensitive electrochemical biosensing of organophosphorus pesticides. *Analyst* **2020**, *145*, 2339–2344. [CrossRef]
54. Zhang, J.; Wang, B.; Li, Y.; Shu, W.; Hu, H.; Yang, L. An acetylcholinesterase biosensor with high stability and sensitivity based on silver nanowire–graphene–TiO_2 for the detection of organophosphate pesticides. *RSC Adv.* **2019**, *9*, 25248–25256. [CrossRef]
55. Song, D.; Wang, Y.; Lu, X.; Gao, Y.; Li, Y.; Gao, F. Ag nanoparticles-decorated nitrogen-fluorine co-doped monolayer MoS_2 nanosheet for highly sensitive electrochemical sensing of organophosphorus pesticides. *Sens. Actuators B Chem.* **2018**, *267*, 5–13. [CrossRef]
56. Yang, Y.; Zhao, Y.; Sun, F.; You, T.; Gao, Y.; Yin, P. Electrochemically synthesized superhydrophilic 3D tree-like Ag microstructure for ultrasensitive detection of omethoate. *Microchem. J.* **2020**, *159*, 105427. [CrossRef]
57. Suwannachat, J.; Saenchoopa, A.; Tun, W.S.T.; Patramanon, R.; Daduang, S.; Daduang, J.; Kulchat, S. An electrochemical AChE-based biosensor for organophosphate pesticides using a modified CuNWs/rGO nanocomposite on a screen-printed carbon electrode. *Food Chem.* **2024**, *434*, 137431. [CrossRef] [PubMed]
58. Lu, X.; Tao, L.; Li, Y.; Huang, H.; Gao, F. A highly sensitive electrochemical platform based on the bimetallic Pd@Au nanowires network for organophosphorus pesticides detection. *Sens. Actuators B Chem.* **2019**, *284*, 103–109. [CrossRef]
59. Zhao, F.; Yao, Y.; Jiang, C.; Shao, Y.; Barceló, D.; Ying, Y.; Ping, J. Self-reduction bimetallic nanoparticles on ultrathin MXene nanosheets as functional platform for pesticide sensing. *J. Hazard. Mater.* **2020**, *384*, 121358. [CrossRef]
60. Wu, Y.; Jiao, L.; Xu, W.; Gu, W.; Zhu, C.; Du, D.; Lin, Y. Polydopamine-capped bimetallic AuPt hydrogels enable robust biosensor for organophosphorus pesticide detection. *Small* **2019**, *15*, 1900632. [CrossRef]
61. Yang, Y.; Yu, D.; Xu, X.; Zhao, Y.; Mi, Y.; Gao, F. One-dimensional bimetallic PdRh alloy mesoporous nanotubes constructed for ultra-sensitive detection of carbamate pesticide. *Anal. Biochem.* **2022**, *652*, 114726. [CrossRef]
62. Jin, R.; Kong, D.; Zhao, X.; Li, H.; Yan, X.; Liu, F.; Sun, P.; Du, D.; Lin, Y.; Lu, G. Tandem catalysis driven by enzymes directed hybrid nanoflowers for on-site ultrasensitive detection of organophosphorus pesticide. *Biosens. Bioelectron.* **2019**, *141*, 111473. [CrossRef]
63. Chen, J.; Guo, Z.; Xin, Y.; Gu, Z.; Zhang, L.; Guo, X. Effective remediation and decontamination of organophosphorus compounds using enzymes: From rational design to potential applications. *Sci. Total Environ.* **2023**, *867*, 161510. [CrossRef]
64. Zhao, C.; Zhao, Y.; Wang, L.; Lu, K.; Guo, W.; Lu, X.; Gao, F. AuNPs and CNTs embellished three-dimensional bloom-like α-Fe_2O_3 nanocomposites for highly sensitive electrochemical pesticides detection. *Microchem. J.* **2023**, *191*, 108762. [CrossRef]
65. Luo, R.; Feng, Z.; Shen, G.; Xiu, Y.; Zhou, Y.; Niu, X.; Wang, H. Acetylcholinesterase biosensor based on mesoporous hollow carbon spheres/core-shell magnetic nanoparticles-modified electrode for the detection of organophosphorus pesticides. *Sensors* **2018**, *18*, 4429. [CrossRef] [PubMed]
66. Yin, Z.F.; Wu, L.; Yang, H.G.; Su, Y.H. Recent progress in biomedical applications of titanium dioxide. *Phys. Chem. Chem. Phys.* **2013**, *15*, 4844–4858. [CrossRef] [PubMed]
67. Bao, J.; Hou, C.; Huo, D.; Dong, Q.; Ma, X.; Sun, X.; Yang, M.; Galil, K.H.A.E.; Chen, W.; Lei, Y. Sensitive and selective electrochemical biosensor based on ELP-OPH/BSA/TiO_2NFs/AuNPs for determination of organophosphate pesticides with p-nitrophenyl substituent. *J. Electrochem. Soc.* **2017**, *164*, G17. [CrossRef]
68. Tao, S.; Guo, Y.; Wang, S.; Xu, F.; Zhou, X.; Guo, Q. A sensitive and stable acetylcholinesterase biosensor with TiO_2 nanoparticles anchored on graphitic carbon nanofibers for determination of organophosphate pesticides. *Anal. Methods* **2022**, *14*, 2396–2404. [CrossRef] [PubMed]
69. Hu, H.; Wang, B.; Li, Y.; Wang, P.; Yang, L. Acetylcholinesterase sensor with patterned structure for detecting organophosphorus pesticides based on titanium dioxide sol-gel carrier. *Electroanalysis* **2020**, *32*, 1834–1842. [CrossRef]
70. Cui, H.-F.; Wu, W.-W.; Li, M.-M.; Song, X.; Lv, Y.; Zhang, T.-T. A highly stable acetylcholinesterase biosensor based on chitosan-TiO_2-graphene nanocomposites for detection of organophosphate pesticides. *Biosens. Bioelectron.* **2018**, *99*, 223–229. [CrossRef]
71. Liu, Y.; Xiao, Y.; Zhang, Y.; Gao, X.; Wang, H.; Niu, B.; Li, W. ZnO-rGO-based electrochemical biosensor for the detection of organophosphorus pesticides. *Bioelectrochemistry* **2024**, *156*, 108599. [CrossRef]
72. Li, Z.; Lu, X.; Liu, G.; Yang, L.; Gao, F. Core-shell ZnO@CoO nitrogen doped nano-composites as highly sensitive electrochemical sensor for organophosphate pesticides detection. *Anal. Biochem.* **2024**, *686*, 115422. [CrossRef]
73. Bao, J.; Huang, T.; Wang, Z.; Yang, H.; Geng, X.; Xu, G.; Samalo, M.; Sakinati, M.; Huo, D.; Hou, C. 3D graphene/copper oxide nano-flowers based acetylcholinesterase biosensor for sensitive detection of organophosphate pesticides. *Sens. Actuators B Chem.* **2019**, *279*, 95–101. [CrossRef]

74. Sun, Y.; Xiong, P.; Tang, J.; Zeng, Z.; Tang, D. Ultrasensitive split-type electrochemical sensing platform for sensitive determination of organophosphorus pesticides based on MnO_2 nanoflower-electron mediator as a signal transduction system. *Anal. Bioanal. Chem.* **2020**, *412*, 6939–6945. [CrossRef]
75. Wang, R.; Li, B.; Li, G.; Shen, Q.; Zou, L. NiCoFeS/rGO nanozyme-mediated multifunctional homogeneous sensing system for ultrasensitive electrochemical assay of pesticides residues in fruits and vegetables. *Sens. Actuators B Chem.* **2025**, *422*, 136664. [CrossRef]
76. Itsoponpan, T.; Thanachayanont, C.; Hasin, P. Sponge-like $CuInS_2$ microspheres on reduced graphene oxide as an electrocatalyst to construct an immobilized acetylcholinesterase electrochemical biosensor for chlorpyrifos detection in vegetables. *Sens. Actuators B Chem.* **2021**, *337*, 129775. [CrossRef]
77. Zhao, Y.; Li, X.; Chen, J.; Lu, X.; Yang, Y.; Song, D.; Gao, F. Porous hierarchical peony-like cobalt-based bimetallic oxides structured by ultrathin nanosheets for highly sensitive electrochemical pesticides detection. *Sens. Actuators B Chem.* **2022**, *352*, 131072. [CrossRef]
78. Shi, Q.; Teng, Y.; Zhang, Y.; Liu, W. Rapid detection of organophosphorus pesticide residue on Prussian blue modified dual-channel screen-printed electrodes combing with portable potentiostat. *Chin. Chem. Lett.* **2018**, *29*, 1379–1382. [CrossRef]
79. Miglione, A.; Raucci, A.; Mancini, M.; Gioia, V.; Frugis, A.; Cinti, S. An electrochemical biosensor for on-glove application: Organophosphorus pesticide detection directly on fruit peels. *Talanta* **2025**, *283*, 127093. [CrossRef] [PubMed]
80. Yu, H.; Lin, P.; Zhou, L. Applications and advances of MOFs-based sensors for pesticide and veterinary drug residue detection: From design to application and challenge. *J. Food Compos. Anal.* **2025**, *148*, 108147. [CrossRef]
81. Castelvecchi, D.; Naddaf, M. Chemistry Nobel for scientists who developed massively porous 'super sponge' materials. *Nature* **2025**, *646*, 522–523. [CrossRef]
82. Hoskins, B.F.; Robson, R. Infinite polymeric frameworks consisting of three dimensionally linked rod-like segments. *J. Am. Chem. Soc.* **1989**, *111*, 5962–5964. [CrossRef]
83. Yaghi, O.M.; Li, G.; Li, H. Selective binding and removal of guests in a microporous metal–organic framework. *Nature* **1995**, *378*, 703–706. [CrossRef]
84. Chansi; R., P.R.; Mukherjee, I.; Basu, T.; Bharadwaj, L.M. Metal organic framework steered electrosynthesis of anisotropic gold nanorods for specific sensing of organophosphate pesticides in vegetables collected from the field. *Nanoscale* **2020**, *12*, 21719–21733. [CrossRef] [PubMed]
85. Liu, F.; Liang, Z.; Zhang, Y.; Qin, S.; Zhang, S.; Zhang, T.; Li, B.; Liao, X.; Li, X.; Yang, S. Evaluation of the potential application of RAPA-based delivery: Structural evolution and biocompatibility of monodispersed nano-ZIF-8@RAPA. *Mater. Today Commun.* **2025**, *47*, 113163. [CrossRef]
86. Wen, L.; Wang, N.; Liu, Z.; Tao, C.-A.; Zou, X.; Wang, F.; Wang, J. Acetylcholinesterase immobilization on ZIF-8/graphene composite engenders high sensitivity electrochemical sensing for organophosphorus pesticides. *Chemosensors* **2022**, *10*, 418. [CrossRef]
87. Wu, Q.; Wang, Y.; Wang, L.; Su, Y.; He, G.; Chen, X.; Hou, L.; Zhang, W.; Wang, Y.-Y. A portable electrochemical biosensor based on an amino-modified ionic metal–organic framework for the one-site detection of multiple organophosphorus pesticides. *ACS Appl. Mater. Interfaces* **2024**, *16*, 55802–55812. [CrossRef] [PubMed]
88. Xue, Y.; Gao, R.; Lin, S.; Zhong, Q.; Zhang, Q.; Hong, J. Regulating the interface electron distribution of iron-based MOFs through ligand functionalization enables efficient peroxymonosulfate utilization and catalytic performance. *J. Colloid Interface Sci.* **2024**, *663*, 358–368. [CrossRef] [PubMed]
89. Gokila, N.; Haldorai, Y.; Saravanan, P.; Rajendra Kumar, R.T. Non-enzymatic electrochemical impedance sensor for selective detection of electro-inactive organophosphate pesticides using Zr-MOF/ZrO_2/MWCNT ternary composite. *Environ. Res.* **2024**, *251*, 118648. [CrossRef] [PubMed]
90. Mahmoudi, E.; Fakhri, H.; Hajian, A.; Afkhami, A.; Bagheri, H. High-performance electrochemical enzyme sensor for organophosphate pesticide detection using modified metal-organic framework sensing platforms. *Bioelectrochemistry* **2019**, *130*, 107348. [CrossRef]
91. Hosseini, M.; Sadat Sabet, F.; Khabbaz, H.; Aghazadeh, M.; Mizani, F.; Ganjali, M.R. Enhancement of the peroxidase-like activity of cerium-doped ferrite nanoparticles for colorimetric detection of H_2O_2 and glucose. *Anal. Methods* **2017**, *9*, 3519–3524. [CrossRef]
92. Abuzeid, H.R.; El-Mahdy, A.F.M.; Kuo, S.-W. Covalent organic frameworks: Design principles, synthetic strategies, and diverse applications. *Giant* **2021**, *6*, 100054. [CrossRef]
93. Song, Y.; Chen, J.; Sun, M.; Gong, C.; Shen, Y.; Song, Y.; Wang, L. A simple electrochemical biosensor based on AuNPs/MPS/Au electrode sensing layer for monitoring carbamate pesticides in real samples. *J. Hazard. Mater.* **2016**, *304*, 103–109. [CrossRef]
94. Wang, L.; Wu, N.; Wang, L.; Song, Y.; Ma, G. Accurate detection of organophosphorus pesticides based on covalent organic framework nanofiber with a turn-on strategy. *Sens. Actuators B Chem.* **2022**, *372*, 132608. [CrossRef]

https://doi.org/10.3390/molecules31040717

95. Wang, X.; Yang, S.; Shan, J.; Bai, X. Novel electrochemical acetylcholinesterase biosensor based on core-shell covalent organic framework@multi-walled carbon nanotubes (COF@MWCNTs) composite for detection of malathion. *Int. J. Electrochem. Sci.* **2022**, *17*, 220543. [CrossRef]
96. Yang, K.; Zhao, H.; Li, N.; Wang, Y.; Sun, B.; Cui, M.; Zhang, C. Facile synthesis of novel porphyrin-based covalent organic frameworks integrated with Au nanoparticles for highly sensitive detection of organophosphorus pesticide residues. *Microchem. J.* **2024**, *203*, 110945. [CrossRef]
97. Liu, Y.; Zhou, M.; Jin, C.; Zeng, J.; Huang, C.; Song, Q.; Song, Y. Preparation of a sensor based on biomass porous carbon/covalent-organic frame composites for pesticide residues detection. *Front. Chem.* **2020**, *8*, 643. [CrossRef]

https://doi.org/10.3390/molecules31040717

MDPI AG
Grosspeteranlage 5
4052 Basel
Switzerland
Tel.: +41 61 683 77 34

Molecules Editorial Office
E-mail: molecules@mdpi.com
www.mdpi.com/journal/molecules

www.ingramcontent.com/pod-product-compliance
Lightning Source LLC
Chambersburg PA
CBHW061447210926
51850CB00008B/1642
9783725878574